The Political Economy of Necessity and the Biolimnological Vulnerability of Lake Victoria

The Political Economy of Necessity and the Biolimnological Vulnerability of Lake Victoria:

A Thermodynamic Perspective

Okbazghi Yohannes

Africa World Press
Trenton | London | Cape Town | Nairobi | Addis Ababa | Asmara | Ibadan

AFRICA WORLD PRESS
541 West Ingham Avenue | Suite B
Trenton, New Jersey 08638

Cover design: Carles Juzang

Library of Congress Cataloging-in-Publication Data

Yohannes, Okbazghi.
The political economy of necessity and the biolimnological vulnerability of Lake Victoria : a thermodynamic perspective / Okbazghi Yohannes.
pages cm
Includes bibliographical references and index.
ISBN 978-1-59221-978-0 (hard cover) -- ISBN 978-1-59221-979-7 (pbk.) 1. Water--Pollution--Victoria, Lake. 2. Victoria, Lake--Environmental conditions. 3. Environmental degradation--Africa, East. 4. Fisheries--Africa, East. 5. Wildlife utilization--Africa, East. 6. Aquatic resources--Africa, East. 7. Capitalism--Africa, East. I. Title.
TD319.E18Y64 2014
333.95'60967827--dc23

2014004635

Dedication

To Keren and Tamara for being the rocks of my life.

Contents

Acknowledgements

No book is ever written without the contributions of so many people. Indeed, this book could not have been written without the enormous insights and intellectual resources I gained from the voluminous studies of those who invested precious time and effort in Lake Victoria. So my thanks go to all those from whose cumulative research products I have greatly benefited. I hope I have done some justice to all of them by citing and including their works in the bibliography.

I am greatly thankful to my colleagues at the University of Louisville, especially to Charles Ziegler, Dewey Clayton, Susan Matarese, Ron Vogel, and Rodger Payne for their moral and intellectual support. I owe Laurie Rodebeck a huge debt of gratitude for finding me the right graduate research assistants. My particular thanks also go to Victor Oliviera, Chris Ledford, Christy O'Toole, Christy Rose, and Josh Applegate for their due diligence with research assistance.

I am also in a debt of gratitude to Kassahun Checole, owner and publisher of Africa World Press, for his enormous generosity of spirit, encouragement, and moral support. My thanks also go to K. Chase for her meticulous editing and invaluable suggestions. I cannot thank Damola Ifaturoti, senior editor at Africa World Press, for his professional proficiency and soothing communications.

My family deserves special thanks. As in all previous works, my wife Tamara has been part of this project from start to finish providing inspiration, guidance, invaluable insights, and prodigious checking of sources. Even when I started to speak in fish tongue, she treated it with grace. I am in her debt for the patience of Job she displayed when my mood alternated between excitement and frustration, between hope and despair. My daughter Keren also merits special thanks for her invaluable suggestions and her frank and witty criticisms. I have handsomely benefited from the wealth of information she brought back from her personal visit to Lake Victoria, especially to the Lake Victoria Fisheries Organization at Jinja, Uganda, where she was treated to a full and complete exposition of their research operation, accompanied by a delicious East African tea reception.

Finally, to state the obvious, for any errors this book might contain, the responsibility is completely my own.

1 | Introduction

If he could see Lake Victoria today, John Speke would probably stare in shock and disbelief. The once clear, life-filled lake is murky, smelly, and choking with algae. The ecological health of Lake Victoria has been affected profoundly as a result of a rapidly growing population, clearance of natural vegetation along the shores, a booming fish-export industry, the disappearance of several fish species native to the lake, prolific growth of algae, and dumping of untreated effluent by several industries. Much of the damage is vast and irreversible. Traditional lifestyles of lakeshore communities have been disrupted and are crumbling. There is a consensus among scientists that, if an accelerated push to save the lake is not made soon, this much-needed body of water will cease to sustain life. (Nancy Chege, 1995)

DEFINING THE PROBLEM

In 1994, the Dutch fish ecologist, Tijs Goldschmidt, wrote a bestseller entitled *Darwin's Dreampond*, in which he lamented the mass extinction of native haplochromine cichlid fishes of Lake Victoria. Goldschmidt squarely places the blame for this sorry ecological catastrophe on human interventions in the aquatic ecology of Lake Victoria in East Africa, motivated by the greed to increase the supply of the highly prized Nile perch at the expense of other species.

Then, exactly ten years later, the well-known Austrian film maker, Hubert Sauper (2004), produced a widely acclaimed documentary entitled

Darwin's Nightmare, featuring the impact of the ugly realities of the globalized fish food supply network on lakeshore communities around Lake Victoria. Reminiscent of the colonial era, the documentary exposes the paradox of abject poverty among fishery-dependent communities in the midst of plentiful fish food. Fishers are producing the best fish in abundance for export to feed mostly rich Europeans, while the actual local producers are condemned to subsisting on rotting Nile perch carcasses. All the while, local producers are being haunted by the ravages of HIV infections, homelessness, prostitution, violent crimes, drugs, and alcoholism. The film shows how homeless kids sniff toxic smoke from melting plastic packaging, discarded by owners of fish-processing facilities; this is their way of escaping the cruelty of social iniquity by getting high, albeit temporarily. Sauper sees a double irony in this situation. On the one hand, cargo planes fly in from Europe loaded with war materiel destined for the various warring factions in the continent, and then they fly out to Europe loaded with fish fillets from the lake. On the other hand, the World Food Program pleads with the international community for $17 million to feed 2 million starving Tanzanians at a time when the country is exporting planeloads of highly prized freshwater foodstuff.

In their different ways, Goldschmidt and Sauper addressed the anti-ecological and antihuman aspects of appropriation of natural resources and human labor as a function of global capitalist production. The book and the documentary together clearly demonstrate the simultaneity of aquatic resource depletion and Lake Victoria's vulnerability to hydro-ecological degradation, as well as the consequent mass immiseration resulting from the commodification and globalization of the lake's hydrobiology.

Of course, the story of Lake Victoria is the story of all lakes in Africa; indeed, in all developing countries. Even though lakeshore communities have always used lacustrine resources, it has been in the second half of the 20th century that inland-capture fisheries in developing countries have grown in prominence in the overall global food ecology.

There are two main reasons for this development: First, thanks to globalization, the ever expanding appetite of affluent global northerners for exotic food sources has made fish the most traded animal species on earth. This has spurred global fish production on unprecedented scales, rising from 12 million metric tons per annum in the 1950s to 143.6 million metric tons in 2006. In monetary terms, $91.5 billion worth of fish and fishery products were globally traded in 2006 (FAO 2008b). The veritable outcome of fish exploitation on this scale is serial depletion of ocean- capture fisheries. According to the FAO, 80 percent of global marine fish species are now either depleted, overexploited, or fully exploited without room for further expansion. In light of this serial depletion of marine-capture fisheries, producers and fish traders have had to migrate to inland waters to close the growing gap in global fish supply.

Second, cash-starved developing countries have targeted inland-capture fisheries as important sources of exportable commodities, in which case the export of fish and fishery products from inland waters has now become a generator of much-needed hard currency. For example, net exports of fish and fishery products from the global south to the world increased from $2.9 billion in 1978 to $27 billion in 2008; exports of fish and fishery products from the global south to the global north now represents 50 percent of total world fish export in terms of value and 60 percent in terms of weight (Napolitano, 2010).

The 304 million natural freshwater lakes in the world (covering 4.2 million km^2) and the 77 million artificial reservoirs and farm ponds (covering 335,000 km^2) are now fully integrated into the global fishery mode of production. This means that the interest of the more than 56 million people who derive their main livelihoods from inland fisheries must give way to hard currency generation (Welcomme et al., 2010).

Thus, globalization of the fishery mode of production has an inordinate impact on the status of the poor, as inland freshwater bodies are invaluable to the livelihoods of subsistence and artisanal fishers. Particularly in Africa, where terrestrial resource depletion, drought, and desertification, as well as human overpopulation are becoming problematic, lacustrine resources take on an added importance in the overall food ecology. Africa is richly endowed with both vast oceanic and lacustrine resources. However, it is the continent's inland water bodies that are crucial to the livelihoods of the poor for several reasons.

Even though a large number of Africans depend on marine fishing, coastal communities have not been able to fully develop capabilities to venture out farther to the deep oceanic fishing grounds. In fact, African marine resources have hitherto been subjected to international predation by long-distance foreign vessels without generating long term benefits to coastal communities. After all, most Africans live far from seas and oceans without access to modern distribution networks of highly perishable seafoods. This reality makes the continent's inland fisheries critical to the overall food ecology.

Next, the vast physical barriers in the continent insulate inland water bodies from foreign predation. Such a situation allows local communities to sustainably use them, unless national elites allow foreign capital to make an inroad into inland fisheries, as has happened with respect to Lake Victoria.

Finally, inland water bodies provide a natural safety net to millions of Africans who have no access to land or urban employment. The open-access character of the lacustrine fishery regime allows the poor to derive livelihoods in the fishing sector, especially during emergencies when household food sources are in short supply due to unexpected shocks or crop infestations. Moreover, inland fisheries serve a variety of purposes in terms of providing employment, additional or alternative sources of income, an inexpensive source of protein, and essential macronutrients, fatty acids, minerals, and vitamins (Welcomme et al., 2010).

The above-stated reasons explain why the state of fisheries in Lake Victoria is of extreme importance to lakeshore communities and why the lake's state transition has alarmed researchers and fishery managers. But, as indicated, the hydroecological condition of the lake must be placed within the context of how integration of Africa and other developing countries into the global network of food production and consumption has profoundly transformed the character of the global fishery mode of production. Africa, as a whole, is richly endowed with lakes. The continent has 670 large natural lakes where fisheries are widely utilized by the poor. The continent's 13 largest natural lakes alone cover 176,000 km^2 in area, which is equal to one-third of the area of the North Sea. Moreover, there are numerous man-made lakes in the continent suitable for fish production. The seven largest artificial lakes have the potential to produce between 188,000 and 2.2 million tons of fish annually; the smaller man-made lakes can add between 30,000 and 65,000 tons a year to the total harvest (Pitcher, 1995).

In West and Central Africa alone, more than 227,000 full-time fishers annually harvest 540,000 tons of fish from inland water bodies, worth $295 million. There are another 400,000 to 500,000 persons involved in the secondary sector, engaged in fish processing and trade, packaging and transporting, and providing other services as well. If the full potential of the inland fisheries in West and Central Africa alone are fully exploited, the annual production could reach 1.34 million tons, with a total value of $750 million (Thieme et al., 2005). The 12 member states of the Southern African Development Community have witnessed a sharp increase in their inland fisheries production, from 168,000 metric tons in 1961 to 635,000 metric tons in 1997.

For the whole of Africa, some incomplete data suggest that the contribution of inland fisheries to total fish production over the past 50 years has progressively increased. In the 1950s, around 600,000 metric tons of fish were annually harvested from African inland waters, which was more than the harvests of the Americas and Europe combined (Fryer and Iles, 1972). In 2000, inland-capture fishery production in Africa was 2.2 million tons, rising to 2.5 million tons in 2008, representing a quarter of the total global inland-capture fisheries.

In fact, real fishery production could be twice the amount suggested by the nominal statistics if catches from minor lakes, rivers and streams are properly included in the computation (Welcomme, 2003; Welcomme et al., 2010). These increases in African inland fisheries' production reflect the urgency for hard currency generation on the part of the ruling elites, as well as the rise in the number of people seeking livelihoods in the fishing sector for lack of opportunities in agriculture or industry (Gordon, 2005; Weyl, 2001; Thieme et al., 2005). To be sure, the exponential growth of demand for aquatic resources is the result of the interplay of deliberate state policy, population pressure, and the rising living standards of the domestic elites, coupled with rising international demand for foodstuffs. In addition, regional increases in aridity in response to hydroecological variability, in conjunction with income differentials between terrestrial and aquatic systems, have naturally made all African lakes magnets for settlers and migrants. For example, in the Kigoma region of Lake Tanganyika in Tanzania, the average artisanal fisher makes $340 a year compared to the national per capita income of $190 (Molssa et al,. 1999).

Most of Africa's productive lakes are located in the Great Rift Valley, where Lakes Victoria, Tanganyika and Malawi are the leading fish producers. These are the first, second and third largest lakes in Africa, respectively, and are the second, fifth and ninth largest lakes in the world, respectively. A focused study on any one of these 3 Great Rift Valley lakes can reveal a lot about the general state of African lakes. After all, as the various terrestrial ecosystems in and around the Great Rift Valley continue to degrade in response to a combination of increased human activity and natural variability, more and more people are bound to congregate around the lakes in search of livelihoods.

These 3 great East African lakes share a number of characteristics defined in terms of tropical locations, geomorphology, limnological complexity, biological diversity, and sensitivity to both natural and anthropogenically induced changes. Because of their sheer sizes, coupled with their fabulous aquatic riches, any one of the Great Rift Valley lakes can serve as an ideal biomonitor of the general state of ecosystem integrity, biodiversity, productivity, sensitivity, and vulnerability. In a way, they are the proverbial "coal mine canaries" of the aquatic ecology. Tradition has that coal miners used to take caged canaries along into tunnels, to serve as biomonitors in the event there were dangerous gas leakages. Canaries can sense imminent danger, so they would stop singing and making noises, or would show signs of stress, if they sensed that something bad was about to happen. This indicated that it was time for miners to exit the tunnels. In the same way, shrinkages in lake size, depletion or extinction of aquatic species, and lacustrine degradation are crucial biological indications that something serious is taking place.

TIME TO TAKE ACTION

It is time for understanding and action. This is the purpose of this volume. Even though the bulk of my study is centered on Lake Victoria, the discussion strays to other satellite lakes in significant ways in order to fully capture all dimensions of hydrobiology and socioeconomics that are germane to the fishery mode of production.

In embarking upon this project, my intention is not to empirically reproduce what has already been produced, but rather to grapple with some epistemological issues that are relevant to understanding the fishery mode of production in and around the lake in question. After all, there is not much that one could add to Goldschmidt's and Saufer's moving depictions of the events that have scarred the hydrobiology of Lake Victoria and consequently thrown the lakeshore communities into utter destitution. But, however powerful the microdescriptions of the hydrobiological state of Lake Victoria and the socioeconomic conditions of lakeshore communities are, neither Goldschmidt nor Saufer can tell us why these circumstances have occurred in the first place. The basic question they have not answered is: Given the inherent nature of capitalism to externalize costs to the environment and poverty to society, while internalizing the benefits to the perpetrators, a situation contextualized by the lower status of the littoral states in the global division of labor, how would the hydroecological integrity of the lake and the welfare of the lakeshore communities be protected?

This volume is dedicated to answering this central question. The methodology employed is materialist. The master narrative is historical. And the orientation guiding the analysis is transdisciplinary, informed by the laws of thermodynamics and evolutionary biology. My epistemological point of departure inheres in the recognition that we will not be able to understand the causal link between the mass extinction of the haplochromine cichlids of Lake Victoria and the mass destitution of lakeshore communities unless we fully grasp how the fundamental logic of global capitalism works. At the very basic level, this requires proper understanding of the organic connection of socioeconomics to ecology. This position presupposes the unity of knowledge.

The internationally renowned American evolutionary biologist, Edward O. Wilson (1998), points out that researchers and commentators frequently commit glaring errors in their assumptions, analyses, and conclusions, not only because they are not properly trained to perceive the natural world as an integrated whole, but also because they are not trained to understand the organic interplay of evolutionary biology and socioeconomics.

The chief source of these errors is the increasing fragmentation of knowledge and the overspecialization of educators and researchers, often forgetful of the transdisciplinary nature of knowledge. Locked up in their separate intellectual compartments, the natural scientists and the social scientists are bound to think of the two separate worlds as if they are wholly unrelated. As Wilson correctly points out, just as a person trained in biochemistry may not know much about trade with China, a person trained in the humanities would not know anything about the phylogenic origins, molecular biology, or cellular physiology and food requirements of fish species.

To remedy this pitfall, Wilson proposes "consilience" as a corrective orientation requisite for the unification of knowledge. This involves the integration of principles from across the disciplines in ways as to capture the essence of knowledge in all its dimensions, in which the inductions from the natural sciences and the humanities are interlinked in crucial ways that reflect the real world. Wilson sees the real obstacles to achieving consilience as embedded in the epistemological confusion, scientism, simplism, conflation, and ontological reductionism of intellectuals, think-tank gurus, and media interrogators, all of whom are purveyors of contemporary scholarship.

This problem is manifestly pervasive in the humanities in general and neoclassical economics in particular. Wilson (1998: 39) puts the problem in these terms:

> Without the instruments and accumulated knowledge of the natural sciences: physics, chemistry and biology, humans are trapped in a cognitive prism. They are like intelligent fish born in a deep shallow pool. Wandering, restless, longing to reach out, they think of the world outside. They invent ingenious speculations and myths about the origins of the confining waters, of the sun, the sky and the stars above, and the meaning of their own existence. But they are wrong, always wrong because the world is too remote from ordinary experience to be merely imagined.

Wilson's epistemological position with respect to the danger posed by the fragmentation of knowledge and the overspecialization of public intellectuals is unassailable. However, he does not tell us how the unification of knowledge can be achieved within the context of the prevailing social order, which owes its growth and expansion to the increasing fragmentation of knowledge itself and to the microsegmentation of capitalist production.

To rectify some of the epistemological shortcomings inherent in conventional scholarship, I take Wilson's insights one step further to elucidate the fact that there are two logics, which move in tandem, central to the organization and evolution of human civilization.

The first is the logic of survival, which is innate to all humans. For all members of the animal world, including humans, survival is the first order of existence. In fact, Wilson (1998) himself acknowledges that humans are governed by the rules of evolutionary biology in their struggle to survive to the next day because "the brain is a machine assembled not to understand itself but to survive." Wilson is by no means suggesting essentialization of the biological instinct of humankind to survive, but rather describing a simple biological fact.

As the chapters in this volume will show, the logic of survival supplies a partial (but only partial) explanation for the aquatic resource depletion and degradation in Lake Victoria. Increases in the number of fishers and their dependents have necessitated intensification of fishing activities, not necessarily to make profit, but rather just to survive. Ottar Brox (1990) has correctly characterized this phenomenon as horizontal growth. As human offspring establish their own households, they compete with their parents to harvest fish in the same lakes. This natural overcrowding is compounded when people migrate to lakeshores, following terrestrial degradation in their biosocial region, and thus end up themselves being fishers competing with the "native" communities for the same resources.

The increasing congregation of people around areas such as Lake Victoria, however, has not occurred in a vacuum, but rather in the context of an evolving social order. In the epoch of capitalism, the primary driver of this social order is the logic of accumulation. Like the biosocial necessity of survival, the logic of accumulation requires an uninterrupted flow of natural resources for its continuous expansion: in this case, the continuous harvesting of fish.

Here, the second logic of accumulation takes place in the context of continuous injection of capital into the fishery sector in the form of modern technologically efficient fishing boats, and modern fishing nets. The process of capital injection into the fisheries sector is continuous until marginal costs equal marginal benefits—or until there are no resources available in the fisheries sector.

This is what Brox (1990) calls vertical growth. Because the necessity of survival and the logic of accumulation interactively move in tandem, the impacts on the lacustrine ecology and human welfare are cumulatively far-reaching. It thus follows that lacustrine depletion, aquatic species extinction, ecological degradation, and human destitution are the unavoidable results of the interaction between the two logics unless we fundamentally change the drivers of both logics simultaneously. To accomplish this requires understanding of the basic fact that unsustainable resource exploitation, social inequality, and poverty occur in the context of the power of market forces and politics.

In the core chapters of this volume, I will show how the biosocial necessity of survival and the capitalist logic of accumulation have in combination led to species extinction and the ecological disintegration of Lake Victoria, as well as inevitable mass destitution. While this first chapter clarifies the conceptual and methodological frameworks, Chapter 2 explores the impacts of demand conditions on the state of Lake Victoria. Chapters 3 and 4 tackle how the insertion of negative externalities into Lake Victoria and human biomanipulation of the lake's species composition have adversely affected supply conditions. Chapter 5 deals with structural issues. The final chapter attempts to show a theoretical way forward by synthesizing and refining the arguments advanced in the preceding chapters.

A PARADIGM SHIFT TOWARD AN ENTROPIC WORLD VIEW

Following the first comprehensive survey of Lake Victoria in 1927 and 1928, British fish ecologist Michael Graham in 1929 sounded the alarm that the commercially fishable species in the lake were in peril of collapse—unless drastic measures were taken to control human predation on them. To this end, he proposed a sweeping state control of fishing power, as well as public control of the importation and distribution of all fishing gear.

Over the past 80 years, the same managerialist theme has been orchestrated over and over again by fishery managers and researchers in their attempts to address the depletion and extinction of fish species in Lake Victoria (Ogutu-Ohwayo, 1990; Balirwa, 1998; Awange and Ong'ang'a, 2006; Ochumba and Kibaara, 1989). However, the exclusive focus on managerialism glosses over the fundamental problems concerning the real causes of the lake's aquatic resource depletion. As the old German saying goes, "a focus on regulation alone, in whatever form, is akin to beating the sack for fear of the donkey."

The problem is much more structural, rooted in the capitalist mode of production, engendering conditions that simultaneously drive the toiling masses to desperation and encourage the rich to invest in modern operations. Excluded from the equitable dispensation of material wealth, the poor have no option other than to increase efforts to harvest more fish to feed the growing members of their families. Likewise, capital accumulators, goaded by the urgency to make more profit, continually update their operations in order to haul ever greater quantity of fish from the lake. Thus, the poor (out of desperation) and the rich (out of greed) have refused to worship the idol of managerialism; instead, they have opted for increasing the horizontal growth of effort and the vertical growth of investment.

The preceding observation suggests that aquatic resource depletion and mass extinction of species in Lake Victoria are the results of flaws in conventional fishery management philosophy. This stems from the commitment of fishery managers and researchers to pseudoscience and avowedly ideological disposition that serves to obscure their class limitations and epistemological poverty. Believers in managerialism usually import concepts and theories from the natural sciences to give their research product seeming concreteness and empirical value. As G.L.S. Shackle (1972: 360) observes, by transforming itself into the science of all sciences, economics has perfunctorily breached the gulf between the humanities and the physical sciences, where "Economics has veritably turned imprecision itself into a science; economics, the science of the quantification of the unquantifiable and the aggregation of the incompatible. It has followed this road at so violent a gallop that much, which is of significance and influence, has been trampled on; much territory has been claimed, which cannot be held."

Indeed, since the dawn of the scientific revolution five centuries ago, the conventional epistemological approach to ecological resources has essentially evolved in tandem with the mechanical philosophy of the Newtonian age. This mechanical worldview presupposes the knowability and domesticability of nature by human beings—thanks to the supposed scientific revolution and the consequent technological progress. In keeping with the tradition of this mechanical world view, neoclassical economists have long been preoccupied with quantification and modeling processes to establish universal axioms of economic growth. In Shackle's (1972: 3) words:

> When the time came to invent economic theory, a number of established, exact and thoroughly explored modes and schemes of thought were ready to hand. They pressed themselves upon the inventors' minds. They gave to economic theory their own shape, derived from other contexts, devised for other questions. The procedure of invention was often to accept such self-suggesting analogy and make the economic questions vivid not to ask what is peculiar and essential in economic questions but to ask what is the essential nature of the world to which those questions belong. Thus economic theory took on a character belonging to the manipulable, calculable, external world of *things*, not a world of the conscious mind in its eternal station on the edge of the void of time, the conscious mind whose being consists precisely in the endless gaining of knowledge. *Knowledge and novelty*...was given a casual and subsidiary role.

Imported from physics to neoclassical economics, the central organizing idea of neoclassical economics is based on Newtonian mechanics, in which all economic agents are expected to obey the laws of conservation because they are assumed to operate in a market defined by homogeneous rationality and perfect information (Georgescu-Roegen, 1971 and 1975; Merchant, 2005; Princen, 2006; Daly, 1990; Daly and Farley, 2003). The equilibrium models that result from the undertakings of economic agents are supposed to determine the present physical conditions of production, from which future physical conditions can be extrapolated mathematically.

In this mechanistic epistemology, economic phenomena are not only predictable, but also reversible, through scientific manipulation and control of economic data and effort, in which case economic events are assumed to operate in ways that return the system as a whole to its previous quality. But this presupposes the relative permanence or stability of these conditions of production, including ecological conditions. Since the biophysical conditions of production are by no means deterministically constant, the knowledge of present conditions cannot assure us of anticipated knowledge of the yet unformed future (Geogescu-Roegen, 1971 and 1975; Shackle, 1972; Daly and Farley, 2003; Smith and Smith, 2001; Cox, 1993; Perrings, 1987; Pitcher, 2001). Indeed, spontaneity, rather than logical ordering, is what underpins all physical and biological evolutions. The interacting components that make up complex ecosystems are not necessarily symmetrically organized, but rather dance on the edge of chaos between order and disorder, between continuity and change.

Additionally, the spontaneous organization of complex ecosystems is nondeterministic, haphazardly surging forward through a series of negations and adaptations without central coordination mechanisms or general equilibrium. Because ecosystems are made up of a myriad of structural components, understanding of one or some components does not mean that we do understand the universe of all ecosystems (Smith and Smith, 2001; Daly and Farley, 2003; Cox, 1993; Pitcher, 2001; Perrings, 1987).

Following World War II, some authors, undiscerning of the inherent limitations of the mechanistic epistemology alluded to above, advanced resource exploitation theory to determine how wild populations could be sustainably harvested (Cox, 1993; Smith and Smith, 2001). Undergirding the theory was the conviction that sustainable yield could be had indefinitely by prior determination of the carrying capacity of the wild stocks in question. The underpinning logic driving this analysis was that, by regulating exploitation of a target population, an equilibrium condition could be maintained between ecological carrying capacity and population density, which would theoretically make sustainable yield a useful fishery management objective.

THE GORDON-SCHAEFER MODEL

Since harvesting above the sustainable yield curve leads to depletion and/or extinction and harvesting below the curve leads to overcrowding and diminution of food resources available to the target population, fishery managers can stabilize the population at the sustainable yield curve through lowering population density by removing a given number of individuals when overcrowding occurs, or by curbing access to fisheries when overfishing occurs. This would require careful monitoring and proper computation of survival and fecundity levels of the species in question. This way, the equilibrium condition between recruitment to the population and harvest of individual species could be maintained. Unless the threshold of the target species is violated, the species harvested could be replaced by increased reproduction of the remaining adult species. After all, if not used by humans, the surplus populations above the sustainable yield curve would naturally die off anyway as a result of intraspecific competition, predation, and starvation. Thus, in theory, the entire harvestable surplus above the threshold level can be taken out from a lake without impacting the status of the target species. After all, higher levels of population growth automatically indicate the need for higher rates of exploitation. This is what constitutes maximum sustainable yield (Cox, 1993; Smith and Smith, 2001; Daly and Farley, 2003).

In the 1950s, based on the aforementioned theory, experts in the fishery sector proposed and popularized the notion of maximum sustainable yield that could be had through scientific fisheries development. The notion was soon universalized as the Gordon-Schaefer model. The premise underpinning this Gordon-Schaefer model was and still is that the total population of a given fish population and the direction of its growth could be predetermined scientifically through advance identification, quantification, and monitoring of stock growth responses. Then, on this basis, harvest of fish could be regularized and managed in a sustainable manner.

The underpinnings of the model rest on the following presupposition: an equilibrium can be established between recruitment and stock levels on the one hand, and between the fishing effort and the standing crop of the species involved on the other, without causing depletion of fish stocks in order to achieve maximum sustainable yield (Clark, 2006; Cox, 1993; Weyl, 2001; Smith and Smith, 2001; Daly and Farley, 2003; Pitcher, 2001). Knowledge of fish population dynamics, determination of fish biomass, distribution of species, the rates at which yields are obtained, and effort levels exerted together were supposed to constitute the framework for fisheries management. Since the natural state of fish stocks is assumed to be knowable by studying the fish population dynamics prior to catching the fish, and the fish ecology is assumed to be stable or fluctuations of environmental factors are scientifically anticipated, a dynamic optimization of fish harvests could be established through quantification and modeling processes. Since the natural state of a given fish stock is the point of departure, this invites regulation of fisher behavior and the instruments of fishery production as a mediating mechanism to maintain the equilibrium relationship between stock recruitment, the number of fishers, and fishing capacity. In this hyperrational choice model, since the root causes of fishery resource depletion are intrinsically linked to economic behavior, the solution requires regulation and monitoring of economic behavior. There are several managerialist derivations from the Gordon-Schaefer model that have been attempted or suggested over the decades (Daly and Farley, 2003; Charles, 2001; Cox, 1993; Smith and Smith, 2001; Clark, 2006).

One approach to fishery management is the fixed quota system in which managers determine the amounts of fish that can be harvested in a given period. Obviously, this requires prior knowledge of the size, age structure, fecundity, and recruitment rates of the target species, as well as the actual or anticipated environmental changes. However, as many cases demonstrate, since the aforementioned factors are difficult to quantify in advance, the fixed- quota approach has proved to be inadequate to solve the structural dilemma in the fishery mode of production. Borrowing from Shackle (1972), even though experts claim to know about the "unformed future" by simply extrapolating its features from the recorded past, the truism is that knowledge about the past is inadequate to help us explain what has not been formed, because reason—scientific or not—is a prisoner of time, incapacitated by circumstances to anticipate the unformed future.

A second approach to fishery management involves regulation of fishing effort. Controlling the number of fishers, fishing days, length of fishing periods, and types and varieties of fishing gear are considered central to maintaining the sustainable yield curve. A variant of this approach to the fishery mode of production is the permit system, which predetermines the maximum amounts of fish that fishers are permitted to remove from the standing crops of the target populations in areas designated by management for control and monitoring purposes.

As well-versed evolutionary biologists (Wilson, 2002; Fryer and Iles, 1972; Cox, 1993; Smith and Smith, 2001) have pointed out, some of those involved in fish ecology have complicated what is otherwise a simple biological proposition by using mathematical axioms. By definition, these axioms are static because they either are based on past experiences or their preoccupation is with the natural state here and now from which all future projections are extrapolated. This glosses over the simple observable fact in evolutionary biology that ecosystems are complex, dynamic, and changing in nonlinear manner, characterized by spontaneity and adaptiveness. The source of this epistemological confusion is the presentation of the economy as a system.

Conversely, the environment on which the economy depends for throughput and sink is defined as a subsystem of the economy. Informing this thinking is what Daly (1990 and 1993) calls the neoclassical preanalytic vision. This vision, which passes as a science, is an ideology, which essentializes the utilitarian conception and rationality of man and justifies the unlimited capitalist accumulation.

In any event, the Gordon-Schaefer model is selective in orientation and reductionist in method because it cannot account for the interplay of human impacts, economic and biological overfishing, biotic succession, climatic shifts, catastrophic stochastic events, and the thermodynamic forces that are continually at work, which determine the evolutionary state of fish ecology. These sets of factors operate interactively on the fish ecology in ways that make predictions difficult, if not impossible, especially under conditions of the prevailing capitalist mode of production, where desperation of the many and greed of the few are its defining features.

Determination of the precise point where recruitment into the stocks equals the rates of mortality due to both harvest and natural conditions becomes illusory because, in the absence of certainty relative to carrying capacity of the environment and the fecundity of the populations in question, the maximum sustainable yield can be much greater than population growth as suggested by various mathematical models (Cox, 1993; Smith and Smith, 2001; Daly and Farley, 2003).

Moreover, this preanalytic vision glosses over two objective facts of life under the prevailing conditions: the necessity of survival, contextualized by poverty and desperation of the dispossessed masses; and the logic of accumulation, driven by greed and profligacy of the rich. These conditions are at the heart of the capitalist mode of production. Admittedly, this observation is hardly novel since Adam Smith had already pointed out the potential contradiction between the indigence of the majority and the opulence of the few, the mother of all contradictions under revolutionary capitalism. The poor, who have no alternative means of survival, are not going to stop fishing and risk starvation. The only things they own are time and their instinct to survive, complicated by the added pressure to feed the growing number of their families. So they will continue to expend all the effort they have to catch increasing amounts of fish of any size.

Proponents of the Gordon-Schaefer model do not seem to understand this necessity of survival because either: they do not bother to ask why people overfish knowing full well that the fish stock is being depleted; or, simply, their class position insulates them from the immediacy of poverty, demoralization, and desperation that the toiling fishers experience daily.

The logic of accumulation operates more or less in the same manner as the necessity of survival with partial exceptions. The cost of labor, subsidies, and the existence of other options determine whether capital stays or exits. In the early phase of fishing, owners of capital heavily invest in high-powered vessels and modern gear to maximize production. Over time, continuous investment in the modernization of fishing operations inevitably leads to overcapacity, which means hauling in more fish resources than the populations of target stocks can support.

Moreover, deceptive reporting of catches makes monitoring and assessing fish stocks virtually impossible. After all, fishery reports from operators are often taken at face value. As Tony Pitcher (2001) correctly notes, for example, even though the official report of global fish harvest in 1997 was 80 million metric tons, the actual collection of fish that year could have exceeded 160 million metric tons; add to the above figure the 27 million metric tons of by-catch that was discarded by operators.

As we shall see in Chapter 2, processing companies in Lake Victoria found it profitable to continue operations despite depletion of adult fish, because they shifted to processing immature Nile perch caught by underpaid fishers in light of the exceptionally high global demand for perch fillet, especially in the European markets. The veritable result has become biological overfishing, which refers to the gradual or sudden reduction in adult fish population below the depensation marker, which refers to a population dropping below the threshold for continued survival or recovery due to predation or other events, leading to the depletion of the population concomitant with the reduction of breeding adults. By definition, removal of a given amount of fish from the ecology by harvesters represents a reduction in the capital of the fish stock concerned (Fryer and Iles, 1972; Daly and Farley, 2003).

As the experience with the Nile perch in Lake Victoria has shown, the expenditure of more and more effort and the deployment of ever more sophisticated modern gear to haul fish means that increase in the fish yield leads to the removal of more capital from the stock. Since fish fecundity is intimately related to the biomass of adult fish in a given habitat, the removal of too many adult fish signifies the opportunity cost of lost eggs necessary to replenish the capital stock. The capture of juvenile fish and by-catch worsens biological overfishing since the fish are caught before they are able to reproduce themselves in order to maintain the original level of fish populations (Fryer and Iles, 1972; Cox, 1993; Smith andSmith, 2001; Daly and Farley, 2003). This makes establishment of the depensation marker impossible.

Fish ecologies and fish population dynamics are also affected by stochastic events, unrelated to overfishing and population density. As intimated earlier, fish ecology is a subset of complex local, regional, and global ecological systems, which produce many unpredictable extreme events. These events may include meteorological variability, change in pH levels, rise in water temperature, and change in the food web. Fryer and Iles (1972), for example, use the case of Lake Chilwa in Malawi to illustrate how external events can negatively affect fish population dynamics. In 1965, Malawian fishers hauled 9,000 metric tons of fish from Lake Chilwa in which tilapiine species accounted for 48 percent of the total. In 1967, however, a severe drought hit the region, resulting in sharp drop in lake level, leading to drastic reduction of catches and death of huge number of tilapias. As a result, in 1968, subsistence fishers managed to scoop a mere 100 metric tons of fish from the lake to which the tilapias contributed a mere 4 metric tons.

Thus far, the discussion invites 3 questions that challenge the Gordon-Schaefer model. First, biologically, how does one exactly pinpoint where the depensation curve lies? Complete information on survival, fecundity, recruitment, reproduction, size and age structure, number of individuals left behind after harvest required to maintain the exact density of the target populations, sex ratio, and new recruits to the harvestable species in exact proportion to the number of individuals removed from the populations is exceedingly difficult to obtain, much less to precisely determine the weight of each variable (Smith and Smith, 2001; Cox, 1993; Daly and Farley, 2003).

Second, environmentally, what are the chances that a model such as the Gordon-Schaefer model will account for the impacts of change in land cover, climate variability, and nutrient fluxes, as well as for the nature of species interactions, uncertainties, and surprises that are inherent to evolutionary biology and thermodynamics? Historical evidence suggests that fishery managers, guided by the Gordon-Schaefer model or its derivations, usually see harvestable stocks not as integral to the environment within which they live and reproduce.

Third, socially, how can one be certain about how much fisher folks and large commercial operators harvest from the standing crops of the target populations, or how can management be sure about what is taken out from a lake is equal to new recruits to the populations? Answering this question obviously requires complete understanding of the social context in which the fishery mode of production occurs. After all, natural resources, defined by an open-access regime, are invariably subject to complex sets of social vagaries inherent to the capitalist mode of production. As stated before, the poor cannot cease and desist from fishing in order to make ends meet, even under conditions of full knowledge that the stock of the fish in question is being threatened by serial depletion. In the absence of alternative means of livelihood, it is not an option for the poor not to continue harvesting. Similarly, owners of big money entrenched in the fishery sector are not bound to refrain from overfishing so long as marginal costs equal marginal benefits; they will exit from the sector only after the stock has been fished down to where marginal costs exceed marginal benefits. To compensate for expenditure in effort, big operators have the natural tendency to improve their technological capacity, coupled with the gross underpayment of labor. Neither the big operators nor the politicians who support them are bound to heed warnings of serial depletion of fishable species.

In light of the preceding considerations, it was no coincidence that researchers soon discovered that the Gordon-Schaefer model was impractical in Africa, precisely because constant catch and constant effort could not predictably be established, since fish production in African inland waters fluctuated from one year to the next in response to climate variability (Weyl, 2001). However, climate variability, although an important determinant, cannot by itself account for all the factors that make the Gordon-Schaefer model unworkable.

In the final analysis, the very logic of capitalist accumulation, which relies on the production of knowledge to justify the logic and the power to maintain it, is decisive. After all, the fishery mode of production is integral to accumulation since the $100 billion global fishing industry, supported by billions of dollars in subsidies, is a critical fulcrum of the trajectory of global accumulation. To be sure, as the statuses of the Peruvian anchovies, the North Atlantic cod, and the orange roughy demonstrate, there is interplay of human predation and natural forces at work undermining the integrity of the fish ecology. These shed light on why the biodiversity, productivity and limnology of Lake Victoria have been adversely impacted.

As the sources cited below collectively demonstrate, the fluctuating status of Peruvian anchovies offers a powerful illustration of how quantification and modeling the interaction of excessive human predation and natural phenomena can be a fanciful exercise under the Gordon-Schaefer model (Cox, 1993; FAO, Peru 2002; Charles, 2001; *Economist*, March 11, 1978; Niquen and Bouchon, 2004; Castillo et al., 2007; Westrup, 1998; Cavides, 1984; Line, 1998). The oceanographic features off the coast of Peru are such that the 300- by 30-mile-wide marine ecosystem is characterized by extraordinary biodiversity and superabundance of species, especially anchovies and sardines. The Humboldt Current, which extends from the middle of the Pacific Ocean to the Peruvian coastline, is responsible for the biodiversity and productivity of this ecosystem. When the Humboldt Current collides with the coastal shelves of South America, nutrient-rich cool waters bubble up to the surface, conveying carbon, nitrogen, phosphorus, and other essential elements to the phytoplankton upon which the anchovies, sardines and other species feed. The anchovies are pelagic phytoplanktivorous species, known for their exceptional fecundity and extensive school formations along the Peruvian coast.

In the 1950s, the abundance of the anchovies was such that fish experts projected that Peru could harvest and process 10 million metric tons of anchovies per annum without affecting the population dynamics of the species. The experts were not off the mark in their prediction, except that the computation was based on the then current abundance of the anchovies without taking into account that the impact of human predation and unanticipated natural events could have on the future status of the anchovies. They could neither determine in advance how many artisanal fishers and giant fishery operators would enter the field, nor could they foresee the effect of severe weather events on the species. What they saw was that the anchovies were prolifically breeding throughout the year along the entire Peruvian coast where planktonic food sources were abundant.

In any event, lured by the exuberant projection of anchovy abundance and the rising global demand for fishmeal and fish oil, fish operators rushed to exploit the anchovies. The rush had both horizontal and vertical dimensions. By 1970, up to 30,000 Peruvians were directly harvesting anchovies, and another 75,000 to 90,000 Peruvians were indirectly involved in the sector.

Along with the rising number of fish operators, big money was plowed into increasingly larger fishing vessels and anchovy processing plants. Before long, 650 giant industrial vessels and an equal number of artisanal boats began chasing the anchovies along the Peruvian coast and beyond, transforming Peru overnight into the biggest anchovy harvester and the biggest exporter of fishmeal and fish oil in the world, accounting for 40 percent of global fishmeal supply. As a result, the catch of anchovies soared from 3 million metric tons in 1950 to 12 million metric tons in 1964. Between 1950 and 2001, Peru exported over 264 million metric tons of fish, of which 209 million metric tons were anchovies. Indeed, in 2001 Peru earned $1.124 billion from export of fish, second only to the mining sector. The human predation of anchovy was so huge in scale that the FAO warned Peru in 1970 that the anchovies were being decimated by overfishing and overcapacity of fishing fleet, since fishmeal plants were processing 16 million metric tons of anchovies and sardines per annum.

Moreover, as if the scale of human predation was not big enough, an unexpected natural event began to take heavy toll on Peru's anchovies. In 1972, the El Niño phenomenon wreaked havoc along the entire Peruvian coast, destroying the food supply of the anchovies. As indicated earlier, the fate of the species is tied to the nutrient-rich waters of the Humboldt Current. The anchovies are known to prosper under weak El Niño conditions and cooler waters, when the normal temperatures are in the range of 14 to 23 Celsius. Modulated by the Humboldt Current, upwelling of such nutrients as nitrates and phosphates enriches the productivity and supply of phytoplankton.

During El Niño occurrences, however, temperatures rise and the upwelling of warmer and nutrient-poor waters undercut the food supply. Whenever the westward trade winds shift direction, warmer waters from the western Pacific Ocean move to the Peruvian coast, trapping the nutrient-rich cool waters underneath. Without these waters reaching the surface, the phytoplankton community collapses. Running out of food supply, the anchovies either die from starvation or make vertical migration in search of food. Not only anchovies are affected by the El Niño phenomenon: seals, sea lions, and sea birds, which feed on anchovies, are also severely affected.

For example, during the 1982-83 El Niño event, as many as 18 million pelicans, boobies, penguins, and other marine mammals (which feed primarily on anchovies) are believed to have died from starvation. Prior to the 1997-98 El Niño, there were 40,000 fur seals and 150,000 sea lions on the Peruvian coast, which fell to 2,800 and 28,000, respectively, following the El Niño event.

Thus, the biological state and productivity of anchovies are highly sensitive to weather events. For example, annual anchovy catch precipitously dropped from 12 million metric tons in 1970 to 1.3 million metric tons in 1973. Since then, the status of the anchovies has continued to fluctuate in response to the severity and duration of the El Niño phenomenon and human predation. Following the 1997-98 El Niño, the harvest of anchovies was barely 1.2 million metric tons. Between 1999 and 2006, the annual catch of anchovies wildly fluctuated between 1.2 million and 6.4 million metric tons.

Even though natural events have been both disruptive and destructive to the anchovies, vested state and private interests have been an important factor in the fate of the species. Demonstrating the importance of the anchovy sector to public revenue, the Peruvian state nationalized 115 fishmeal factories, along with 1,400 fishing vessels. This state interest alone made the anchovies the single most important species for hard currency generation. Even when scientists warned against the potential collapse of the anchovies, the state resisted meaningful protection measures. In 1978, for example, the Pinochet regime fired the director of the country's Marine Research Institute for recommending a ban on anchovy harvest until the species sufficiently recovered. Insofar as state interest is concerned, it has proved difficult to give up anchovy exports, since these exports still continue to play an important role in hard currency generation. In 1977, for example, the harvest of 1.5 million metric tons of anchovies produced 300,000 metric tons of exportable fishmeal, accounting for 10 percent of Peru's total export earnings.

The event surrounding the state of the anchovies is not new. Rather, it is a familiar one we have heard many times. The story of the North Atlantic cod fishery comes to mind. Because the cod was so abundant in the 1950s, Canadian scientists never thought that the cod was ever exhaustible. What the experts did not know was the fact that small-scale inshore cod harvesters and occasional migratory fishers were limited by the state of their technology and social needs to use the cod resources sustainably. As a result, the cod continued to flourish in the North Atlantic region. Then lured by the story that the region had inexhaustible cod resources, distant fishing vessels came from as far as Russia and Germany, armed with on-board processers and freezers to exploit offshore cod stocks. The giant factory ships harvested, processed, froze, and stored the fish, day in and day out, without ever coming onshore. Overcapacity of fishing fleets, overestimation of cod abundance, economic incentives from heavy subsidies, and discarding of small cod made any computation of fish mortality unfeasible. Under assault from national inshore harvesters and international offshore bottom trawlers, catches of cod off of Newfoundland reached a peak of 800,000 tons in 1968.

By 1975, however, the North Atlantic cod populations showed signs of near exhaustion, compelling Canada and the United States to assert economic sovereignty over the 200 miles of their territorial waters in order to keep foreign fish harvesters off the area, whereby the cod were expected to recover. However, recovery of the cod fishery was slow as indicated by the catch of only 139,000 tons in 1978.

The matter was made worse when Canadian operators, supported by generous public subsidies, saw the answer to the sharp decline in cod fishery in the modernization of their fishing fleet, so that they could venture farther out into the ocean in order to chase more of the cod. Even though inshore fishers complained to authorities about the implications for their livelihoods of introducing new fishing technology, the government ignored their pleas and, instead, poured more resources into the construction of modern stern trawlers, armed with dragger nets. The government rationalized its protrawling policy by citing the importance of keeping jobs.

Dragger nets are a football field long and can sweep the ocean floor. In the process, dragger nets not only sweep up entire schools of cod and every other living species, but they also destroy fish eggs. The draggers became particularly deadly when descending on large aggregations of cod during spawning period. In any case, with the aid of these baglike nets, Canadian fishers were able to land 250,000 tons of cod per year in the 1980s.

However, in 1992, the entire cod fishery suddenly collapsed, prompting the Canadian government to impose a moratorium on fisheries off Newfoundland. When small-scale commercial inshore fishery was permitted in 1998, it was found that the cod populations were far from recovery to support fishery at any level. Then, in 2003, Canada closed cod fisheries in the area indefinitely. The lives of Newfoundlanders were devastated; 42,000 fishers lost everything, and all the draggers and cod processing plants became idle. The Federal Canadian government spent over $2 billion on social welfare and on retraining the displaced fishers for other jobs (Population Council, 2005; Smith and Smith, 2001; Cox, 1993; Canadian Green Peace, n.d.).

The state of the orange roughy fishery is another textbook case of how the logic of accumulation drives certain species to serial depletion and/or extinction. The orange roughy is a rare fish species. Generally found around seamounts, which are underwater oceanic mountains and plateaux between 500 and 1,500 meters deep, the orange roughy is found off the continental shelf in the vicinity of Australia and New Zealand, the western Indian Ocean, and the North Atlantic Ocean. The average span of maturity for the species is in the order of 20 to 40 years and the life expectancy of the species can be up to 150 years. Because of its low fecundity, the species is extremely vulnerable to depletion in the face of unsustainable human predation. Because orange roughy tend to travel together and congregate around seamounts when feeding and spawning, they are ready-made target for trawlers who could scoop tens of thousands of individual orange roughy in one swoop.

The discovery of the orange roughy was accidental when an Australian fisher suddenly descended upon a group of the species in the 1970s. However, the overexploitation was conscious and deliberate; fishers quickly deployed their destructive fishing trawlers to scoop the orange roughy, in utter disregard of the ecological and biological consequences of their actions. The result was a voracious appetite for the new fish.

The roughy was particularly sought after in the United States, which began importing 19 million pounds of the fish a year, accounting for 90 percent of global orange roughy harvest. With such demand, the fishers could not resist widening and deepening the hunt for the species. In less than four years, the orange roughy populations near Australia and New Zealand were almost wiped out. To make up for lost import of the species from the Australian region, trawlers promptly migrated to the western Indian Ocean, where the U.S. market now depended for the supply of orange roughy; even with Namibia accounting for two-thirds of the import, serial depletion saw the import of the fish fall to just 2 percent by 2002 (Smith and Smith, 2001; World Wildlife Fund Report, December 2003).

These three cases illustrate the observation that the existing quantification modeling processes cannot adequately capture the consequences of the interplay of human predation and natural events on the fish ecology. In fact, in light of the conceptual and empirical cracks found in the Gordon-Schaefer model, a number of authors have tried various derivations of the same model, hoping to promote sustainable fisheries. However, the simulated derivations, too, have faltered and the notion of sustainable fisheries proved to be problematic for many fishing communities.

To rescue the model, fish experts in such global institutions as the UN Food and Agriculture Organization (FAO) now claim the ecosystem approach to fishery management to be the right model to adopt. The presumed novelty of this approach inheres in the contention that fisheries cannot and should not be isolated from the general ecology in which fishing activity occurs. Because the phrase "ecosystem sustainability" is thrown virtually into every official document, the concept has today taken on a semireligious connotation. In a nutshell, the underpinnings of the ecosystem approach presuppose the urgency to develop strategies for an integrated management of biological resources, water, and land resources in ways that would promote conservation, equity, and sustainable uses.

In support of this contention, Anthony Charles (1994 and 2001) attempts to ground the sustainability of fisheries in an overarching framework of "integrated sustainability" of which ecological, socioeconomic, community, and institutional sustainabilities become its organic constituents; stock assessment would guide fish managers to determine when and how much to harvest from a given population of a given species. Ecological sustainability in the fisheries sector requires maintenance of fish stocks at levels that allow fish species to flourish in their natural conditions. This criterion imposes restraint on harvesting in favor of conservation. Continual stock assessment, based on precautionary principles in order to forestall the consequences of uncertainty, would guide the approach to ecological sustainability of fisheries. As Charles puts it, since sustainability is defined as an ecosystem's ability to maintain itself in the face of continuous demands made on it, concerns over its integrity are understood in terms of 3 operative notions: vulnerability, sensitivity, and resilience. Vulnerability refers to potential or actual exposure of people or biotic resources to high degrees of stresses and shocks such as prolonged droughts and floods; sensitivity is the measure of a system's ability to respond to external perturbations; and resilience measures a system's ability to bounce back after an unfavorable event occurred. Thus a vulnerable ecosystem is one that exhibits high sensitivity and low resilience, while a holistically stable ecosystem displays low sensitivity and high resilience.

Socioeconomic and community sustainability call for the imperative to understand the material needs and cultural environment of fishing communities, as well as the participation of fishing communities in the decision-making processes regarding the distribution of benefits and costs, conservation, representation, and access rights. Institutional sustainability is supposed to address issues of structure, agency, financial support, regulation, and policy direction. Based on the analysis and assessment of these four facets of sustainability, fishing communities, and fisheries management can develop integrated strategies to deal with resource system complexity, prepare adaptive management steps to cope with ecological uncertainties, accompanied by the installation of an appropriate fishing regime via the definition of access rights, and provide a comprehensive planning of economic diversification in order to reduce the pressure on fish stocks by providing fishermen alternative employment opportunities and incomes.

A variant of the ecosystem approach is the sustainable livelihood approach, which seeks to reconceptualize fishery development within an overarching framework of natural resources management, rural poverty reduction, geographic and occupational mobility of people, and the social context that encourage opportunistic exploitation of resources (Weyl, 2001). Given the fluctuations in fish stocks, which result from microregional variations in climatic conditions that influence biological interactions and stock dynamics independent of fishing effort, proponents of the sustainable livelihood approach begin with the recognition that the equilibrium assumptions about Maximum Sustainability Yield (MSY) are bound to be violated.

This view, as we shall see in Chapters 3 and 4, is supported by how climate-induced changes in primary and secondary production in Lake Victoria has led to substantial fluctuations in fish stock abundance and spatial distribution. It stands to reason that the Gordon-Schaefer model cannot account for climate-induced fishery fluctuations, nor can it capture the dimensions of geography, occupational mobility, and sociocultural conditions of the people involved in fisheries. Moreover, it will be difficult to locate the equilibrium point where returns from fisheries is supposed to equal the cost of harvesting the stocks, since signals of stock abundance/scarcity may be distorted by stochastic events and public subsidies or by faulty data about stock assessment and fishing effort (Weyl, 2001).

PROBLEMS WITH APPROACHES AND SOME SUGGESTIONS

Although the ecosystem and livelihood approaches are a significant improvement on the traditional Gordon-Schaefer model, they are still wanting for several basic reasons.

First, the methodology of stock assessment is fraught with uncertainties, surprises, miscalculations, and erroneous computations, concerning the point at which a given stock can be harvested sustainably. The concept of MSY presupposes that factual determination, based on quantification modeling processes, about the position of the equilibrium point, can be made prior to action in which social demand would equal natural stock supply. Put differently, sustainable harvest is the net removal of fish from a given stock without affecting the natural reproduction and maintenance of the stock's populations. In theory, harvest rates on the sustainable yield curve would be matched by equal growth rates in the populations of the stock, and this would be known through stock assessment. However, as Daly and Farley (2003) have pointed out, no one can be certain about the definitive populations of the species in question, much less about its reproduction and mortality rates. Moreover, such external interventions as climate change, habitat fragmentation and destruction, diseases, predator/prey relations, and changes in trophic profile can render any stock assessment meaningless, which means that the equilibrium point may have long changed while harvesters are fully deployed to extract the resource.

Second, the fundamentals of capitalist market expansion and the logic of accumulation are assumed away as exogenous. The capitalist system is conveniently naturalized and therefore unalterable, which makes it immune to contestation; only our approaches to resources are presumed to be capable of social manipulation. Proponents of the ecosystem approach are purveyors of the dominant procapitalist ideology, an orientation that makes one oblivious to the historical reality that capitalism is a system committed to an open-ended accumulation, which, in turn, depends on a total dedication to an open-ended maximization of exploiting natural resources. Without prefiguring the structural framework in which integrated ecosystem sustainability can be entrenched, the theoretical proposition in favor of ecological sustainability or sustainable livelihood is an exercise in futility.

Third, the approaches are reductionist in that they fall short of comprehending the complex interconnections and dynamic interactions between the lacustrine ecology and the biophysical processes in terrestrial ecosystems on the one hand, and between the economy and the ecology at the local, regional, and global levels, on the other. Analysts who subscribe to the ecosystem approach tend to assume that the lacustrine ecology is self-contained, forgetting the lakes are by definition a natural sink for terrestrial ecosystems. Without simultaneous changes in activities in the aquatic and terrestrial ecologies, lacustrine ecological sustainability cannot be achieved. Both matter and energy are continually exchanged between biophysical processes in terrestrial subsystems and processes of lacustrine production. Since changes in the material transformation associated with the exchange of matter and energy is inevitable, both the patterns and accumulation of entropic changes will determine lacustrine sustainability. Those who use the sustainable livelihood approach, seeking alternative source of employment and income for fishers in agriculture, are equally guilty of forgetfulness.

Terrestrial ecosystems are also undergoing rapid entropic degradation, owing to several factors: overpopulation, overcultivation, overgrazing, deforestation, climate change, and creeping desertification. Therefore, the terrestrial ecosystems in the countries surrounding Lake Victoria under investigation also urgently require relief from intensive activities.

Finally, it is important to note that local ecosystems cannot be abstracted from the global ecology. Only a global unit of analysis can help us understand these interconnections. It is here where the thermodynamic perspective is helpful.

Over the past two centuries, a number of thinkers have attempted to offer a biophysical theory of economics. But it has been only in the last four decades that this effort has been treated with more systematic considerations. Nicholas Georgescu-Roegen and Herman Daly (among others) are generally credited with integrating what had been hitherto scattered ideas about the complex linkages between the economy and the environment into a coherent theory of ecological economics. Georgescu-Roegen (1971) begins his analysis by rejecting the neoclassical presentation of the economy as a pendulum movement of production and consumption as a circular flow of exchange values between firms and households. In his view, production processes, like all biological processes, are subject to the laws of evolution and, consequently, subject to the entropy law and dissipation. Natural resources flowing into the production system are transformed into commodities and wastes, suggesting simultaneous depletion of low entropy resources and generation of high entropy wastes leading to the exhaustion of sinks.

Along similar lines, Daly argues that understanding the economy as a subsystem of the earth's ecosystem is critical to grappling with issues of conservation and sustainability. As a subsystem of the total environment, the economy is entirely dependent on the environment for the inflow of matter and energy and for the outflow of its wastes. Physical expansion of the economy would entail more demand for throughput to increase production and for more sink to dispose of the generated wastes. Since the environment is finite and nongrowing, it behooves societies to establish a sustainable scale of the economy at which the demand of the economy for flow through and sink should be equal to nature's capacity to renew itself and to absorb the wastes. In other words, short of establishing a steady state economy, one which neither depletes natural resources nor pollutes the environment beyond its ability to regulate and control the biogeochemical cycles, risks, ecological collapse (Daly, 1990 & 1993).

The seminal idea here is that the dynamic relationship between the economy and the environment can be explained by reference to the laws of thermodynamics. Energy is the force, driving both the economy and the environment. The first law of thermodynamics states that matter or energy can neither be created nor destroyed; it can only be transformed from useful and available to perform work to nonuseful and unavailable to perform further work.

Thus, the second law provides extremely useful insights into the functions of ecosystems by highlighting the irreversibility of energy since it is released to the environment as heat once work is done. This means that a decreasing amount of energy is available to support populations at each successive trophic level, with important implications for predator/prey relations (Jorgensen and Szirezhez, 2004). According to this second law, entropic processes are bound to entail continuous reductions of low-entropy throughputs and corresponding increases in high-entropy wastes. The difficulty in reversing the dissipated energy to low-entropy makes the entropic process unidirectional. If the extraction and import of low-entropic resources from the environment into the production system is faster than the capacity of nature to regenerate itself and if the export of high-entropic wastes to the environment is beyond the environment's ability to process and assimilate the wastes, increases in entropic dissipation are accelerated (Jorgensen and Szirezhez, 2004; Georgescu-Roegen, 1971).

The crucial point here is that the global physical system is thermodynamically closed in the sense that the global environment exchanges energy in the form of solar radiation but does not exchange matter with the solar system. However, the economy is open as a subsystem of the global environment, exchanging both energy and matter. The economy receives throughput from the environment and exports waste to the environment. As Perrings (1987) notes, the first interaction involves exactions on the environment entailing reduction or depletion of virgin resources, and the second interaction involves general degradation or insertions of pollution into the environment. It thus stands to reason that, since the economy entails continuous material transformations, ecological sustainability becomes a function of the difference between the magnitude of exactions on virgin resources and the quantity of insertions of high entropic conditions into the environment.

CONCLUSION

From the general review of the literature, five conclusions thus emerge that are relevant to understanding and explaining the state transition of Lake Victoria, as well as the present status of the Nile perch and the fate of the haplochromine cichlids.

First, taking too much out of the lake by both subsistence fishers and industrial operators and biomanipulating its species compositions in order to commercially privilege bigger species over other biotically critical species appear to have impeded the biological trajectory of the lake's evolution.

Second, the metabolic transformation of the urban ecology has become an important source of insertions of discharges into the lake. Burgeoning lakeshore towns and industrial establishments simply dump their untreated wastes into the lake, which have become major sources of cultural eutrophication.

Third, heavy exactions on terrestrial resources within the wider basin of Lake Victoria, connected with growing human biomass and the rapid expansion of chemical-dependent agrocommercial plantations, have substantially modified the basin's terrestrial ecology, leading to habitat fragmentation, deforestation, range land degradation, soil erosion, siltation of water channels, and the resulting sedimentation of the lake itself.

Fourth, regional and global climate change has impacted the heat budget of the lake, not only significantly influencing the rainfall regime of the area, but also compounding prolonged thermal stratification and hypolimnotic deoxygenation of the lake.

Fifth, compounding the aforementioned problems is the ubiquity of poverty of knowledge and/or class limitations of the producers and enforcers of knowledge. It is a sorry commentary on the lack of understanding on the part of fishery managers that, like all ecosystems, Lake Victoria is veritably governed by the laws of thermodynamics and evolutionary biology. The lake represents complex and dynamic hydroecological characteristics, not all easily quantifiable and manipulable by subjectively juggled models. It appears that fishery managers and researchers, their minds colored by self-interest and uncritical commitment to the growth ideology, are unable to appreciate the uncertainties that arise from the complexity and variability inherent to evolutionary biology and the operation of the laws of thermodynamics. The seminal point to bear in mind here, when examining the state transition of Lake Victoria, is the fact that the necessity of survival (central to the livelihood of the poor) and the logic of destructive creation (integral to the greed of capital accumulators) are, in the final analysis, at the root of Victoria's entropic disintegration.

2 | Lake Victoria: The Demand Conditions

INTRODUCTION

Lake Victoria, the size of Switzerland and lying on the equator between the two arms of the Great Rift Valley—the Eastern Rift and the Albertine Rift—is the second largest body of freshwater in the world (after Lake Superior in North America). The lake is indeed vast—its length spans 412 km from north to south and 355 km from east to west. The lake has a surface area of 68,800 km^2 and a catchment area of 251,000 km^2. The lake's catchment area extends over five countries: Kenya, Tanzania, Uganda, Burundi, and Rwanda, a situation that magnifies the problem of collective action in terms of fashioning policy coordination and joint action. The lake has a shoreline of 3,500 km long plus 3,000 islands squatting throughout the lake. The shorelines are fringed by thick papyrus that extends into vast swamps, providing crucial buffering services to the lake (Crul, 1995; Hughes and Hughes, 1992). However, its long, rugged, and broken shoreline is accessible to fishers from almost every direction, a situation that makes the lake vulnerable to overexploitation.

Having a maximum depth of 80 meters and a mean depth of 40 meters, Lake Victoria is generally considered very shallow when compared to Lakes Tanganyika and Malawi. Under relatively stable climatic conditions, rainfall and evaporation rates largely determine the water balance of the lake. This is because the lake receives 85 percent of its water budget directly from rainfall, while it correspondingly loses roughly 85 percent of its water to evaporation.

Even though the lake annually supplements its 2,760 km^3 of water storage by 7.9 billion cubic meters (bcm) from the Kagera River (which rises in the highlands of Burundi and Rwanda), and by another 7 bcm from six Kenyan rivers, water inputs from rivers play a minor role in the hydrological balance of Lake Victoria. Indeed, if one divides the water volume of the lake by the river inflows, its flushing time comes to 140 years. This is compared to Lake Tanganyika's 7,000 years and Lake Malawi's 750 years (Crul, 1995; Collins, 2002; Waterbury, 2002). Because the hydrological state of Lake Victoria is determined by the difference between precipitation and evaporation, the lake is hypersensitive to climate change. This situation, in turn, determines the lake's hydroecological health and productivity. During its 750,000-year history, Lake Victoria has been subjected to desiccation on three occasions, the last one believed to be 14,600 years ago (Johnson et al., 2000).

Exhibiting extraordinary levels of endemism, Lake Victoria has one of the richest faunas in the world, second only to Lake Malawi. Given the lake's short history, the diversity and richness of the faunas of endemic haplochromine cichlid fishes are unparalleled anywhere in the world. The lake experienced explosive speciation and extraordinary adaptive radiation of these endemic cichlids in just 12,400 years. Believed to have descended from a common phylogenetic tree or from a few ancestors, the haplochromine cichlids differentiated themselves over time through adaptive radiation, undergoing revolutionary morphological and behavioral transformations. These transformations endowed the haplochromine cichlids with different traits, sizes, shapes, colors, structures, and other features, which allowed them to colonize a variety of ecological niches and to specialize in different food sources.

In the context of expansive ecological niche segregation and extraordinary trophic specialization, the hydroecological conditions of the lake offer the explanation for this remarkable sympatric speciation of the haplochromine cichlids (Goldschmidt, 1994; Fryer and Iles, 1972; Witte et al., 2007). Before human intervention undermined the ecological conditions of the lake, and human biomanipulation of its hydrobiology made many species of fish vulnerable to predation, there were more than 500 endemic haplochromine cichlids in Lake Victoria. The cichlids are distinguished from one another by their feeding habits—the food they eat or the ways they digest their food—and/or by their shapes and colors. They are recognized as the detritus eaters, mud biters, algae grazers, leaf choppers, snail crushers, plankton feeders, zooplankton eaters, mollusk crushers, fish eaters, insect eaters, snail scale scrapers, snail engulfers, and prawn eaters (Graham, 1929;

Worthington and Worthington, 1933; Goldschmidt, 1994; Fryer and Iles, 1972; Witte et al., 2007). In addition, Lake Victoria hosts 59 noncichlid species of which 29 are endemic to the lake (Rabi, 1996; Yohannes, 2008). Because they specialize in different feeding habits, coexistence and coevolution have been possible.

Lake Victoria is shared by Tanzania (49 percent), Uganda (45 percent) and Kenya (6 percent). In addition to the annual $300 million worth of fish export by the riparian states, in conjunction with the $600 million that the lakeshore communities annually gross from fishing, Lake Victoria is sensitive and vulnerable to economic activities and disturbance regimes that originate in its wider catchment area. These variables take a heavy toll on the lake's biodiversity. A full third of the three littoral states' combined populations of 114 million live in the lake's watershed area with annual income ranging from $90 to $270; the basin also accounts for a third of the combined GDP of the three littoral nations (LVFO, 1999; UNEP, 2009).

Given the amounts of exactions on and insertions into the lake, carried in and around the lake by both governments and communities, the threats to the long-term sustainability are not difficult to understand. Indeed, as indicated in the Chapter 1, it is understandable why Tija Goldschmidt (1994) and Hubert Sauper (2004) are troubled by the extent to which the hydrobiology of the lake and the socioeconomics of the lakeshore communities are subjected to the ravages of degradation and destitution. Even though both Goldschmidt and Sauper raise profound questions, the answers to those questions remain elusive for several reasons.

In addition to suffering from epistemological limitations, the questions are not informed by history. Epistemologically, it is important to note that capitalism, as a dynamic social system, surges forward through destructive creation; alienation and commercialization of nature is inherent to capitalism. Furthermore, the transformation of what is alienated from nature can have exchange value only through the application of labor. It is here that hydrobiology and socioeconomics meet dialectically.

The aforementioned point can have meaning only if we understand the history of the political economies of the nations surrounding Lake Victoria. Indeed, the history of the political economies of the nations in question is one of commodity production for export. The system of commodity production introduced in the Lake Victoria basin was, and still is, based on a series of primary commodities. These commodities have been grounded in continual alienation of terrestrial resources, specifically land and natural vegetation. The cash crops produced by millions of laborers and peasants have not been for domestic consumption, but rather for export, to satisfy the needs and wants of affluent global northerners. The present penetration of

Lake Victoria by external capital and its consequent commodification simply affirm the point already made above.

Finally, during the colonial era, the colonizers chiefly relied on indigenous intermediaries to effectuate the alienation of nature, the exploitation of labor, and the collection of taxes, as well as to enforce the will of the colonizers. That role is now being enforced by the national elites who play a mediating role between the global class of capitalist accumulators and local producers. In this sense, contemporary national elites of today are just as dependent on the global capitalist accumulators for nourishment as the traditional intermediaries were on the colonizers. A key difference between the old and the new intermediaries is that the present elites control and use the state for purposes of direct personal wealth creation and accumulation, as well as for receipts from external patrons in exchange for the legitimating functions they perform on behalf of the global neoliberal consensus. This point needs some clarification.

Several recent studies have pointed to a distinction between administrative corruption and state capture (Hellman et al., 2000; Harrison, 1999; Hanlon, 2004). Administrative corruption involves payments by members of civil society to officeholders for public services, which any member of society is entitled to receive without paying anything for those services. Payments to health officers, school headmasters, the police, and judges fall under this category. In contrast, state capture involves either corporate control of the legislative and regulatory machinery of the state, or the total takeover of state institutions by the upper echelon of the national elites. The elites then use them for direct personal wealth creation and accumulation, as well as for dispensation of tangible benefits to political clients, supporters, and partisans. This objective reality makes the national elites pirate state bourgeoisie. I am not using the term "pirate" polemically, but rather as an objective qualifier of a class that owes its power, wealth, and continued existence to the monopoly control of the state. Under conditions of pirate capitalism, the state becomes not a facilitator and regulator of accumulation, but rather the direct instrument of accumulation on behalf of the class that "owns" it.

In light of the aforementioned reasons, the simultaneous commodification and globalization of Lake Victoria is simply a continuation of the colonial mode of production. Form and appearance have changed, but content and essence remain unchanged. Indeed, the symmetry of patterns between the colonial mode of production and the neocolonial economic interest articulation is remarkable. To substantiate these points, I have revisited the history of the political economy for each country. The purpose is not to delve into what happened in yesteryears, but rather to make the larger point that what is happening today is the direct product of what happened before.

KENYA: THE POLITICAL ECONOMY OF COMMODITIES AND THE FISHERY MODE OF PRODUCTION

Situated between Somalia and Tanzania on the Indian Ocean, Kenya occupies the heartland of East Africa. The country's total area is 580,300 km^2, of which 11,200 km^2 is occupied by bodies of water. Kenya's Indian Ocean coastline stretches for 536 km. Much of Kenya's eastern and northern regions are semiarid to arid, with little rainfall and high temperatures. The highly prized land in the country is found in the highlands, where colonists developed relatively modern infrastructure to serve the modern agrocommercial plantations. The colonists had a long vision to make the settler colony a permanent abode for white Europeans. Thus, Kenya was relatively far ahead of its neighbors. In fact, no other country in East Africa displayed more promise for independence than Kenya. The country was thought to possess a modest population size, comparatively good infrastructure, relatively well-developed commercial arteries, abundant water resources from its five water towers, and rich wildlife of mountain areas of much attraction to foreign visitors. All of these factors were considered sufficient preconditions for a successful takeoff along the lines of modern capitalism. Moreover, the national elite's hospitality to foreign capital, their beneficent attitude toward the resident citizens of European and Asian origin, and their extrovert adoption of an open macroeconomy as a tested road to evolutionary development all stood in stark contrast to the introvert politics and interventionist policies of its neighbors.

These qualities were regarded as prudent steps to attract foreign private capital flows with the potential to reinforce the virtuous cycle of growth. However, five decades later, one would be hard-pressed to find a trace of the country's early promise and of the much touted capitalist road to development. As a measure of performance, a quick glance at some key macroeconomic indicators presents a bleak picture. In 2008, the unemployment rate stood at a staggering 40 percent and the annualized inflation rate was running at 25.5 percent. Today, the Kenyan government survives year to year by the skin of its teeth, largely with the help of donor countries as its twin difficulties (budget and trade deficits) have taken on a permanent fixture. In 2008, the government collected $6.6 billion in revenue, and yet its expenditures amounted to $8.2 billion; its external debt stood at almost 6.8 billion; the value of the country's exports was under $4.9 billion against almost $10.8 billion in imports (CIA World Book of Facts, 2009).

After five decades on the capitalist road of growth, Kenya now shows very little in terms of providing the basic necessities of life to its citizens. In 2008, three out of four in Kenya's workforce were still engaged in subsistence agriculture, a sector that contributed a mere 26 percent to GDP (UNEP, 2009). In 2007, 16.6 million Kenyans were categorized as extremely poor; another 18.9 million were barely above this classification, merely able to afford a dollar per day (*Africa News*, April 27, 2007). Moreover, 2 million Kenyans in the semiarid and arid regions of the country are permanently on famine relief, and during drought years their number rises to 5 million (Aquastat, Kenya, May 2006). The Kibera slum in Nairobi, with over 800,000 inhabitants cramped into a tiny space and riddled with stark destitution, prostitution, crime, filth, and diseases, epitomizes the level of poverty in the country. In 2004, the government conservatively put the number of slum dwellers in the country at 4 million (UNEP, 2009). Others have observed that up to 70 percent of Nairobi residents live in slums or slum conditions (Khal, 2006).

> What went wrong? The answer to this question lies in the salient features of Kenya's political economy. Over four decades ago, Alan Waters (1972) perceptively observed that the study of Kenyan political economy must be approached with a sharper focus on how the control systems that were developed during colonialism survived into the postcolonial order with minor modifications or adaptations.

To come to the point without much ado, Kenya's nondevelopment and structural poverty are explained by a confluence of four salient factors: the character of capitalism inherited by the nationalists and simulation of the inherited control systems; the retention of the inherited division of labor; a teeming population; and the dissipation of ecological resources in response to a mix of anthropogenic agency and natural processes.

Although this is not the occasion to delve into colonial histories, it is important to put a few historical facts in perspective in order to establish a foundational framework for understanding the discussion that follows. As it is already too familiar to the historian, colonial and postcolonial politics in Kenya fundamentally revolved around land alienation and its consequent contradictions. Colonial land alienation was governed by the 1902 Crown Land Ordnance, which, for all practical purposes and intents, excluded Africans from the Kenyan highlands and segregated them by ethnicity into reserves. Crown land could be freely granted or leased only to white Europeans. A third category of reserves comprised nature reserves, a network of parks and game reserves for the psychic pleasure of the settler population and potentially for their material consumption through excision or degazetting.

Against the backdrop of this ordinance, land alienation, which began in 1903, reached its high-water mark by the 1920s with the creation of the "white highlands"—constituting 11,571 square miles of modern farms under unassailable European monopoly, encircling the Rift Valley (Morgan, 1963). The crisis of colonial accumulation during the Great Depression and World War II engendered powerful objective material conditions with which the colonial administration was forced to come to terms.

In the 1930s, the settler commodity production virtually collapsed, following the crash of world commodity prices. African laborers who worked the white plantations suddenly became unemployed and thus were unable to meet their tax obligations to the colonial state. After all, the colonial state used 50 percent of the tax it collected from Africans to subsidize European commercial production (Anderson and Throup, 1985). Depleted of fungible resources, the very survival of the colonial state was now at stake. Previously, exclusion of Africans from commercial production of important crops was rationalized by the logic of insulating white farmers from indigenous competition. Given the financial difficulties facing the state, African peasant production now presented itself as a default option, and consideration of settler interest had to give way to state interest.

In this context, Africans were allowed and encouraged to engage themselves in commercial production without necessarily endangering the long-term interest of white farmers (Anderson and Throup, 1985; Throup, 1985). Simultaneously, with the promotion of commercial production in the native reserves, the colonial state strengthened its marketing board to insure its grips on the flow of revenue from Africans. The stimulation of African cash crop production in the 1930s readily lent itself to the emergence of African petty commodity producers, who expanded their economic engagement into setting up and/or buying shops, tracks, land, and related facilities over time. More important, in order to circumvent the state marketing board, these petty accumulators created their own subterranean distribution and marketing channels, whereby they enhanced their ability to accumulate more wealth.

The emergence and growing strength of what Anderson and Throup (1985) call African "protocapitalists" had two major social and political implications. First, in buying land and producing cash crops in the native reserves, the petty accumulators grew bolder to increasingly challenge the authority of traditional chiefs. Second, conscious of their interest and with sufficient resources at their disposal, the African "protocapitalists" began to question the hitherto unassailable dominance of white farmers.

Indeed, these two social consequences of the commercialization of African commodity producers would supply the fuel for the anticolonial movement. Contextualized by the Mau-Mau rebellion in the 1950s, the land issue came to the fore and could no longer be postponed. There was a growing anxiety among colonial administrators that the Mau-Mau anticolonial insurgency might undo the entire colonial architecture, resulting in not only Great Britain losing Kenya, but also in jeopardizing the long-term vision of replacing the formal colonial empire by an informal empire.

Even worse, the anticolonial nationalism could take on a radical peasant revolution along the lines of "red" China. Such a scenario could lead to mass exodus of white settlers and all British subjects of Asian origin. In the midst of such fears, containing and marginalizing Kenyan nationalism became a priority for the colonial officers. The means for doing so was addressing the land question frontally, accompanied by progressive cooptation of the awakened trouble-making class of agitators and wealth accumulators, transforming them into pliant collaborators.

To this end, two parallel strategies were deployed. The first called for the urgent need to ameliorate the land hunger in the overcrowded African reserves and among the landless masses. In the African reserves, the colonial administration transformed communal land ownership into individual freeholds through consolidation and registration of existing land rights. Although the official rationale was to give farmers the incentive to become more efficient producers by moving up the production ladder from subsistence farming to commodity production for the market, the hidden purpose was to create a well-to-do class of landowners with a stake in the new status quo and thus avoid falling victim to the appeals of nationalism, whose primary goal was to expropriate the expropriators and redistribute the land to the indigenous people.

Along the lines of this plan, several schemes were conceived to marginally excise land from the white highlands for resettling Africans. These schemes were finally consolidated into what came to be known as the "million-acre" scheme. This meant that lands from white farmers close to the reserves would be settled by Africans. On the eve of independence, some 350,000 acres were allocated to some 25,000 smallholders under the schemes. More important, Africans, whether in the reserves or on the newly created schemes, would grow cash crops in order to generate functional satisfaction from increased income (Morgan, 1963; Jones, 1965; Harbeson, 1971; Throup, 1985). Just as the principle of race was used to alienate land during the colonial period, the structure of the new African resettlement schemes would be guided by the principle of ethnicity.

The second strategy banked on cooptation of African petty capitalist accumulators, followed by their integration into the white highlands under the rubric of "multiracializing" the plantation system. These petty capitalist accumulators had some capital of their own and would be assisted financially to buy out European farmers. To expedite their transformation as landowners, Great Britain, the World Bank, West Germany, and the Commonwealth Agricultural Development Corporation contributed to the pot some $30.2 million in grants and loans. On the eve of independence, 125 projects, covering 100,000 acres, were transferred from white owners to some 5,000 African owners as a prelude to gentrifying an important stratum of Africans (Morgan, 1963; Jones, 1965; Throup, 1985; Harbeson, 1971).

Embedded in the notion of promoting a multiracial society, an argument was orchestrated that radical tampering (as proposed by the nationalists) with the colonially developed plantation system would result in mass exodus of white farmers and the subsequent reversion of Kenya's modern mixed agriculture into subsistence farming, thereby ruining the whole economy and with it the future of the country (Jones, 1965; Harbeson, 1971). In order to accentuate the relevance of the external markers of identity to perpetuating the continuity of the colonial order in modified form, the colonial planners scrupulously carried forward ethnic considerations as prime factors in the allocation of agrarian resources. Even former Mau-Mau fighters were given land. Since the Kikuyou were most affected by land alienation and were seen as the main drivers of the anticolonial nationalism, they appeared to have gained the most from the agrarian realignment, capturing some 45 percent of the new land allocations (Jones, 1965, Harbeson, 1971).

Thus, by skillful manipulation of the land issue, the British officers managed to defuse the gathering storms of anticolonial nationalism and to thwart the potential radicalization of the land question—something that would become a source of ideological division among the new class of African elites after independence. Faced with a fait accompli, the conservative wing of the Kenya African National Union (KANU), led by Jomo Kenyatta, legitimized the land settlement by officially accepting it. The conservative nationalists now saw settlement of the agrarian question prior to independence as instrumental to depoliticizing postindependence economic development (Harbeson, 1971; Throup, 1985). The fear was that politicization of the land issue might usher in a class-oriented peasant uprising, posing a potential threat to the modified neocolonial control systems and to the regime itself.

Kenya's anticolonial nationalism was championed and led by an eclectic group of people. Coming—as they did—from the churches, the teaching profession, and petty trade, these strata of the petit bourgeoisie all shared one fundamental attribute in common: they were bourgeoisie in aspiration. They had also internalized Kwame Nkrumah's biblical recitation that the first step to economic prosperity (whether personal or national) was conquest of the "political kingdom." After all, the entire investment of the nationalists (who saw their embourgeoisement as tied to the monopoly control of the state) was now in politics.

The historical dilemma facing the nationalists was how to transform themselves into the national bourgeoisie in a country where there was no national capital. One option was to expropriate the accumulated wealth of the alien appropriators and correspondingly dismantle the inherited institutions of domination and exploitation. The conservative nationalists, however, made a conscious (albeit naive) decision that self-transformation into a national bourgeoisie was feasible through collaboration with the resident citizens and transnational corporations.

In the calculus of the conservative nationalists, indigenization of accumulation could be achieved by using the levers of state power to transfer resources from the nonnative resident citizens to themselves, mainly through forming partnerships with and shareholding in the transnational companies. The Industrial and Commercial Development Corporation, for example, was created to provide loans to Kenyan businessmen and bureaucrats to gain access to the transnational corporations (Currie and Ray, 1987). Moreover, control over fiscal and trade policies and receipts of foreign assistance would facilitate the elevation of the ruling class to the status of a national bourgeoisie. To this end, a number of parastatals proliferated to assist in the Kenyanization of capitalist accumulation. Thus, the preliminary act of the first generation of Kenya's politicians was to integrate themselves into the existing power structure rather than to dismantle the received institutions of law and order.

The rival factions within the postliberation forces found themselves entangled in a struggle over the political spoils as a way of obtaining access to and control of the plantation economy. The struggle within the national elites eventually resolved itself in the formation of a de facto one-party system and the subsequent political marginalization of the radical elements who had espoused wholesale dismantlement of the plantation economy in order to fully satisfy the land hunger under a form of petit bourgeois peasant socialism (Currie and Ray, 1987).

Secure in their monopoly control over the trajectory of Kenya's political evolution, the victors under Jomo Kenyatta then turned their attention to the urgent task of adapting themselves to the neocolonial mode of primitive capital accumulation for purposes of speedy personal enrichment. This step, on one hand, required material gratification of loyal clients and the cooptation of potential "counterelites" into the state apparatus as partners in the spoils of capital accumulation, while creating a mass power base in the peasantry on the other.

Kenyatta and his hangers-on were conscious of the centrality of the land issue to the intraelite struggle. To contain this struggle within manageable bounds, it was of paramount importance to create a peasant class of smallholders wrapped in the illusion of landownership. In keeping with this objective, smallholder cash crop production assumed a hegemonic role in the reproduction of primitive capital accumulation. Smallholder peasant cultivation served, not only to depoliticize the land issue and thereby create political legitimacy for the regime, but also to maximize the transfer of surplus value to the bureaucratic bourgeoisie from a highly segmented peasant class (Currie and Ray, 1987). The creation of the smallholder peasant class removed the threat to the inherited large-scale farms from those who called for egalitarian land redistribution. Furthermore, it concealed the fundamental contradiction between the 13 million Kenyans who held 1-2 hectares (ha) of farmland and the 3,433 large estates, which together controlled more than half of Kenya's arable land (Currie and Ray, 1987).

The crystallization of class formation in Kenya in favor of the conservative nationalists eventually insured the retention of Kenya's specialization in the production and supply of primary commodities to global markets. Tea plantations, coffee growing, horticultural production, exploitation of fisheries, tourism development, and dependence on foreign aid all became the focus of the national elite as vehicles for rapid personal wealth creation and accumulation. Constituting the main pillars of Kenya's economy, these sources of primitive wealth accumulation share common characteristics: they are all prone to rental predation, structural oversupply and related global price volatility, and they are vulnerable to the vagaries of weather and politics.

A synopsis of each of these pillars of primitive accumulation should provide foundational understanding of Kenya's place in the global division of labor. The overarching purpose of the British colonial undertaking in Kenya (as in all other colonial cases) was to develop a sustainable complementarity between the metropolitan economy and the colonial economy. The logic underpinning the colonial enterprise was that the crisis of accumulation in Britain could have been contained or postponed indefinitely if two conditions were met: if there were a steady flow of primary commodities to fuel the production and reproduction of capitalism; and if enough external market outlets were found to ease home market saturation. Kenya's specialization in the production and supply of coffee and tea was the initial outcome of this colonial enterprise.

Beginning in the 1860s, French Catholic missionaries were pioneering in the experimentation, introduction, and propagation of cash crops in East Africa. The desire to base the evangelization of Africans on a concrete material foundation was the motivation of the Holy Ghost Fathers. Earnings from cash crops were deemed essential for sustaining the self-supporting project of evangelization. The experiments in cash crops included coffee, tea, vanilla, cotton, and related crops, and the Holy Ghost Fathers were not disappointed in their primary enterprises (Kieran, 1969). They did effectively contribute to the construction of a system geared toward specialization in primary commodity production from which the East African states have yet to wean themselves.

At any rate, the Catholic missionaries carried their success in coffee production from Tanzania with their missionary work to Kenya where the missionaries first introduced coffee in the 1890s. Enticed by the prospect of steady revenue, the colonial administration bought 2,000 seedlings in 1904 from the missionaries for purposes of experimental demonstration. Spurred by the growing interest among settlers and the colonial administration, the missionaries then increased growing and selling coffee seedlings so rapidly that they were soon offering up to 100,000 seedlings to a single buyer in one swoop (Kieran, 1969; Waters, 1972; Freeman, 1988). Thus coffee proved to be the answer to the crisis of colonial accumulation, and an important bait to attract European settlers.

One veritable result of the rapid ascent of coffee as the lucrative source of cash was the exclusion of native Kenyans from growing coffee, thereby institutionalizing the race-based division of labor. The simultaneity with which the alienation of native land and the appropriation of indigenous labor were achieved in the coffee sector was remarkable. With Africans excluded from the sector, coffee growing and marketing became an exclusively European monopoly. Reliant on cheap indigenous labor, the coffee operations evolved into giant estates with powerful influence on colonial politics. Using their dominant position, coffee plantation owners vigorously and successfully resisted the colonial administration's tentative moves to tax coffee exports, as well as to block potential native competitors from participation in coffee production well until the 1930s and 1940s (Waters, 1972; Freeman, 1988).

Upon independence in 1963, however, the new government modified the inherited control structure over coffee production in two minor ways. First, it universalized the formal segmentation of the coffee sector into estates and smallholder farms; the colonial administration had already allowed a limited number of Africans to plant coffee trees under strict control. For example, the colonial administration saw the number of Africans growing coffee increase from 62,000 in 1957 to 193,000 at independence (Freeman, 1988). Now the postcolonial state simply expanded the role of smallholders in the coffee sector until 60 percent of total national coffee production was accounted for by some 600,000 smallholder farmers but under even more stringent control systems. The ministry of agriculture and the Coffee Board of Kenya (a successor to the colonial Coffee Marketing Board) were soon empowered to use the issuance or revocation of licenses to regulate and control the planting, growing, producing, and marketing of coffee; coffee trees deemed illegal would henceforth be arbitrarily uprooted. Under the new system, growing coffee was not a right but a privilege, since the pirate state bourgeoisie determined the identity and number of growers as well as the amount of land under coffee production. Moreover, as a condition of having the privilege of growing coffee, smallholder farmers must belong to coffee cooperative unions. The Kenya Coffee Planters Cooperative Union was then given monopolistic power to purchase all coffee beans from the coffee cooperatives at prices set by the Coffee Board of Kenya (CBK).

Second, in addition to numerous levies and taxes by local governments, the new pirate state bourgeoisie imposed an export tax on coffee growers in order to insure a steady flow of revenue to the government. After all, in 1966, Kenya exported 49,000 metric tons of coffee out of the total national production of 52,000 metric tons (Waters, 1972; Freeman, 1988). Fully in control of the state, the new ruling class could and did use these tax revenues to enrich themselves. The institutional mechanism for extracting surplus value from coffee growers was the parastatal CBK. Thus, the manner in which the control systems in the coffee sector of the colonial era have survived into the postcolonial period without compromising its salient features exemplified the remarkable continuity of the colonial mode of alienation of nature and appropriation of labor. The distinction of analytical importance between the "old" state and the " new" state was that the role of the colonial state was limited to creating favorable conditions for private monopoly capital and providing enabling services to coffee plantation owners, whereas the postcolonial state inserted itself in the economy through parastatal institutions—not only as a facilitator of accumulation, but also as an instrument of personal wealth creation, accumulation, and patronage distribution on the basis of kinship, ethnicity, political loyalty, and social sycophancy.

In the end, neocolonialism and crony capitalism worked well off one other to perpetuate Kenya's role in the global division of labor as a producer and supplier of raw coffee. The coffee sector has remained one of the mainstays of the ruling elite as the yearly average of foreign exchange earnings from coffee between 2003 and 2008 was almost $129 million. The coffee sector is still regarded as one of the areas where the state bourgeoisie can extract surplus value and increase the number of coffee growers as a way of containing the unemployment crisis. Even the 2002 "national poverty reduction strategy," drawn with the help of the World Bank and the International Monetary Fund (IMF), targeted the coffee sector for increasing foreign exchange earnings and employment through privatization and foreign investment. It bears to note that the Mwai Kibaki government came to office promising 500,000 jobs a year, a promise that entailed a price tag of $4 billion in foreign assistance (Kamau, 2004).

Since Kenya formally accepted the structural adjustment program as way forward, growing and exporting cash crops was regarded as the sole engine of export-led "development." The subtext of this logic has always been to enable Kenya to earn hard currency by exporting commodities and then repay its debt. The problem is that the 600,000 smallholder coffee farmers, who account for 60 percent of Kenya's total coffee production, lack enabling services to remain self-sustaining in terms of income. Smallholder farmers -- bedeviled by the combination of a lack of credit facilities, poor infrastructure, high cost of inputs, low prices and global oversupply of coffee, and heavy taxation -- have been made increasingly poor, despite increased acreages under coffee cultivation. For example, the coffee area under active cultivation increased from 102,400 ha in 1980 to 170,000 ha in 2004. Yet yields of coffee beans fell from 8,919 hectograms per ha to 3,794 hectograms per ha over the same period. This reality is reflected in the fact that national coffee production progressively plunged from 130,000 metric tons in 1988 to 56,000 metric tons in 2003 (Damianopoulos, 2005; Kamau, 2008). So the logic underpinning the "national poverty reduction strategy" is that rehabilitation of the coffee sector through promoting further liberalization and adopting investor-friendly legislation would enable the pirate state bourgeoisie to contain both the crisis of accumulation and the crisis of unemployment in the coffee sector.

Ironically, it was since the official adoption of liberalization that Kenya's coffee production began to decline. In fact, the Kenya Planters Cooperative Union (KPCU) (representing all smallholder coffee farmers) rightly pointed out that it was liberalization that had destroyed the coffee sector. Under the guise of promoting technical efficiency and competitive innovation, the pirate state bourgeoisie issued licenses to capital-rich commercial millers. Enjoying superior access to capital, credits, and political connections, the new coffee millers outcompeted the KPCU and the latter lost 70 percent of its market share to the commercial millers. After all, smallholder coffee growers could not directly sell what they produced; according to the cooperative framework, growers must sell their green coffee beans to the KPCU, which is regulated by the Coffee Board of Kenya with total monopoly control over price. Because of this convoluted arrangement, Kenyan coffee growers have been receiving just 40 percent of what they produced (*Africa Analysis*, July 27, 2001).

The colonial and neocolonial experiences in the coffee sector are replicated in equal proportions in the tea sector. In Kenya, tea's arrival was not far behind coffee, when white settlers started cultivating tea in 1903. Tea cultivation was, from the start, an exclusive settler preserve. By the early 1930s, Kenya was an important supplier of tea to the British home market. After independence, the pirate state bourgeoisie transformed the tea sector into a mainstay of Kenya's economy by encouraging smallholder cultivation. As a result, national tea production steadily increased from 18,000 metric tons on 24,448 ha in 1963 to 315,000 metric tons on 131,000 ha in 2006; Kenya accounts for 9 percent of global tea production and for 20 percent of the world tea market.

The tea sector provides livelihoods directly or indirectly to 3 million Kenyans (FAO, May 2008a; SOMO, 2008). Today, Kenya is the third largest tea producer in the world and shares the distinction of being the first or second largest tea exporter in the world with Sri Lanka. In 1996, Kenya beat Sri Lanka in terms of both the amount of tea produced and the amount exported when it produced 244,200 metric tons of tea compared to 244,000 metric tons for Sri Lanka (UNESCO Courier, Dec. 1997).

The yearly average of hard currency flowing to Kenya from the export of tea between 2003 and 2008 was over $636 million, endowing the tea sector a premier status among exportable commodities. The tea sector is divided into large-scale corporate operations and smallholder farms. The plantation sector (accounting for 40 percent of total production) is dominated by about 70 multinational corporations, which grow, process, blend, package, and market their own tea brands. Together, the corporations own 39 tea factories in Kenya; Unilever, James Finlays, and Brooke Bond are among the largest of these foreign companies. Combined, the tea multinationals provide employment to more than 50,000 Kenyans; Unilever by itself has more than 18,000 of those laborers on its 13 tea plantations (Attwood, 2008; UNEP, 2009).

Although the tea companies provide housing, food, water, and health care to their laborers, these benefits are a far cry from being adequate. A recent study by a Dutch research organization on multinational companies (SOMO, 2008) found that the company houses are extremely overcrowded and in poor condition with leaking roofs, inadequate water supply, and poor sanitation services. Low wages, poor working conditions, lack of job security, lack of negotiating power, application of biocides without protective gear, fractures from falling, back pain, respiratory illnesses, and discrimination on the basis of ethnicity and gender are common features of the plantation system. Women who failed a pregnancy test are not hired and, if they resist sexual advances by their supervisors, they are assigned to dangerous plucking areas.

For purposes of extortionate rates of exploitation, the plantation owners use the piece-rate system; there are no hourly wages. The piece rates are determined by the companies, they are not a result of negotiation. As a result, the income of workers can fluctuate widely, depending on the seasonality of tea harvesting and the health and strength of laborers. Moreover, the plantation owners increasingly rely on contingent laborers in order to deny them pension, insurance, health, and other benefits. The definition of "temporary" laborer is so arbitrary that laborers may remain classified as such, even after working in the plantations for up to six years (SOMO, 2008).

Furthermore, external markers of identity (such as ethnicity and gender) are widely used in terms of hiring and promoting. Such practices block social solidarity among laborers since they serve to foster interethnic tension and even violence. In the wake of the 2007 contested national elections, for example, tea plantation workers at Unilever operations went at each other on the basis of ethnicity. In January 2008, 600 young men, armed with machetes, bows and arrows, invaded one of Unilever's tea estates in Kericho, near the Mau forest, and killed 14 laborers on account of being ethnically different. To avert mass slaughter, Unilever used its trucks to transport 3,000 of its laborers to their ethnic enclaves where they could find security among their kindred (Attwood, 2008).

The 600,000 farmers in the smallholder tea subsector are organized into cooperatives under the direction and supervision of the Kenya Tea Development Agency (KTDA) for purposes of taxation, price, and quality control. All smallholder tea growers must sell their raw tea to one of the 56 processing facilities of the KTDA, usually below cost of production. The exploitative character of this convoluted arrangement is reflected in the fact that the small farmers, on average, received $0.21 per kg of tea leaf between 2000 and 2007, while the average world price for Kenyan tea during the same period was $1.72 per kg (SOMO, 2008).

In addition to having to pay taxes and levies, the small tea producers in Kenya are handicapped by constantly rising costs of inputs, poor infrastructure, high costs of fuel, and nonrepresentation in the key parastatal institutions (SOMO, 2008).

Like all primary commodities, tea is subject to structural oversupply either as a result of increased yields on existing acreage or as a result of new entrants into the production and marketing of tea. For example, tea production in China increased by 9.5 percent in 2006 over the previous year; the increase of tea production in Vietnam was even more dramatic, rising by 133,000 metric tons in 2006 over the previous year (FAO, May 2008a). The veritable result is depressed global prices for tea and loss of market share. In 2005, Kenya exported 98 million kg of tea worth $186 million to Pakistan. That amount fell to 80 million kg worth $155 million in 2007 as a result of competition from Sri Lanka, Malawi, Rwanda, and Uganda. In fact, Kenyan tea exports to Pakistan in the first half of 2008 declined by 43 percent to 27 million kg (Inter Press Service, Sept. 23, 2008).

The consequence of competitive overspecialization among cash crop producers is structural oversupply. In order to hang in the market, the competitors are bound to continually expand production, setting a self-reinforcing process in motion. In light of global oversupply of tea, Kenya accumulated 367 million kg in 2004, carried over from previous years (Butler, 2005). In the hope of stimulating tea consumption at home, the Kenyan government launched a national campaign promoting tea drinking as a healthy thing; the campaign raised national tea drinking by 10 percent. However, raising the national tea consumption of tea by 10 percent in a market whose historical absorption capacity normally was a mere 12,000 metric tons of tea would not make a dent in the country's unsold stock of tea (Butler, 2005).

Since the crisis is structural, such palliative measures are not going to solve the crisis of national and global oversupply. Global tea production and marketing are dominated by giant multinationals, which are present at every stage of the production and supply networks. The multinationals both grow tea on their own plantations and buy raw tea from small producers. In the context of structural oversupply, the corporations possess enormous power to freeze out small producers if and when they fail to the standards and prices set by the companies. For example, Unilever annually produces 35,000 metric tons of tea on its own plantations in Kenya, Tanzania, and India, and purchases another 290,000 metric tons a year from other producers; the company then markets its blended tea in over 100 countries. Likewise, James Finlays grows 45,000 metric tons of tea a year on its plantations in Kenya, Uganda and Sri Lanka (SOMO, 2008; FAO, 2008a). More important, the global corporations have total monopoly over blending, packaging, and marketing tea. These functions account for 80 percent of the retail value of tea and is retained by the actors in the downstream supply chain. Historically, the deviation in the value of downstream functions from upstream activities has been remarkable. Whereas the producer prices have declined by 44 percent in real terms over the past 30 years, the retail value of tea has remained relatively stable (SOMO, 2008).

Negative ecological externalities are another associated legacy of specialization in tea production. Forests in Kenya have historically been converted to tea cultivation or logged for timber to process the tea. Since tea leaf is grown monoculturally on slopes, tea cultivation fosters soil erosion and nutrient depletion. In order to replenish the nutrients, the massive application of agrochemicals is required. Residues of these agro-chemicals and eroded soils inevitably enter the natural water systems, polluting and silting rivers, streams, and lakes (SOMO, 2008).

The export sector of horticultural production has, in recent decades, become the latest addition to the sources of primary accumulation. Some 2 million Kenyans are engaged in the sector producing cut flowers and foliage, fruits, and such vegetables as chilies, okra, French beans, avocados, and baby cabbages.

The annual foreign exchange revenue between 2003 and 2008 from the export of horticultural commodities was over $655 million. The export sector is, however, dominated by some 200 licensed companies and 70 licensed shipping agencies. Three large companies combined control 55 percent of Kenya's horticultural exports.

The growing global significance of horticultural production is exemplified by the fact that exports of horticultural products from Kenya steadily rose from over 20,000 metric tons in 1982 to almost 85,000 metric tons in 1996, with Europe being the largest market for Kenyan horticulture. In 1994, 85 percent of Kenyan horticultural goods went to four countries: the United Kingdom (30 percent), the Netherlands (29 percent), France (16 percent), and Germany (10 percent). Indeed, Kenya has become the leading supplier of green beans to the European Union (EU), with France and the United Kingdom absorbing 43 percent and 36 percent of Kenyan green beans, respectively. Moreover, Kenya is the leading supplier of roses to the EU with 35 percent market share; it bears to note here that Homegrown (one of the three largest firms in the sector) barely managed to export 17 metric tons of cut flower in 1982, whereas it exported 10,000 metric tons in 1997 (Dorsey, 1999; Barrett et al., 1999; Mead, 1998; The East African Standard, March 26, 2008). Between 1993 and 1997, export of horticultural products from Kenya grew by 20 percent a year. In cut flower alone, Kenya exported 35,736 metric tons to Europe in 1997, earning $100 million in hard currency, which accounted for 60 percent of total cut flower export from Africa. In 2005, Kenya exported 80,000 metric tons of cut flowers to Europe, raking in $350 million (PANOS, Nov. 17, 2006; Mead, 1998).

The chief driver behind the growing industrialization of the horticultural sector has been the desire to expand the portfolio of cash crops in order to strengthen the accumulation process. Under the rubric of diversification, horticultural production in okra, chilies, French beans, avocados, papayas, cut flowers, and other fresh products has become very attractive to both multinational companies and petty commodity traders, not only competing with coffee and tea as a foreign exchange earner but also surpassing them. In the cut flower industry alone, the three largest firms (Unilever's Sulmac division, Oserian, and Homegrown) employ over 20,000 people today, mostly women (Attwood, 2008). Tea, coffee, and horticulture together now account for 55 percent of Kenya's total exports (Blomfield, 2008). In fact, in 2007, horticultural exports dwarfed all traditional sources of hard currency when the sector raked in $1.12 billion, of which cut flower accounted for 64 percent, compared to $972 million from tourism and $63.9 million from tea export (Kabukuru, 2009; *The East African Standard*, March 26, 2008).

The floriculture industry in Kenya is dominated by roses, carnations, alstroemeria, and statice. By 1998, 1,300 ha of well-watered and fertile land was devoted to floricultural production around Lakes Naivasha and Thika, areas dominated by Homegrown, Sulmac and Oserian. There are also more than 50 flower farms in the Lake Naivasha basin, in addition to several hundred smallholder farms with much smaller flower plots (Hughes, 2001; Mead, 1998). Kenya Flowers Council (an association of flower growers) boasts that the floriculture sector provides direct employment to more than 70,000 people of whom 60 percent are unskilled women, while benefiting another 1 million from flower-related activities (PANOS, Nov. 17, 2006). Workers, however, told a different story since they work a 10-hour shift without compensation for overtime work and are forced to pick freshly sprayed flowers without protective gear. In the past, flower fields were sprayed the previous night in order to allow the chemicals to dissipate before workers were deployed to the fields the next morning to pick the flowers. Now, however, the flowers are sprayed in the morning; then two hours later, workers pick the flowers to supposedly preserve freshness, regardless of health risk (PANOS, Nov. 17, 2006).

As in the other cash crop sectors, the horticultural sector is subject to extortionate rate of exploitation, price volatility, and domination by large firms and receiving retailers, who use their financial and organizational resources to determine the quantity, quality, and prices of horticultural products. Smallholders do not have land large enough to scale up their operations and lack access to credit needed to modernize their production. In the floricultural subsector in particular, smallholders are completely shut out because cut flower operations are capital-intensive. Thus, concentration of horticultural production resulting from the increasing mechanization of operations by ever fewer firms undermines the logic of diversification as a poverty alleviation strategy. Additionally, the horticultural sector is subject to predation by intermediaries who pack and export the horticultural products. And finally, commercialization and globalization of the horticultural sector perpetuates rather than ameliorates poverty since the sector is geared toward the satisfaction of the psychic wants of affluent Europeans and North Americans.

Foreign supermarkets determine the amounts, quality, timing, and prices of horticultural products, depending on domestic demand conditions and the number of their international suppliers (Barrett et al., 1999; Dorsey, 1999; Hughes, 2001). Producers who do not meet the requirements set by retailers are shut out of the market. Additionally, the fact that horticultural products, by definition, are perishable goods compounds the vulnerability of producers to the volatility of the psychic wants of consumers and their appetite for exotic tropical produce. It bears to highlight here that supermarkets have increasingly assumed an unassailable dominance in the global supply network relative to horticultural products. This supply chain now comprises producers, agents, shippers, exporters/importers, and supermarkets. In this neatly structured pyramid, the supermarkets determine the quantity and quality of products, regulation of production, price, and timing of supply. Although a complex vertical integration in the supply chain is in place, the supermarkets do not own the farm operations, and thereby shift all risks associated with production to the grower. Meanwhile, however, they control what is produced, how it produced, how much must be produced, and when the products must be delivered (Barrett et al., 1999). Growers of horticultural products who do not meet the set requirements are frozen out from the market. Moreover, flexibility is central to supermarket decisions needed to cope with changing moods of consumers, hence the supply chain is governed by the operative principle "just in time" supply of fresh produce. These onerous conditions on growers militate in favor of concentration and specialization of horticultural production and therefore, against a sustainable smallholder economy.

Finally, there is the dismal environmental and social record of floricultural producers. Capital-intensity in floricultural production means heavy reliance on synthetic fertilizers, toxic pesticides and herbicides that, in combination, deplete natural soil nutrients, pollute natural water systems, and pose serious health risk to workers. In the 1990s, Kenyan floricultural laborers were applying 300 metric tons of methyl bromide a year to kill pests and weeds. Methyl bromide is known as an ozone-layer-depleting chemical. In addition, it kills every living thing in the soil, which renders the soil totally barren and requires massive application of synthetic fertilizers to grow anything. The reliance on this product is also expensive since, in the late 1990s, Kenya was using 5 percent of its hard currency to import methyl bromide used by flower growers and other cash crop producers. The expanding irrigation schemes around Lake Naivasha have also resulted in appropriation of more land belonging to indigenous communities, which have long been used to grow food and graze livestock (Madeley, 1998).

A perceptive reader may legitimately raise a question as to whether the discussions on coffee, tea, horticulture, and other related issues have a bearing on the state of Lake Victoria and the fishery mode of production. Here are some pertinent answers that should help to situate the current state of Lake Victoria and its satellite water bodies in broad macrostructural conditions.

First, the progressive expansion of agrocommercial plantations has taken a heavy toll on nature. A series of vast forests, woodlands, grasslands, and wetlands have been cleared to make way for these plantations. This has not only contributed to the incremental modification of microclimatic conditions in the Lake Victoria basin but also facilitated soil erosion, leading to progressive sedimentation of Lake Victoria.

Second, by their very nature, commercial plantations are chemical-dependent. Residues of reactive nitrogen, pesticides, herbicides, and other trace elements have been inserted into Lake Victoria for more than a century. Agrochemicals have thus become integral to the cultural eutrophication of Lake Victoria, which, in combination with other factors, may have triggered state transition of the lake.

Third, concentration on the appropriation of nature as the primary source of wealth generation (both before and after decolonization) introduced a powerful social force in the competition for land and water resources. As a result, the best land, defined in terms of fertility, and proximity to water, infrastructure, and markets, has been concentrated in the hands of the few, mainly to grow cash crops for export. The mass population has been socially marginalized, and has been pushed off to even more marginal lands. Unable to compete with the national elite for good land and water, the majority poor have continued to depend on clearing forests for cultivation, charcoal production, construction material, and overgrazing grassland. This has worsened soil erosion throughout the basin, further contributing to the sedimentation of Lake Victoria. Other dispossessed peasants and livestock keepers have migrated to the lakeshores of Lake Victoria to eke out a living from fishing, worsening the crisis of overfishing.

The social dimension of competition for land and water resources in Kenya and its potential impact on the state of the country's natural capital shows no sign of changing. Elite concentration on coffee, tea, and horticulture as the major pillars of primitive capital accumulation is bound to be driven by objective material conditions. Note that, upon liberation, the nationalists found themselves unable to rapidly hold control of the urban economy which was dominated by multinational corporations or by Asians who were imported to Kenya by the colonial administration. While the Europeans and Asians (capitalizing on their early-mover advantages) prospered under colonialism, the natives were relegated to utter dispossession and exclusion. Even in the three decades following independence, Asians (representing under 1 percent of the country's total population) still owned 75 percent of Kenya's medium and large manufacturing firms, while indigenous African ownership was limited to a dismal 5 percent (Vandenberg, 2003).

This objective material reality posed an insurmountable difficulty for those who sought to penetrate the urban economy. In consequence, the necessity to alienate more of nature and labor forwarded itself as an alternative. Under the guise of indigenization of nature (particularly land), the national elite managed to buy out some settlers from the plantation economy. Once having done so, it was not difficult to distribute the land resources to members of the national elites in proportion to their political standings in the new system. Furthermore, having acquired a foothold in the agricultural sector, it was not difficult to aggrandize personal holdings through the acquisition of more land through the dubious instrumentation of crude state power. Between 1978 and 1999, for example, the Maasai of the Mau forest lost more than 1 million acres of their virgin land to politically connected and powerful land speculators, tourism developers, and large-scale commercial coffee and tea plantation operations. It bears to note here that the Maasai had already lost 15 million acres of their prime ancestral land to European settlers and companies during the colonial era (Olol-Dapash, 2002).

The traditional forest-dwelling Ogiek people have not fared better than their Maasai neighbors. President Daniel Arap Moi opened more than 148,000 acres of virgin forest to commercial logging companies and tea plantations. Even though the Ogiek obtained an injunction from the court in 1997, proscribing further alienation of Ogiek forested land until the court ruled President Moi disregarded the court order and publicly announced a government plan to open up another 412,000 acres of land from 14 forested areas in the name of promoting rural development (Fitzhenry, 2002).

To be sure, the regulation and control of the production, marketing, and pricing of primary commodities have strengthened the material and financial position of the new pirate state bourgeoisie vis-à-vis the populace. The rewards have been immediate and palpable. Preserving and perpetuating the inherited division of labor has served the pirate state bourgeoisie well. For example, between 1967 and 1987, the world market share of Kenya's coffee and tea doubled, whereas exports of finished goods as a proportion of total manufacturing declined from 20 percent to 10 percent over the same period. In 2001, tea alone raked in 35 billion Kenya shillings (KSh) in foreign exchange earnings compared to KSh27 billion from all manufacturing exports combined (Vandenberg, 2003).

The primary sources of primitive capital accumulation (discussed earlier) contain deep contradictions since they hinge on continual alienation of nature and appropriation of labor. In consequence, further expansion of the pillars of primitive accumulation, requiring more and more appropriation of water and land resources, has created a structural situation pitting the pirate state bourgeoisie against the broad masses.

It bears noting here that only a small fraction of the country's land area of 569,250 km^2 is suitable for productive agriculture. Even of this amount, the land is subject to increasing fragmentation in response to population rise and is vulnerable to extreme weather events such as recurring droughts, floods, aridity, and desertification, compounded by soil erosion, extensive deforestation, and water quality deterioration. In 1950, the per capita land availability in Kenya was 9.8 ha, which progressively declined to 1.7 ha per person by 2005; this will further decline to 0.3 ha per person by 2050 (UNEP, 2009). Of the 57.6 million ha of land devoted to agrarian and livestock production, only 9.4 million ha (17 percent) are characterized as having moderate to high fertility; the remaining 48.2 million ha (83 percent) are classified as either arid or semiarid and can support only limited farming, livestock production, and pastoral economy (UNEP, 2009). In 2003 the country had a mere 1,030 km^2 under irrigation. Even though it may be possible to bring more land under irrigation, this would require drawing massive amounts of water from such as Lake Victoria.

The country's population has risen from 8 million in 1964 to 40 million today. If the present demographic trends persist, Kenya's population will reach the 51 million mark by 2025; 43 percent of the population are under age 14 and the median age is 18.7 years (UNEP, 2009). Providing food, shelter, and jobs to such a young population is going to be a formidable task. To provide shelter to the burgeoning new urban dwellers, the government will need to increase the number of new housing units from the current 35,000 a year to 200,000 units a year (UNEP, 2009). It is against these structural conditions that the expansion of primitive accumulation to the country's wetlands and lakes for resource extraction and appropriation must be understood.

Having reached the outer limits of primitive accumulation in the cash crop economy, Kenya's pirate state bourgeoisie have increasingly been looking to the country's wetlands and natural water systems to expand the scope of accumulation. As a result, the struggle for hydrobiological resources in Kenya has intensified. Even though allocation of resources in the 1960s to develop the fishery sector was minimal, the pirate state bourgeoisie have made their determination known that the fisheries mode of production (centered primarily on Lakes Victoria and Turkana) would be integral to the country's overall development trajectory.

The Lake Victoria basin, which represents only 8.4 percent of Kenya's territory, is home to more than 40 percent of the country's population (Coche and Balarin, 1982). Since the average land availability per person in this area was 0.32 ha as early as 1981, the economic transformation of the basin was considered contingent on the fisheries development of Lake Victoria. In 1964, of the total catch of 20,122 metric tons, Lake Victoria's fisheries accounted for 12,000 metric tons and marine capture fishery contributed another 4,652 metric tons, while the remainder came from other minor lakes and rivers (Rhodes, 1966).

In spite of the fact that the 1964 fish production made a very small contribution to overall food security, optimism was abounding. Indeed, the

1965-70 fisheries development plan envisioned a tripling of fish production to 60,000 tons a year by 1970, with freshwater fisheries accounting for the additional 25,000 metric tons and marine capture fisheries for another 15,000 metric tons per annum. Over the long horizon after 1970, the planners saw astronomical possibilities of increasing fish production; the projected annual production from Lake Turkana was in the order of 50,000 to 225,000 metric tons; for Lake Victoria it was in the range of 8,000 to 40,000 metric tons; and marine capture fishery was expected to add between 65,625 and 260,000 metric tons per annum to the total national production (Rhodes, 1966; Cohen et al., 1996).

While the projections for Lake Turkana and marine capture fisheries were off the mark by huge margins, fish production in Lake Victoria exceeded all expectations. Of course, the newly discovered productivity of Lake Victoria was not the result of natural evolution in the abundance of fish species, but, rather, a result of human biomanipulation of the lake's species composition having to do with the relocation of the Nile perch and four exotic tilapias from other aquatic systems to Lake Victoria. The success of the biomanipulation project was reflected in the steady growth in fish harvest in the Kenyan portion of Lake Victoria alone from 25,000 tons in 1978 to 46,000 tons in 1981 and then to 175,000 tons in 1989 in which the Nile perch by itself accounted for 60 percent of the total harvest. For example, the haplochromis and the Nile perch traded positions in terms of contribution to the overall commercial fishery. The share of the Nile perch to total landings rose from just 1 percent in 1977 to 68 percent in 1983, whereas the share of the haplochromis dropped from 32 percent to less than 1 percent over the same period.

Correspondingly, the number of Kenyan fishers in Lake Victoria grew from 16,900 in 1971 to 31,000 in 1981, and the number of canoes deployed to the lake increased from 14,918 to 45,667 over the same period. In contrast, marine capture fisheries stagnated as 11,000 fishers harvested only 5,336 tons in 1980. In fact, of the 194,000 metric tons of national fish production in 1995 in which harvest from Lake Victoria represented 94 percent, marine fishery accounted for a mere 3 percent. Overall, though, Kenya's fishery mode of production became one of the major pillars of accumulation as 798,000 Kenyans came to depend on the sector for direct and indirect livelihoods by 1995 (Coche and Balarin, 1982; Jansen, 2007; Ogutu-Ohwayo, 1990; Ikiara and Odink, 2000; Yohannes, 2008). In 2000, Kenya harvested 202,639 metric tons of fish, of which 192,798 metric tons came from Lake Victoria; in fact, 93 percent of all fish landings in Kenya throughout the 1990s were from Lake Victoria in which the Nile perch constituted 57 percent of total fish landings (Mkombo et al., 2007). Thanks to the dominance and popularity of the Nile perch in foreign markets,

Kenya's foreign exchange earnings from Nile perch progressively increased from just $400,000 in 1980 to $60 million in 2003 (Njiru et al., 2008).

Even though Kenya's jurisdiction over Lake Victoria is limited to just 4,100 km^2 or 6 percent of the lake's surface area, its portion of the lake is considered highly productive, reflected in the fact that the country regularly accounted for 35 percent of the total Lake Victoria fish landings over the decades. However, Kenya's dominance, reflected in the growing number of fishers and vessels deployed to the lake, has been a major contributor to the decline of Nile perch fishery. Indeed, Nile perch landings on the Kenyan portion of the lake fell from the peak of 122,781 tons in 1991 to 61,416 in 1998. By 1995, the share of Nile perch had already fallen to 47 percent of total landings compared to 44 percent for the *O. niloticus*. As a result, the catch per unit effort dropped from 180 kg per boat per day in 1989 to 80 kg by 1999 (Mkombo et al., 2007).

Next to Lake Victoria, the Rift Valley lake on which the elites had pinned their hope for vast fish harvests was Lake Turkana. Indeed, as indicated earlier, in the mid-1960s, FAO experts had estimated that Kenya could harvest between 50,000 and 225,000 tons of fish annually from Lake Turkana alone, something that has come to naught.

The 4th largest lake in Africa by volume with 237 km^3, Lake Turkana is 260 km long and 30 km wide on average; it has maximum depth of 114 meters and mean depth of 31 meters. Lake Turkana is the largest lake in northwestern Kenya's arid region. For the 300,000 people in the region, the primary source of livelihood used to be livestock production. However, owing to the growing desiccation of the grazing land, the communities have increasingly resorted to fishing on the lake as a viable alternative. But this alternative is losing ground to growing hydroecological change. For example, the number of fishers grew from 150 in 1962, operating four nonmotorized boats, to 7,000 fishers with 200 nonmotorized boats in 1980, and then fell to fewer than 2,000 fishers and 150 canoes in 1988 (Kolding, 1995). Although there are 40 species of fish in Lake Turkana, only the Nile perch, tilapia, and catfish are commercially harvested. Traditionally, the annual fish catch in Lake Turkana fluctuated between 15,000 and 30,000 metric tons a year; due to increased alkalinity, the catch has fallen drastically in recent years. The annual fish catches from the lake fell from 12,384 metric tons in 1980 to 5,239 metric tons in 1999, and the average annual catch between 1999 and 2003 was far below 4,000 metric tons (Awange and Ong'ang'a, 2006; Export Processing Zones Authority, Kenya, 2005).

Lake Turkana receives 80 percent of its water from the Omo River arising in southwestern Ethiopia. But the Omo River is in the process of being harnessed for purposes of modern hydroelectrification. The Ethiopian Electricity Company is building the Gibe III hydroelectricity dam, some 600 km away from Lake Turkana at the cost of $1.7 billion; upon completion, this second largest dam in sub-Saharan Africa will hold 11 billion cubic meters of water, the equivalent of two wet years' flow of the Omo River.

This development may translate into a death sentence for Lake Turkana, which has already been shrinking in size and becoming more saline. Since the British Cambridge Expedition to the lake in 1931, the length of the lake has already decreased by 96 km and its average width has been reduced by half (Worthington and Worthington, 1933). Moreover, the delta of Lake Turkana has expanded by 380 km^2 due to sediment deposition associated with anthropogenic destruction of the lake's catchment area (Ollago and Odada, 2007). The ruling elite in Nairobi have so far remained silent on Ethiopia's decision to harness the Omo River—probably willing to sacrifice the hundreds of thousands of people who depend on the lake for survival. The reason for this silence is that Kenya has agreed to import between 200 mw and 400 mw of hydroenergy from Ethiopia annually; after all, Ethiopia's plan entails exportation of 900 mw of electricity (half of the Gibe III's installed capacity) to Sudan, Kenya, Djibouti, and Yemen (BBC News Monitoring Africa, April 3, 2009).

Lake Naivasha is Kenya's other source of commercial fishery, even though results have as yet to meet national expectations. Located in Kenya's eastern Rift Valley 190 km south of the equator, Lake Naivasha is the country's second largest body of freshwater after Lake Victoria. The lake is 140 km^2 in area and has a catchment area of 3,200 km^2. The River Malewa accounts for 90 percent of total surface inflows while the Rivers Gilgil and Karatti contribute the remaining 10 percent of riverine inflow (Hughes and Hughes, 1992; Britton et al., 2007).

Prior to colonization, there had been no known fishing activity of commercial significance in Lake Naivasha since the lake is characterized by extreme poverty in native fish species. Beginning in 1925, the colonial administration subjected Lake Naivasha to a series of biomanipulation through numerous introductions of exotic fish species in the hope of making commercial fishery a viable source of income and revenue. In 1959, commercial gillnet fishery began with successful establishment of some introduced fish species. Indeed, during the period 1959 to 1973, commercial fishery saw relative success as maximum annual fish catches reached 1,150 metric tons. However, contrary to predictions that a sustainable annual harvest could stabilize around 900 metric tons, mean annual catch stagnated around 280 metric tons between 1974 and 1988. Between 1989 and 2000, mean annual catch actually declined to 150 metric tons (Hickley, 2004; Britton et al., 2007).

By year 2000, the 120 commercial fishing boats deployed to Lake Naivasha had driven the fish populations to almost extinction, prompting the fishery department to close the lake to fishery for the entire year of 2001. When reopened in February 2002 with only 43 licensed boats, it was discovered that the benthivorous common carp (Cyprinus carpio), native to Asia, had invaded Lake Naivasha. The common carp is believed to have escaped sometime between 1997 and 2000 from fish ponds near the River Malewa and then transported to Lake Naivasha where it has established itself throughout the lake. In March 2002, 37 carp, each 220 mm in length and weighing 0.4 km, were caught in the lake (Hickley et al., 2004; Britton et al., 2007). The common carp now constitutes 97 percent of the lake's fish biomass, dominating both the fish populations and the catch composition. Until the early 2000s,, the fish catch had primarily been composed of black bass, *Oreochromis leucostictus*, and *Tilapia zillii. Barbus*, the Louisiana red swamp crayfish (*Procambarus clarkii*), and *poecilia* reticulate of South American origin have also been minor players in the fish ecology of Lake Naivasha (Hickley et al., 2004; Britton et al., 2007).

But now, as the common carp has proliferated and dominated the fish ecology, the relative importance of other species has sharply declined. For example, of the 193 metric tons hauled from Lake Naivasha in 2006, the common carp comprised 80 percent of the catch (*Business* Daily, Oct. 16, 2007). The common carp can grow to 1.5 m in length with maximum weight of 37.5 kg. The species is known for high fecundity since the average female carp lays between 360,000 and 599,000 eggs compared with 3,000 eggs for black bass and 2,000 eggs for tilapias (*Business Daily*, Oct. 16, 2007).

The common carp is generally regarded as unfriendly to the fish ecology because of its feeding habits. As a bottom lake feeder, the carp collects sediments and organisms before spitting out the sediments and retaining its prey. The species uproots aquatic macrophytes, which reinforces the resuspension of sediment particles, resulting in increased turbidity of lake water. Furthermore, the carp releases phosphorus during feeding, enhancing algal growth. Since the common carp is hardy, flexible in its habitat choice, and highly tolerant of a variety of environmental conditions, it has all the advantages in its competition with other species. Since *Tilapia zillii* in particular heavily feeds on macrophytes and depends on water clarity as a visual feeder in order to properly nourish itself, increased turbidity of the lake has extremely hampered its ability to recover. After all, since *T. zillii*'s food in Lake Naivasha consists of 67.7 percent of aquatic macrophytes, carp destruction of macrophytes is damaging to the food ecology of other species (Toowoomba, 2000).

Moreover, the dominance of carp has begun to affect the lake's vegetation ecology. In the past, the aquatic macrophytes and the introduced red Louisiana swamp crayfish traded positions: in periods when crayfish populations declined, aquatic macrophytes recovered and thrived. But now, even when crayfish populations are in flux, the macrophytes may not fully regenerate since the common carp destroy them (Hickley et al., 2004). Thus, resuspension of sediments in combination with macrophyte destruction seems to pose complications to the trophic profile and geochemistry of the lake, as indicated by the apparent correlation between the sharp decline in *Tilapia zillii* and *Micropterus* in recent catch composition, on the one hand, and the growing dominance of the common carp, on the other.

Perhaps the greatest threat to the fishery mode of production in Kenya comes from the hydroorientation of the pirate state bourgeoisie. As in the cash crops' economy, the fisheries sector has been viewed in Kenya as critical to the export-led growth of accumulation. In keeping with this objective, Kenya has long pursued the establishment of industrial fish processing factories. In 2005, there were 17 industrial fish processing companies, all engaged in the fish export subsector. Historically, Tanzanian and Ugandan harvesters landed their fish catches in Kisumu to be processed for export until Tanzania and Uganda later emulated Kenya and built their own export-oriented fish-processing factories.

Unlike in the cash crops' economy, the transformation and domination of the fishery mode of production in Kenya is an indigenous affair in the sense that the pirate state bourgeoisie have been instrumental in the intermediation of fish production for export. Just as the European colonists regarded tropical cash crops as the primary source of revenue for the colonial state and as complementary to the reproduction of capitalism in the metropol, the Kenyan elites have hitherto treated the fishery mode of production as just another source of primitive accumulation via the export of the highly prized Nile perch. However, the rise in fish production and foreign exchange earnings have come with unsustainable costs both to the lake and lakeshore communities.

Indeed, the latest focus on the fishery mode of production has exposed the fundamental contradictions in Kenya's political economy by bringing to the fore the crises of accumulation, legitimation, and ecology. Deluded by the temporary boom in Nile perch production and export, the state-driven vertical growth in investment introduced a dangerous element of competition into the fishery mode of production between the toiling fishers and industrial exploiters. This has resulted in a radical transformation of the Lake Victoria-based fishery mode of production from one based on multispecies to another based on three commercial species. Following the sharp reduction in the relative abundance of some species and the extirpation of other species, commercial fishing in Lake Victoria now depends on the Nile perch, *Rastrineobola argentea*, and the exotic tilapiine. In the early years of the new century, in the Kenyan portion of Lake Victoria, the perch, the *R. argentea,* and the Nile tilapiine represented 58 percent, 30 percent, and 10 percent of the commercial catch respectively (Abila, 2003; Awange and Ong'ang'a, 2006).

The lacustrine resource depletion resulting from the vertical expansion of capitalist accumulation is compounded by the expansion of horizontal growth of the fishing population. Since the 1970s, the number of Kenyans who depend on the fishery sector for their livelihood has ballooned to 800,000 (Abila, 2003). What emerges from the vertical and horizontal expansion of fishing is a pageantry of wealth for the few and a grim picture of poverty for the many. Since the 1980s, the Nile perch-dominated fishery mode of production transformed the Kenyan portion of Victoria into a focus for foreign capital investment. Indeed, Kisumu has become the lake's capital of fish processing plants. In year 2000, for example, 38,000 Kenyan fishers, operating 11,200 boats, hauled 207,000 metric tons of fish, 93 percent from Lake Victoria, with a market value of $125 million, to keep the fish factories going (Awange and Ong'ang'a, 2006). Yet, food security of lakeshore communities has been eroded. In fact, the Nyanza region bordering Lake Victoria has been named "the poverty belt" since the poorest of Kenya's population are concentrated along this belt. In the port city of Kisumu itself, the rate of poverty increased from 48 percent in 1994 to 64 percent in 1997. For the Nyanza region as a whole, the level of poverty in 1997 was 58 percent (Awange and Ong'ang'a, 2006).

As the competition between the commercial processing factories and artisanal fishers intensified, Nile perch populations in the Kenyan portion of Lake Victoria have also begun to fall drastically. It is noteworthy to mention that Nile perch catches in Kenyan waters first rose from 19,000 metric tons in 1977 to 253,000 metric tons in 1991, and then fell to 53,000 metric tons in 2000. In fact, in light of this sharp reduction in perch supply, six fish processing factories folded up their operations and the remaining ones were forced to cut their utilization capacity by more than half (Awange and Ong'ang'a, 2006).

Furthermore, 70 percent of the Nile perch catches are now juvenile. In the early 1980s, the Nile perch caught by harvesters were between 50 and 100 kg by weight, which drastically fell to between five and ten kg by the mid-1990s. Correspondingly, the average mean mesh sizes of gill nets, necessitated by the imperative to catch smaller fish, dropped sharply from 12 inches in 1981 to 6 inches in 1996. With the decline in Nile perch populations, the expenditure of effort also substantially increased. The catch per unit of effort fell from 180 kg per boat per day in 1989 to 80 kg per boat per day by 1999 (Abila, 2003; Yohannes, 2008). Moreover, although the biomanipulators put their estimates of the commercially maximum sustainable yield for Nile perch at 39,200 metric tons per annum on the Kenyan portion of Lake Victoria, between 1997 and 2002, Kenya took out

90,000 metric tons of perch annually from the lake anyway, largely accounted for by juvenile perch (Abila, 2003). Thus, since the beginning of the 1990s, the vicious cycle of economic overfishing and biological overfishing was set in motion. As more and more harvesters (both industrial and artisanal) decreased the mesh sizes of their gill nets in order to catch ever smaller Nile perch to maintain volume of harvest (since bigger ones were fished out) young Nile perch have been caught before reaching sufficient maturity to reproduce themselves and maintain the Nile perch populations above the critical depensation curve.

This fundamental contradiction between the necessity to survive on the part of the toiling masses and the logic of accumulation to strengthen the position of capital could be resolved only when the prevailing social order is contested and replaced by a social order that prioritizes just distribution over accumulation, development over growth, and promotion of domestic welfare over export of fish to feed the affluent global northerners. As of this writing, the Nile perch-based fishery mode of production in Kenya had virtually collapsed, reflected in the fact that Kenyan export of perch fillets to the European Union sharply fell from 5,200 metric tons in 2005 to 1,200 metric tons in 2008 (Josupeit, 2009).

The predicament of Kenya's fishery mode of production is aggravated by the fact that the fisheries in Kenya's other Rift Valley lakes have also collapsed. These satellite lakes have long been undergoing processes of resource depletion, pollution, and desiccation. As a result, fish harvests from these lakes have progressively diminished over recent decades.

The crisis in the fishery mode of production in Kenya is occurring not in isolation from but in conjunction with the general crisis of dependent capitalism resulting from the multiplication of contradictions and the intensification of the class struggle. To mask the class character of the contradictions, Kenya's pirate state bourgeoisie have continually resorted to exploiting the external markers of identity by conflating the idiom of opposition to the dependent capitalist state with ethnic-based animosity toward the Kikuyu. Jomo Kenyatta presented the opposition to his land policy by the liberal wing of the national elites, led by Odinga Odinga, as originating with the Luo because Odinga happened to be Luo. The consequent conflation of Kenyatta's regime with Kikuyu identity served to transform ethnicity as a central organizing principle of Kenyan politics. Kenyatta's successor, Arap Moi, took the imitation to another level. The construction of his patron-client system reflected the elevation of the Kalenjin collectivity in Kenyan politics.

The pattern of elite rule in Kenya has thus come to be embedded in the logic of ethnic mobilization whenever demands are made on the ruling class by internal forces or by international donors to change direction. In 1991, for example, sensing that Kenya was set for an implosion in the face of popular opposition to the Moi regime, the multilateral financial institutions and donor countries suspended $1 billion on which Moi's survival hinged. To mollify donor concerns, Moi allowed a façade of multiparty formation and elections. However, while partially opening the system for multiparty elections, he unleashed Kalenjin and Maasai vigilantes against the Kikuyu and Luo as the standard- bearers of popular opposition to the Kalenjin-dominated regime. Multiparty elections were held in 1992 and Moi marginally won by 36 percent in the midst of fractionalized opposition and bloody skirmishes. Even though 1,500 Kenyans lost their lives and more than 300,000 people were displaced by the violence, Moi's international patrons restored $800 million in balance of payment support as an expression of satisfaction with the outcome of the simulated elections (Khal, 2006). Again, in the post-2007 election violence, more than 1,500 Kenyans lost their lives, and another 300,000 were displaced.

It all began in the Rift Valley and western province by the Kalenjin and Luo mostly directed at the Kikuyu identified with Kibaki -- who was accused of stealing the election from the Luo Raila Odinga -- and his Kalenjin partner, William Ruto. As of this writing, Kenyan politicians and their ethnic supporters appear to be preparing for another bloody showdown in future elections. The Kikuyu paramilitary gang, known as Mungiki, was fanning narrow nationalism, embedded in cultist and millenarian ideology. Kalenjin militia elements were also rearming themselves for a potential showdown (Wrong, 2009).

TANZANIA: THE POLITICAL ECONOMY OF COMMODITIES AND THE FISHERY MODE OF PRODUCTION

Situated on the western shores of the Indian Ocean between Kenya and Mozambique, Tanzania has an area of 947,300 km^2 in which its water bodies cover 61,500 km^2. The country's coastline stretches for 1,400 km, giving the country economic sovereignty over 223,000 km^2. The country is twice the size of California.

A general review of Tanzania's political economy suggests similar conditions as those prevailing in Kenya. Indeed, a quick review of its economic status reveals a disheartening picture. In 2005, Tanzania ranked 162nd out of 177 countries on the human development index; the yearly per capita income of its 37 million population was $300; agriculture accounted for 85 percent of exports and for 80 percent of rural employment; and foreign aid represented 30 percent of its total revenue (European Union, March 2007; Rotberg, 2007).

Even though the way Tanzania came into being was slightly different from its neighbors, the country was nonetheless inducted into a category of nations assigned to specialize in cash crop production. In rapid succession, coffee, cotton, cashews, tobacco, and tea became the major pillars of the colonial economy and have continued to be the mainstays of Tanzanian economy since these five cash crops still account for 85 percent of Tanzania's foreign exchange earnings. The ruling elites rely on smallholder-grown cash crops to sustain themselves in power and, as in Kenya, the elite compel smallholder farmers to sell their farm products to the marketing parastatals, which have monopoly control over prices.

Coffee has reigned for a long time in Tanzania as the premier cash crop. Between 1880 and 1890, Catholic missionaries started growing coffee in the Morogoro region around Mount Kilimanjaro, and from there the plant spread in many directions. Since the early 1960s, the importance of coffee has steadily been growing. Today, almost 1.8 million Tanzanians are directly involved in coffee production on 260,000 ha of land, and some four million people derive significant portions of their income from the coffee sector (Tanzania Coffee Board, 2007). Unlike Kenya, 90 percent of coffee producers in Tanzania are smallholders. According to official accounting, there is much room for further expansion of coffee cultivation since, of the 733,800 ha of land considered suitable for optimal coffee production, only 260,000 ha are under coffee production. In addition, there are more than 4.8 million ha regarded as marginally suitable for coffee cultivation (Tanzania Coffee Board, 2007). Given these estimations, the room for further expansion appears virtually unlimited.

In any case, the coffee sector in Tanzania continues to experience consistent declines in output, owing to a mix of factors including low world price, low inputs, aging coffee trees, and spread of diseases, exacerbated by an extortionate rate of exploitation of coffee farmers. These problems are reflected in the fact that annual coffee production has shown consistent decline from 53,000 tons during the 1976 to 1985 period to 44,400 tons during the 1993 to 1997 period, despite increases in acreages under cultivation (Tanzania Coffee Board, 2007).

Cotton was natural to Tanzania long before the arrival of European settlers. However, it was German settlers who started experimenting with commercial cotton production on a plantation level beginning in 1904. When the plantation system proved to be futile, smallholder producers stepped into the void, a condition that made the sector primarily a smallholder affair in which 5 million Tanzanians are directly or indirectly involved in cotton production. By 1966, annual cotton production reached 80,000 tons, even though production began to experience a steep decline in the 1970s and 1980s (Baffes, 2002).

Customarily, producers are always blamed for decline in quantity and quality of cotton. Ali Ngongolo, General Manager of Tanzania's cotton marketing board, for example, once observed that Tanzania's annual cotton production of around 400,000 bales could have been 1 million bales if growers were good farmers. As he put it:

> What is needed is discipline to grow cotton. We have the potential to produce one million bales a year. Our main problem is that some of the farmers don't understand why care is important and others do not want to get involved in anything with any extra cost. They don't understand that the money they have to pay for fertilizer is paid back many times over by higher yields. ... If we let the farmers grow the cotton without any minimum standards, they might spoil the good name of Tanzanian cotton. (Wallenren, 1997)

Beginning in the early 1990s, Tanzania took a series of measures to intensify extraction of surplus value from cotton growers. In the 1994-95 season, farmers put 400,000 ha under cotton cultivation, which allowed the government to earn $126 million by exporting 440,000 bales (Wallenren, 1997). In 2005-06, farmers produced 575,000 bales on 500,000 ha of which 402,000 bales were exported (Kearney, 2006b). Today, cotton cultivation is undertaken by some half a million Tanzanians, the average holding being 1.5 ha (Baffes, 2005). Although the government plans to raise current production to 1.5 million bales a year by 2015, cotton production has been in decline due to ecological deterioration and soil nutrient depletion. The average yield of Tanzanian cotton is 0.4 metric tons per ha compared with 0.6 to 2.0 metric tons per ha in other African countries. In the view of observers, annual production of 300,000 metric tons of cotton seed would be needed to make the sector sustainable (Obonyo, 2007).

In spite of the fact that productivity is diminishing due to nutrient depletion and ecological degradation, the Tanzanian regime, donor countries, and lending institutions still view the cotton sector as a priority for poverty alleviation by stimulating the sector with private investment. Presumably, Tanzania still has over 500,000 ha of land deemed suitable for cotton cultivation, the utilization of which can bring annual cotton production to over 1 million bales, generating $167 million in hard currency (Kearney, 2006).

In order to lure foreign capital flow into the cotton sector, Tanzania enacted an investor-friendly land reform in 1997, granting persons the ability to freely acquire and sell land, cushioned by exemptions from value-added taxation, import duties, and the right to repatriate up to 100 percent of profits after taxes are paid (Kearney, 2006). Such plans, of course, betray both illusion and desperation. Donor countries and lending institutions have a vested interest in stimulating Tanzanian cotton production, in part to enable the country to earn hard currency in order to service its debt, and partly to induce global structural oversupply in order to lower the price for manufacturers. It bears to note that, in March, 2009, 160,000 bales of cotton from the 2008-09 harvest season were dumped in storage as the world price collapsed to $0.40 a pound from the historical average of $0.72 (Ihucha, 2009). In fact, perturbed by the potential implication of such a price collapse for the cotton industry, the minister of agriculture directly appealed to farmers not to abandon their cotton fields until the government found ways to redeem the industry (Ihucha, 2009).

The cashew nut sector was another addition to the sources of primary accumulation in Tanzania. The sector is a smallholder affair. After starting slow, cashew tree planting gathered steam in the 1950s and 1960s. As a result, exports of cashew nuts increased from 7,000 metric tons in the late 1940s to 42,000 metric tons by 1960, and then to 145,000 metric tons by the 1973-74 season (Sijaona, 2002). By 2002, there were 280,000 households growing cashew trees on 400,000 ha, the average holding being between 1 and 2 ha; dominated by smallholders, large-scale cashew nut operations are limited to just 2,000 ha (Sijaona, 2002).

As in the other cash sectors, cashew nut growers complain about low return on their labor investment, endless taxes, and oversupply, all of which lend themselves to poverty and low production. In the 1986-87 growing season, for instance, Tanzania produced a mere 16,000 metric tons. Since then, production gradually rose until it reached 122,283 metric tons in the 1999-2000 season (Sijaona, 2002). As of 2007, annual cashew nut production was 90,000 metric tons. Even though the country's potential processing capacity is over 100,000 metric tons, 95 percent of the cashew nuts produced in Tanzania is exported raw (Sijaona, 2002; Obonyo, 2007).

Tobacco ranks fourth in the portfolio of Tanzania's cash crops. Tanzania produces three types of tobacco: burley, flue-cured, and fire-cured. Despite international antismoking campaigns, the demand for Tanzanian tobacco has been rising. In response, tobacco production in Tanzania increased from 16,000 metric tons in 1990-91 to 40,000 metric tons in 2006-07 (Obonyo, 2007). The area under tobacco cultivation in 2007 was 47,000 ha, worked by 50,000 smallholder farmers. More than 1.5 million Tanzanians make a living on tobacco production (Africa Research Bulletin, May 9, 2008).

Sisal is another source of primary accumulation. It is a perennial plant, known for its strong fiber used in making rope, baler twine, and even carpet. Sisal was first introduced into Tanzania by German settlers in 1892, and rapidly grew into a plantation system. By 1913, Tanzania was producing 20,000 tons of sisal fiber a year. Indeed, Tanzania was the leading sisal fiber producer in the world until the 1970s, producing 200,000 tons a year with 24 percent of world market share (Pass Limited, 2001).

As in the case of all cash crops, the sisal industry fell on hard times due to oversupply, arrival of new competitors, and production of synthetic materials, which have forced Tanzania to relinquish its premier position as a supplier of sisal to global markets. Now, Tanzania is fourth in the ranking of sisal producers. In 2000, the sisal estates together were producing sisal on 46,120 ha, yielding 20,000 metric tons of fiber, way down from 304,680 ha in 1967 when the estates generated 233,000 metric tons of fiber. By the same token, sisal fiber export correspondingly fell from the 1970 peak of 217,000 metric tons to 6,500 metric tons in 2000 (Pass Limited, 2001). As of 2007, there were 82 sisal estates, producing 34,000 metric tons of sisal fiber (Obonyo, 2007.

Tea was first introduced into Tanzania in 1902 by German settlers, whose tea estates were under exclusive European monopoly. Commercial production on a large scale began in 1926 for both domestic and international consumption, even though export of Tanzanian tea was still modest. For example, in 1934, the settlers managed to export just 23 tons of tea (Baffes, 2003). After World War II, the area under tea cultivation expanded, rising from 2,723 ha in 1945 to 7,336 ha in 1960 (Baffes, 2003). In the mid-1960s, however, the Afrosocialist regime recognized smallholder tea producers as vital to realizing the fundamental aims of romantic peasant socialism, a necessary prerequisite for self-sufficiency, enhanced production, and easing unemployment. As a result, spatial segregation of the tea sector into large estates and a smallholder subsector emerged.

To ease the transition from a plantation system to smallholder production, the World Bank provided Tanzania two loans totaling $8.5 million to enable 14,000 smallholder farmers to plant 8,300 ha with tea trees between 1967 and 1970 (Baffes, 2003). Subsequently, smallholder tea cultivation began to prosper until the adoption of a structural adjustment program and the consequent imposition of liberalization on the country's macroeconomy reversed the gains made by small tea farmers. The share of smallholder tea cultivators had progressively reached 29 percent by 1985, when 28,700 farmers had put 9,000 ha under tea production. Thereafter, small tea cultivators saw their gains quickly evaporating in the face of giant multinational competitors. In fact, by 1998 the share of smallholder tea cultivators had fallen to 5 percent of total output.

The sharp decrease in the smallholder role in the sector was due to lack of financial and credit support, exacerbated by the price squeeze on small farmers by the marketing parastatal and a range of taxes and levies. For example, smallholder productivity decreased from 300 kg per ha in 1975-76 to 198 kg per ha in 1999-2000, whereas the yields on commercial estates jumped from 1,200 kg per ha to 2,115 kg per ha, due to mechanization and massive agrochemical application (Butler, 2005). Moreover, whereas smallholder tea growers are forced to take their produce to parastatals for processing and marketing, the multinational tea companies are vertically integrated, controlling the entire tea operation from planting and harvesting, to processing and marketing, thereby preempting state intrusion and squeeze.

Of the three dominant tea-processing factories, Brooke Bond is the largest tea producer in Tanzania, accounting for 9,500 metric tons of Tanzania's 25,000 metric tons of total tea production in 2004 (SOMO, 2008). Although tea raked in $25 million in foreign exchange earnings in 2004 (making it the fifth biggest hard currency earner), the profile of tea in the portfolio of cash crops has been steadily rising in relation to the other traditional chief foreign exchange earners: cotton, coffee, and cashew nuts. In 2006, Tanzania produced 31,378 metric tons of tea and exported 24,142 metric tons, generating $32.9 million (SOMO, 2008). The tea sector directly employs 50,000 families and another 2 million indirectly eke out a living from the sector. Given these significant employment figures, the government views the tea industry as one of the vital sources of primary accumulation. In the imagination of the regime, Tanzania can potentially put more than 100,000 ha under tea cultivation, capable of producing 1.3 billion kg, valued at $1.4 billion a year (Pass Limited, 2001). In 2008, there were 22,722 ha under tea cultivation, almost evenly divided between estates and smallholder farms. The government's plan is to raise smallholder cultivation by 5,300 ha and that of large estates under multinational companies by 40,000 ha, raising annual tea production to 80,000 tons, the expected hard currency earnings being over $86 million a year (Pass Limited, 2001; Tea Board of Tanzania, 2008).

An area that exhibits a glaring weakness in terms of primary commodity production is sugarcane production. The country has consistently experienced a deficit in the sugar sector. In 2007, Tanzania produced just 115,000 metric tons of refined sugar, requiring importation of 200,000 additional metric tons to cover the deficit (Obonyo, 2007).

The tourism sector is another source of accumulation for the pirate state bourgeoisie. Of course, tourism is intimately connected to the country's nature reserves. Before the 1998 bombing of the American embassy in Dar es Salaam, more than 600,000 tourists visited the country every year. In 2004, 582,000 tourists paid a visit to the country, bringing with them $746 million in hard currency. Tourism now contributes 5 percent to GDP and directly employs 200,000 Tanzanians (World Bank, 2008).

The mining sector, officially contributing 3.5 percent to GDP, is still one of the principal sources of accumulation. Under conditions of liberalization since the 1980s, the mining sector has seen flurries of activities by both global corporations and artisanal miners. In 1998, for instance, the mining sector saw a growth rate of 27.4 percent, resulting from massive foreign investment. In 2003, the amount of gold extracted by large corporations alone was 45,000 kg (World Bank, 2008). In addition to large-scale operations, the regime encouraged small-scale operations in order to increase its own receipts. Mineral resource exporters are allowed to use the proceeds to pay for imports, theoretically for capital or intermediate goods. This permissive policy has lent itself to the rise of people employed in small-scale mining operations to over 500,000 by 1995 and to more than 1 million by 2007; this sharply contrasts with the 30,000 people employed in large-scale mining operations (World Bank, 2008).

However, liberalization of the mining sector has unleashed powerful forces that the regime cannot control. In 1995, for instance, 90 percent of the 550,000 miners were unlicensed and therefore not reporting anything to the regime. In 1998, according to official accounting, Tanzania earned $22 million from the export of precious metals; in contrast, the estimated value of precious metals smuggled out of the country in the same year was a whopping $136 million (smuggled gold alone was between $45 million and $200 million) (World Bank, 2008).

The dark side of specialization in the production and supply of primary commodities is the commodification and exploitation of child labor. One out of three children in rural Tanzania, between the ages of 10 and 14, work on commercial coffee, tobacco, sisal, or tea plantations, being paid half that of an adult laborer for an 11-17 hour workday. On some sisal plantations, children represent 30 percent of the workforce (Global Exchange, June 1, 2008). In the Geita mining district, 12.5 percent of the workforce is comprised of children. As of 2007, there were 3,000 child laborers in the Mererani Tanzanite Mines, nicknamed "snake boys" because they work in deep narrow mines (World Bank, 2008). Children who work in quarries and mines suffer from dust inhalation, silicosis, exposure to mercury, and dangerous injuries of all sorts.

Like boys, young girls are also heavily used in mines. In fact, women constitute 25 percent of miners, directly engaged in all forms of activities for one third of a man's wage (World Bank, 2008). Young girls, who leave the rural areas on empty promises of jobs in the city, also fall victim to urban predators after finding out that domestic servitude amounts to 17-18 hours per workday. They escape urban domestic servitude, but often, only to make a transition to prostitution. By some estimates, 25 percent of urban prostitutes in Tanzania are former domestic servants (Global Exchange, June 1, 2008).

All the sectors so far discussed share a number of characteristics that, in combination, have made Tanzania's environment and division of labor unsustainable.

First, the country's political economy -- primarily rooted in the sections just described -- has taken a heavy toll on the country's water bodies. Residues of synthetic nitrogen, pesticides, and herbicides from commercial plantations along with toxic pollution from mining operations have been regularly inserted into Lakes Victoria, Tanganyika, Malawi, and other satellite lakes, accelerating widespread sedimentation and cultural eutrophication. Particularly in Lake Victoria, Tanzania's plantation system and gold mines have become major contributors to the state transition of the lake.

Second, cash crops, fish production, mining, and tourism are all susceptible to stochastic external shocks. In 1999, for example, fish exports from Tanzania dropped by 50 percent after the EU banned importation of fish from Lake Victoria, citing poor sanitation conditions. In addition to having lost export revenue, one third of the 200,000 persons employed in the fisheries sector lost their jobs; many more saw their earnings slashed by one third. A number of fish-processing factories either closed their operations or were operating at 20 percent of capacity (World Bank, 2008).

Tanzania's rentier economy is also susceptible to variability in natural conditions. As a result, Tanzania's effort at rapid modernization has been hampered by important natural conditions. Of the country's total area of 945,000 km^2, only 4.23 percent is suitable for rain fed agriculture. In 2005, the country had 1,840 km^2 under irrigation. In spite of this gross structural handicap, Tanzania depends on the sector for 27 percent of the GDP, 85 percent of export, and 85 percent of employment. This structural anomaly manifests itself in the ratio of exports to imports; in 2008, the country's monetary value of exports was $2.49 billion as against the value of imports at $5.9 billion.

Since Tanzania abandoned its experimentation in Afrosocialism in the mid-1980s, the country has been much touted as a model example of structural adjustment program. The chief sponsors of the country's neoliberal policy are quick to attribute the modest gains in the macroeconomic growth measured in terms of GDP to market reforms and liberalized trade. The microconditions of the mass population, however, tell an entirely different story. A recent survey indicates that the country is far behind in terms of providing clean drinking water, basic sanitation, health care, and general poverty alleviation; 80 percent of the population is still involved in subsistence agriculture and all of the vagaries associated with precarious existence. In fact, the number of Tanzanians who survive on less than a $1 a day increased by 1 million to 12.7 million since 2001. In 2007, 90 percent of the country's 40 million people had no access to electricity and only 7 percent had a flushing toilet; even in Dar es Salaam, 42 percent of the residents had no piped water (Inter Press Service, Dec. 4, 2008). All this persists despite the fact that the country's GDP grew by 7 percent during the same period because of foreign aid, which accounted for 40 percent of government budget (Inter Press Service, Dec. 4, 2008).

In order to buttress the neoliberal growth model against popular backlash, "poverty reduction through growth" has become a highly orchestrated mantra of the government and its global sponsors. In April 2007, the World Bank agreed to provide Tanzania a loan of $190 million to support its growth and poverty-alleviation programs. The stated purpose of the loan was to strengthen the performance of the economy by accelerating institutional, social, and economic reforms (*Africa News*, April 25, 2007). In October 2008, the African Development Bank also agreed to funnel $156 million to Tanzania in budgetary support for purposes of growth and poverty reduction programs through 2010. The primary aim of this loan was to improve the business climate by undertaking further deregulatory, legal, and procurement reforms (*Africa News*, Oct. 30, 2008). Yet the hefty prize was the $698 million that Tanzania was slated to receive from the U.S. Millennium Challenge Corporation over a five year period (BBC News Monitoring Africa, Sept. 19, 2007). The familiar reference justifying this provision of aid is "poverty reduction" through "growth," in effect meaning improving the business climate by investing in infrastructure, transport, energy, water supply, and diversification of cash crops. However, instead of growth in the welfare of its citizens, Tanzania has seen its indebtedness grow to over $5.3 billion by the end of 2008.

It is in the context of the intractable levels of structural poverty and overdependence on cash crops that Tanzania's fishery mode of production must be examine. Tanzania has increasingly been looking to its inland water bodies and coastal waters for relief. The country's earnings from fish exports steadily increased from just $8.3 million in 1992 to over $189 million in 2007. The yearly average of foreign exchange earnings from the export of fish between 2003 and 2008 was $147.5 million.

Corresponding to intensive activities, the fisheries sector has grown by over 6 percent a year between 2000 and 2007 (World Bank, 2008). In the context of the fact that the country's GDP was $12.78 billion in 2006 and the per capita of 39 million Tanzanians was $308, obtaining 27 percent of their protein intake from fish, the role of fisheries in the economy is very crucial. In 2005, the contribution of fisheries to GDP was $321 million. Primary employment in the fisheries sector in 2005 was over 200,000, of which 149,946 were artisanal; another 2 million were engaged in secondary and tertiary fishery-related activities such as boat building, net mending, and fish processing and marketing. The statistics stated above do not accurately reflect the total number of beneficiaries from the freshwater fishery regime since some 200,000 more persons regularly catch fish for household consumption (World Bank, 2008).

Tanzania's fisheries sector is defined by three regional fishery regimes. Based on the Indian Ocean, the first fishery regime comprises mainly artisanal fishers engaged in demersal and pelagic fishing for subsistence, as well as coastal commercial prawn harvesting. Under this regime, the number of persons catching fish simply for their own consumption can be as high as 120,000.

The second fishery regime is based on offshore pelagic marine fishery in the exclusive economic zone based on the collection of lobsters, sharks, snappers, kingfish, shrimps, rays, sardines, and sea cucumbers, handled and processed by two factory ships, 21 processing plants, and 26 freezer vessels operating in the Exclusive Economic Zone (EEZ). Between 1988 and 2003, marine-capture fishery production was stable at 50,000 metric tons per annum (European Union, March 2007). For all practical purposes, the industrial fisheries subsector in Tanzania's marine offshore fishery is dominated and controlled by foreign trawlers, whose numbers steadily rose from less than 10 in 1998 to more than 170 in 2004, indicating intensified government policy to accelerate marine resources extraction. In so doing, however, Tanzania has become complicitous in helping to accelerate marine resource depletion and degradation of marine ecology with little benefits accruing for the country either now or in the future. In 2004, for example, the foreign vessels collected $300 million worth of fish, of which only $3.4 million went to the Tanzanian regime. Furthermore, even though the estimate of maximum sustainable yield in the Tanzanian EEZ is in the range of 280,000 tons to 350,000 tons a year, foreign vessels have been removing between 400,000 tons and 450,000 tons per year (World Bank, 2008).

Corrupt licensing practices, compounded by gross inability to monitor the operations of the foreign vessels, have placed Tanzania's marine ecology in serious jeopardy. The problem is compounded by the fact that Tanzania shares the western Indian Ocean fisheries with Kenya, Madagascar, Mozambique, Mauritius, Seychelles, the Comoros, and Somalia. Even though the region's fishery potential is estimated at 4 million metric tons per annum, the fishery is prone to overexploitation by both national and long-distant fishing vessels, largely attracted by the species richness of the region, which has more than 2,200 fish species, representing 15 percent of total world marine species.

In the decade leading up to 2004, 61 countries reported to have harvested fish in the western Indian Ocean. Foreign vessels accounted for most of the valuable fish catch because they enjoy comparative advantage in technology and powerful gear that allow them to go into deeper areas of the ocean (Elst, et al., 2005). Moreover, 60 million of the region's population, of whom 3 million are dependent on marine fish resources, live within 100 km of the shores of the Indian Ocean, with greater dependence on marine resources. The combined effect of national and long-distant fishing is serial fishing, which expresses itself in increased harvest of previously commercially less important species. For example, in the 1950s, the composition of the catch involved 85 species, which rose to 150 species in 2000. Because of foreign vessels, the annual catch of orange roughy in the region declined by 85 percent in a matter of two years since its discovery in 1999 (Elst, et al., 2005). In the case of Tanzania, 21 percent of the country's population lives along the coast of the Indian Ocean. Of the active fishers, 95 percent are artisanal while 5 percent play a minor role in industrial marine fishing (Elst, et al., 2005).

The third fishery regime in Tanzania is based on inland water bodies, centered on Lakes Victoria, Tanganyika, and Malawi, which together account for 86 percent of total catches from inland water bodies. Indeed, covering 6.5 percent of the country's total area, Tanzania's inland waters provide 80 to 90 percent of total annual fish production, reflected in the fact that catches from inland waters increased from 294,782 tons in 1993 to more than 300,000 tons in 2003 compared to 40,000 metric tons of marine capture (World Bank, 2008; Magnet, et al., 2000). In 2000, there were roughly 200,000 fishers directly dependent on inland fisheries for their livelihoods, who collectively operated a fleet of 25,000 small craft, principally dugout canoes and plank boats; in 2002, 80,000 fishers were operating from 22,000 landing sites on Lake Victoria alone (Magnet, et al., 2003; European Union, March 2006). Thus, in terms of employment, supply of protein to the majority of Tanzanians, and hard currency earnings for the pirate state bourgeoisie, the freshwater fishery regime is critical to the overall economy of the country.

As in Kenya and Uganda, Lake Victoria has been the most reliable cashcow for Tanzania. The country's fish catch from Lake Victoria alone rose from 146,000 metric tons in 1988 to 231,600 metric tons in 1990, which remained fairly stable at that level throughout the 1990s (LVFO, 1999-2015). The fulcrum of the rising yields of fish harvest from Lake Victoria has been the perch, whose catch steadily rose from less than 1,000 metric tons in 1980 to 124,000 metric tons in 1986 (Jansen, 2007). By the mid-1990s, 3,500 craft were deployed to chasing the perch and three-quarters of the perch caught were directly delivered to fish-processing factories (Gibbon, 1997). Of the total fish catch of 334,000 metric tons in 2002, Nile perch from Lake Victoria accounted for 92,000 metric tons; more important, of the export earnings of $91.1 million that year, Nile perch accounted for $77.1 million (E. Wilson, 2004).

However, the situation on the Tanzanian portion of Lake Victoria is no different from what prevails in the lacustrine jurisdiction of the other littoral states in terms of resource depletion. In fact, the share of Nile perch in the total landings fell from 80 percent in 1995 to 40 percent by 2000, while that of the *R. argentea* increased to 49 percent of the total catches. In fact, catches per net dropped by 67 percent between 1997 and 2000. Even though the MSY was put at 81,947 tons per annum on the Tanzanian side of Lake Victoria, Tanzanians continued to haul 138,324 tons a year, suggesting the capture of immature Nile perch (Mkombo, et al., 2007). This has prompted the government to blame illegal harvesting and smuggling as the main source of the problem. By some estimates, more than 36,000 metric tons out of a total annual harvest of 175,000 metric tons from Tanzania's portion of Lake Victoria had been smuggled to neighboring countries. Since 70 percent of Tanzania's harvest from lake Victoria consists of Nile perch, smuggling fish of this quantity would represent a significant loss to fish-processing factories. Moreover, the collapse of Lake Victoria fishery has put the livelihoods of 550,000 Tanzanians in jeopardy (Yohannes, 2008).

Presumably, to put an end to both illegal fishing and smuggling, in 2005 the government deployed three speedboats (aided by newly acquired surveillance planes) to its portion of Lake Victoria. As the problem persisted, in May 2008 Tanzania's president ordered an investigation into why catch of the Nile perch dropped sharply. He gave the order when he was actually inaugurating the launching of a giant fish factory with a daily processing capacity of 60 metric tons, which promised to generate 600 jobs. To make his point, the president used statistics supplied by the fisheries department. According to his presentation, in addition to rising so-called illegal fishing, the number of fishers rose from 55,985 in 2000 to over 98,000 in 2006 while their fishing vessels increased from 15,434 to 29,732 over the same period (*Africa News*, May 14, 2008). The major concern of the pirate bourgeoisie is, of course, loss of hard currency since export of perch products has been declining. For example, Tanzania's shipment of perch fillets to the European Union fell from 24,000 metric tons in 2005 to just 6,100 metric tons in 2008 (Josupeit, 2009).

UGANDA: THE POLITICAL ECONOMY OF COMMODITIES AND THE FISHERY MODE OF PRODUCTION

This landlocked country in East Africa has a total area of 241,000 km^2 in which its inland water bodies cover 44,000 km^2. The history of Uganda's political economy parallels that of its littoral neighbors, although some minor particulars distinguish Uganda's in that the initial focus of the colonizers was on rubber and cotton plantations. The exhilaration over potential accumulation of wealth in rubber and cotton plantations was little short of intoxicating. To cash in on the arrival of the automobile, 55 rubber companies were rushed into existence in the decade prior to World War I to promote the development and production of rubber plantations in tropical Africa; of these 22 were deployed to East Africa. Rising prices for rubber companies and the production of rubber in Tanganyika by German companies were important considerations for the deployment of the British companies to Kenya and Uganda (Munro, 1983). Even though the outcome of the rush into rubber exploitation was in general disappointing, some British firms did manage to establish some rubber plantations as well as to gather wild rubber.

The story of cotton production in Uganda was entirely different. The British Cotton Growers Association (BCGA) had long warned of the danger of overdependence on American cotton supply. After all, during World War I, the United states accounted for 14 million bales of the annual world production of cotton of 20 million bales; the U.S. was also consuming 55 percent of its own cotton. In effect, the U.S. had a stranglehold on the world supply of cotton, especially in times of bad harvests (Himbury, 1918). From the perspective of the BCGA, any supply disruption owing to bad harvests or political reasons might cause the British textile industry to collapse, affecting the entire economy. According to the association and its supporters, 10 million Britons had their livelihood tied to the cotton trade. In 1913, of the 411.2 million pounds total of British manufactured exports, one-third or 127.2 million pounds was accounted for by cotton goods; domestic cotton trade accounted for another 31.8 million pounds (Himbury, 1918). Since Great Britain was annually importing 80 percent of its cotton from the United States, transforming Uganda into a major cotton producer and supplier of cotton to Lancashire was the answer to overdependence on America.

To realize this vision, the BCGA collaborated with Catholic missionaries to introduce modern cotton cultivation in Uganda by distributing free cotton seeds and farm implements to indigenous Africans. When peasants were not lured to cotton cultivation, the poll tax and coercion were used in increasing proportions to compel the peasants to grow cotton; in order to generate enough cash to pay the tax and in hopes of limiting the severity of coercion, the peasants began to grow cotton in increasing numbers. The poll tax rates were progressively modulated in such a manner as to compel the peasants to put more and more acreage under cotton cultivation; otherwise, they might cultivate just enough cotton to cover the poll tax. Even local chiefs on their own cotton fields (C. Taylor, 1978; Bowles, 1975; Jamal 1978). The result was that cotton emerged as a premier cash crop; equally important, cotton was peasant grown, absorbing 80 percent of rural labor.

The empire builders now concluded that they had finally found a colony that could shield Lancashire against potential supply disruption from America. Raw cotton exports from Uganda now increased by leaps and bounds. In 1904, Uganda managed to export only 54 bales of raw cotton, which increase to 4,000 bales by 1908, to 20,000 bales by 1912, and to 40,000 bales by 1914. Given this geometric progression in exports, the projection was that Uganda could export 400,000 to 500,000 bales of high-quality cotton annually (Himbury, 1918). In 1919, Uganda exported 40,000 bales of cotton to Britain, which rose to 63,594 bales in 1922 (Taylor, 1978). Indeed, peasant cotton production was so successful that it created its own contradiction in which it manifested itself in the new European planters seeing peasant cotton cultivation as a major obstacle to European domination over primitive land accumulation and monopoly control over the labor market in the interest of capital.

As World War I drew to conclusion, the European planters had been accelerating their efforts to grab more and more land, which led to a sudden rise of land value in Buganda and the eastern province, where most of the plantations were concentrated in the nmidst of flourishing peasant cotton production. This was indicated by the fact that, in a matter of one year, the number of European plantations increased from 138 in 1918 to 223 in 1919, having proprietorship over 188 square miles of land (Taylor, 1978). The European planters understood well that, without further alienation of land and the appropriation of local labor, expansion of the plantation system was going to be exceedingly difficult, if not impossible. So, the planters began to lobby the colonial state to discourage peasant cotton production and, instead, assist them to establish cotton plantations, grown on alienated native land and worked by indigenous labor.

Of course, the Ugandans had no say in the matter; the struggle was between the European planters in Uganda and the textile barons in Lancashire, with the colonial government in the middle trying to reconcile the conflicting demands. The people in Lancashire were satisfied with peasant-grown cotton. Moreover, they feared that the transition from peasant production to the plantation system could be long and disruptive. Furthermore, the Lancashire barons strongly felt that peasant cotton production would in the long run give them flexibility. Given these considerations, the BCGA lobbied hard against the planters, placing the colonial administrators in a quandary.

The general role of the colonial state and its immediate self-interest vested in peasant commodity production were two additional factors that complicated the situation. The primary function of the colonial state was to create universal conditions of accumulation, whereas the planters' self-interests were immediate and particular. The colonial state could demonstrate and sustain its legitimacy and relevance to the colonial mode of production only if the long-run trajectory of accumulation was set on the right course. Furthermore, having burrowed its way into the economic sanctum of the peasants, the colonial state had to worry about the impact of disrupting peasant cotton production on its ability to keep taxes coming to its coffers from the peasants. After all, in 1920, 90.3 percent of state revenue came from cotton production, while coffee and rubber contributed just 8.7 and 2.5 percent, respectively (Taylor, 1978). The sympathetic official comprehension of planter demand could not override the material necessity of reliable revenue generation.

The 1920-21 depression appeared to have pulled the rug from underneath the planters' feet as most of them were financially bleeding; most of the plantations were on the verge of collapse and, in fact, more than 60 estates would be abandoned by their owners for lack of financial support from the state. So, pursuing a pro planters policy meant funneling state money into the plantation sector in order to rescue their owners for dubious returns (Taylor, 1978). In light of these considerations, there appeared to have been a convergence of views and interests between the colonial empire builders in Uganda and textile barons in England on the necessity to maintain peasant commodity production. Moreover, the objective conditions, having to do with the 1920-21 trade depression in which the value of cotton was reduced by 50 percent, eventually militated against the cause of the planters, and the colonial administration opted for continuation of peasant-grown cotton without necessarily abandoning the European planters. In fact, the planters were encouraged to set up plantations in western Uganda, leading to the expansion of the colonial mode of production to hitherto unexplored regions.

In the end, peasant commodity production and the European plantation system evolved into what Christopher Youe (cited in Taylor 1978) called a "dual economy," marked by complementarity rather than by contradiction insofar as the general conditions of accumulation went. While actively encouraging peasant commodity production in Buganda and the eastern province, the colonial state vigorously assisted the development of plantations in western Uganda. The preponderance of peasant cotton production was exemplified by the rise in acreage under peasant cotton cultivation from 162,350 acres in 1919 to 572,800 acres in 1924 (Taylor, 1978). Ultimately, resolution of the contradiction in favor of peasant cotton production served to profoundly shape the trajectory of the colonial mode of production in Uganda to the extent that smallholder production continued to define the character of post independence Ugandan political economy.

European settlers became ancillary to the accumulation trajectory. Even though the number of settler-owned estates gradually rose from 170 in 1924 to 219 by 1933, covering 101,944 acres, and primarily devoted to production of coffee, rubber, tea, and tobacco, their role in the overall mode of production remained complementary to peasant commodity production, rather than a substitute for it; by 1933, the number of European settlers in Uganda stood at 1,854 (Keane, 1935).

In any case, the promise of high-grade fiber of medium-staple cotton was such that the acreage under cotton cultivation continued to increase with the full support of the British colonial administration; by the 1930s, cotton production reached 57,000 metric tons a year. Representing 90 percent of total export, in 1933, the value of cotton export was almost 2.7 million pounds sterling (Keane, 1935; World Bank, Supplement 2007).

To insure the continuous production of cotton and the centralization of surplus, the colonial administration created a cotton marketing board. In theory, the purpose of the board was to collect the surplus funds from cotton exports in order to pay growers so that they would not abandon cotton production; whenever world prices fell, peasants would draw some money from this government-run fund. But by 1948, the fund accumulated 5 million pounds, which the government used for other purposes (Bowles, 1975). This was to mask the level of exploitation. For example, between 1847 and 1952, the government paid cotton growers just 40 percent on average of the proceeds that the government received from the export of cotton; this meant that the marketing board and the government retained 60 percent of the proceeds from cotton production (Bowles, 1975).

The determination to leave cotton production as essentially a peasant-based undertaking had a spillover effect insofar as proprietorship of other cash crops was concerned. The important lesson the colonial administrators drew from the 1920-21 trade depression was the danger of overdependence on cotton as the only source of revenue; to reduce vulnerability of the state to price collapse, diversification into the cash crops was actively pursued. Smallholder farmers in Uganda were now not only permitted to grow coffee but also the colonial government took an active role distributing free robusta seedlings to peasants in Buganda. The growing importance of native-grown coffee was indicated by the rise of area under peasant cultivation from just 2,000 acres in 19191, when European estates had 25,000 acres under coffee production, to over 13,000 acres by 1924 (Taylor, 1978). In fact, native Ugandans were allowed to grow Arabica coffee, which was, until the 1930s, a settler preserve. Peasants and native cooperatives in Bugisu district around Mount Elgon planted 3,520 acres with Arabica coffee and quickly proved successful (Keane, 1935). Because of native participation, the value of coffee exported in 1933 stood at more than 210,000 pounds sterling (Keane, 1935).

By the 1950s, coffee surpassed cotton as the premier cash crop. In order to maximize surplus extraction from the cash crops, the colonial state established marketing boards to receive the crops from the producers and directly export them. In addition to the poll tax, manipulation of prices became an essential component of state revenue. The marketing boards set producer prices lower than the prevailing world market prices; the difference was stashed away in state coffers. Between 1945 and 1952, for example, the cotton and coffee marketing boards captured 33 percent of the potential income of cotton growers and 45 percent of the potential income of coffee growers (Jamal, 1978). Over the period between 1945 and 1960, the cotton marketing board withheld 40 million pounds while the state collected another 43.4 million pounds in taxes from cotton exports (Jamal, 1978). The amount collected by the coffee board between 1945 and 1952 was 12.9 million pounds.

To be sure, the colonial heritage has been instrumental in the definition of Uganda's post-colonial economy. The first far-reaching impact of colonialism on the country's economy mode of operation, the Ugandans were directly integrated into global capitalism simply as producers and suppliers of commodities, which they would neither consume nor from which they would derive income to raise their living standards. In this sense, they skipped a crucial element of their social evolution inasmuch as they were directly integrated into the global capitalist mode of production. As the colonists devoured the best land, water, and human resources to feather their nests, the natives simply marched fast toward the darkening clouds of what has become a permanent condition of structural poverty.

The second consequence of colonialism stems from the first in that social classes had been misinformed. The so-called indirect rule that the British exercised as a theory of administration has profound implications for social formation in Uganda. It was a potent methodology of divide and rule. By institutionalizing ethnic fragmentation, the colonists perniciously assumed that they would be able to thwart the emergence of national solidarity, in which the Ugandans would have been able to forge a class-based resistance to colonialism with concrete national scope and national purpose.

Indirect rule meant passing the cost of pacification and exploitation to local chiefs. Chiefs were appointed to perform executive, judicial, and security functions in relation to the colonial subjects. As part of their functions, they also insured cash crop cultivation and collection of taxes on behalf of the colonial state. In return, the chiefs enjoyed all the benefits that their office presented. Using their status as the only standing power between the state and the peasants, they obtained the services of peasants to work their land to grow cotton and other cash crops. In addition, because of their wealth and contacts with missionaries, the chiefs were the first to send their sons to schools, who would in due course become civil servants and, later, politicians (Gartreli, 1983). When the tidal waves of decolonization proved irresistible following World War II, it was the educated sons of chief and their supporters who carried the banner of the independence movement. The important colonial legacy was that, since indirect rule valued ethnic markers of identity and since the main centers of economic resources were local, the nationalist movement had to carry the baggage of ethnicity and parochialism. Even the departing British colonial officers realized the peril posed by the external markers of identity when they pondered how to transform Uganda from a formal possession of Great Britain to an informed sphere of British dominance. Suddenly, the British preoccupation became how to prevent the eruption of the contradictions that were kept under the rug through the coercion of the producers and cooptation of intermediaries.

In a speech before the Royal Imperial African Society in 1957, Sir Andrew Cohen (then governor of Uganda) warned that the forces of "tribalism" were more powerful than those of nationalism in Uganda. He put the conundrum in these terms:

> Tribalism can be an obstacle to progress, but equally tribal institutionshave normally been the channels through which progress has been achieved. any policy which ignores tribal loyalties and tribal institutions in Uganda will fail. The tribe has for many years been the unit of administration at the district level in the eastern, northern and western provinces. (Cohen, 1957)

Governor Cohen actually projected a triangulated struggle emerging in Uganda: between the central government and the various "tribal" institutions; between the 6 million Africans and 60,000 Asians; and finally, between the Africans and the 10,000 or so Europeans (Cohen, 1957).

To insure that Uganda remained within the informal British Empire as an important supplier of primary commodities, Governor Cohen proposed a two-track strategy. The first involved speeding up the induction of "progressive" Ugandans into civil service and develop a framework in which to anchor an Anglo-Ugandan political partnership after independence. The term "progressives" was reserved for those Ugandans who (in the conception of the colonists) had become "detribalized" and yet expected to be quiescent "internationalists," fully accepting the British vision of informal empire. These "detribalized progressives" were expected to implement and run the hastily hatched Westminster model of rule to give the political system a façade of legitimacy while retaining Uganda's division of labor as a producer of cash crops intact.

The second strategy was to give Uganda's political economy a native character, with particular emphasis put on commodity production. Highlighting the importance of Uganda's division of labor, Governor Cohen reminded his audience that "Uganda is the biggest coffee producer in the British Commonwealth and the biggest cotton producer among colonial territories. … Peasant farming will continue to be the country's main industry for as far ahead as anyone can foresee" (Cohen, 1957). Insofar as Afro-Asian relations in Uganda were concerned, Governor Cohen was prescient to the degree that the quasimonopoly of commerce by some 60,000 Asians in a country of 6 million Africans could be a fault line of contention as the country stumbled forward to independence. According to his accounting, there were only 15,000 African traders with an annual turnover of between 15 million and 20 million pounds (Cohen, 1957). The fear was that the struggle for political liberation might spill over to the economic sphere as producers might seek economic emancipation. After all, the cooperatives movement had been gathering steam for quite sometime. In fact, by midcentury, there were 1,300 cooperative societies in Uganda, which sought autonomy in terms of managing and marketing what they produced without intermediaries or marketing boards (Cohen, 1957).

Governor Cohen was also prescient in his estimation of the road Uganda was destined to take, namely, the road of economic growth and ensconced in primary commodity production. Indeed, upon independence, cotton production continued to rise until it peaked at 86,000 metric tons in 1974 (World Bank, September, 2007). Although coffee was catching up with cotton because of its rising prominence among coffee growers, cotton still continued to command between 25 and 40 percent of foreign exchange earnings until the 1970s, when it suddenly collapsed, accounting for just 3 percent of all cash crop exports. In 1985, 199,000 ha were planted with cotton, generating just 13.5 million pounds in export revenue (Byrnes, 1990). On the advice of the World Bank, the Ugandan regime embarked on a $28.5 million rehabilitation program through liberalization beginning in 1993; of the total cost of the project, the World Bank group furnished $13.8 million to speed up the effort. In response, average yearly cotton production rose from 35,000 bales in 1993-94 to 120,000 bales in 2000-01, worked by some 400,000 producers, each cultivating 3.2 ha of cotton; the area under cotton production correspondingly rose from 55,000 ha to 119,000 ha over the same period. With further assistance from the United States, Uganda managed to boost production to 150,000 bales in 2004-04 (World Bank, September 2007).

The coffee sector dramatically rose to prominence among Uganda's crash crops. The area under coffee production grew from 58,000 ha during World War I to over 272,000 ha by 2005. More than 500,000 smallholder farmers produce coffee, each farmer cultivating between 0.5 and 2.5 ha on average. In addition to the direct producers, another 3.5 million families make a living from coffee-related activities (UCDA, 2009; Nakkazi, 2007). Concomitant with the installation of the neoliberal regime in the second half of the 1980s, particular attention has been devoted to coffee production in order to secure much-needed currency. This is reflected in the fact that the area under coffee cultivation increased from 224,000 ha in 1980, to 264,000 ha in 2004, and to 272,000 ha in 2005, and coffee yields rose from 6,036 hectograms per ha to 7,045 hectograms per ha over the same period (Damianopoulos, 2005; Nakkazi, 2007). The result was a boom in coffee production as Uganda earned $264 million in 1988. In 1989, 191,700 ha were under robusta cultivation, while another 33,000 ha were under Arabica coffee production (Byrnes, 1990). In 1998, coffee export fetched $296 million or two-thirds of the $536.5 million of agricultural exports that year (Aquastat, Uganda, June 2006).

As with other coffee growers in neighboring states, Ugandan coffee growers have historically been subjected to a mix of three sources of vulnerabilities: extortionate taxation by local and central authorities, the coffee wilt disease, and external stochastic shocks. According to the Uganda Coffee Development Authority (UCDA, 2009), in 2008, the coffee wilt disease destroyed 150 million of the 300 million robusta trees. The coffee wilt disease is a fungus, which blocks circulation of nutrients and water from the plant roots to other parts of the tree. Deprived of important nutrients and water, the coffee tree wilts and dies. The coffee wilt disease was first identified in Uganda in 1993, and since then the disease has proved resilient, eroding both the livelihood of small farmers and government earnings. On average, Uganda has been losing $50 million annually as a result of the wilt disease infestation. For instance, Uganda earned just $84 million from coffee exports in 2002.

However, because the coffee sector is the mainstay of government revenue, the regime has given extraordinary attention to the importance of fighting the wilt disease by introducing new tree varieties and by encouraging the planting of more trees. The Uganda Coffee Development authority (UCDA) has been empowered to develop fighting strategies, which include distributing free coffee seedlings to farmers. Since 2002 alone, more than 85 million new coffee trees have been planted (UCDA, 2009). In response to such intensified efforts, the sector does exhibit spurts in recovery from time to time, reflected in the fact that Uganda exported 3.2 million bags of coffee in 2008, earning $300 million (UCDA, 2009). In order to insure control, the government abolished coffee cooperatives in 1994 under liberalization and placed coffee growers directly under the Ugandan Development Authority.

In response to aggressive state intervention, low prices, and the wild coffee disease, growers have been compelled to shift to other cash crops, something that made the regime apprehensive. The apprehension stems from what Bunker (1985) called the potential defection of farmers to subsistence crops. Under circumstances of unchecked infestation, small farmers might abandon commercial crop production in favor of subsistence food, which the regime could not control for purposes of tax collection and foreign exchange generation (Bunker, 1985).

Thus, in order to insure control of commercial crop production, the government aggressively intervened in commercial agriculture in ways that placed the shift from robusta coffee production to other cash crops under government direction and supervision. For example, the regime helped more than 3,000 coffee growers in Buganda to convert their holdings from coffee to cocoa production, while that many more were being aided to do the same in 2002 (*Africa Analysis*, August 23, 2002). Cognizant of the trend, the government has been actively distributing free cocoa seeds to farmers in 15 districts, with a plan to help 500 coffee growers each year switch from coffee to cocoa production. In 2008, Uganda exported 13,000 metric tons of cocoa and earned $26 million with stated plans to boost production to 50,000 metric tons a year. To this end, the regime proposed plans to plant 35 million cocoa trees over seven years ending in 2015; as of 2009, Uganda had 250,000 ha of land for certified organic cocoa production (*New Vision*, January 21, 2009). Of course, the government was reacting not to the concern of coffee farmers, but rather to its own coffers and the international market conditions. In mid-2002, Uganda was collecting just 46 cents from one kilo of coffee on the world market, of which only one cent went to the coffee farmers. By contrast, one kilo of Ugandan cocoa was going for $1.70 on the international market (*Africa Analysis*, August 23, 2002).

Paradoxically, while assisting coffee growers in Buganda to shift to cocoa cultivation, the regime was simultaneously encouraging more coffee planting in other regions of Uganda. Indeed, the regime was implementing plans in 2002 to extend state support to 15,900 coffee growers in the country, of whom only 168 growers were in Bugunda, with a view to increasing coffee production from 3.5 million bags in 2002 to 12 million bags by 2006 (*Africa Analysis*, August 23, 2002). The 2006-2015 coffee production campaign is meant to expand the area under coffee production in northern and northeastern regions of the country (UCDA, 2009).

In addition to the onerous burden of various taxes, Ugandan coffee growers deal with global structural oversupply of coffee. Just as Uganda has been preoccupied with how to expand areas under coffee cultivation and boost productivity, other countries have equally been busy doing the same. For example, Vietnam (a latecomer to the global coffee market) has been expanding its area under coffee production. As a result, supply of Vietnamese coffee rose sharply from 84,000 tons in 1990 to 950,000 tons in 2001, making Vietnam the second biggest exporter of coffee after Brazil; the resulting oversupply led to the lowest world price for coffee in three decades. To put the matter in historical perspective, in 1976-77 Uganda earned $588.5 million in hard currency from 2.45 million bags of green coffee compared to the $339 million it earned in 2006 from 2.5 million bags of exported green coffee beans (Nakkazi, 2007).

Tea cultivation was another source of primary accumulation in Uganda, which soon followed the introduction of coffee. In the late 1940s, tea interest groups identified that Uganda had 200,000 ha of land suitable for commercial tea cultivation, stretching from the West Nile to the western highlands and from the northern crescent of Lake Victoria to the southern regions (butler, 2005). Historically, the tea sector has been dominated by large estates. However, the expulsion of resident citizens of Asian and European origin in the early 1970s hurt the tea sector in profound ways. For example, tea exports plummeted from 22,000 tons in 1974 to 1,700 tons in 1981. In 1989, tea cultivation occupied 21,000 ha, which produced a mere 4,600 tons. In the late 1980s, the tea sector was dominated by three corporations, while 11,000 smallholders put 9,000 ha under tea production (Byrnes, 1990). After devastation from the civil wars, Uganda's tea sector is now fully rehabilitated under a privatization regime and multinational corporate domination. As in Kenya, Unilever and James Finlays are dominant players in Uganda's tea sector. In 2004, Finlays alone accounted for 26 percent of the 35,000 metric tons of Uganda's total tea production (Butler, 2005). In 2006, Uganda earned $36 million from almost 33,000 metric tons of tea export (SOMO, 2008).

By the late 1990s, floriculture became the new addition to the trajectory of primary accumulation in Uganda. In 1998, Uganda exported 6,000 metric tons of cut flowers, valued at $23 million (Aquastat, Uganda, June 2006).

A dark side of specialization in cash crops has been the commercialization and dehumanization of children. According to the International Labor Organization (ILO) (Cassidy, 2006), 7.9 million child laborers (whose ages ranged between 10 and 17) were involved in all facets of commercial agriculture in 2006. In the tea plantations in the Tooro District in western Uganda alone, there were 40,000 child laborers earning just 30 cents a day; one out of five children were said to have been orphaned by AIDS and compelled to carry the burden of being the head of household (Cassidy, 2006).

The contemporary fundamental contradictions in Ugandan political economy are located in the country's dependency on cash commodity production and exportation. Indeed, in a matter of less than a decade,

Uganda's primary commodity production began to crack from its own internal contradictions. By 1969, the country's level of unemployment had become formidable. Revenue from primary commodity production was not enough to go around and meet the basic needs of the population and/or was squandered by the pirate state bourgeoisie. In fact, Milton Abote recognized this reality and saw two pathways out of the crisis of legitimation: either partial delinking from the global capitalism or further deepening of integration into it. Abote toyed with the first option by calling for Ugandan version of socialism. In late 1969, Abote adopted the "Common Man's Charter," containing 44 articles spelling out the basic elements of Ugandan socialism. Subsequently, Abote called for nationalization or controlling interest in 85 firms, equalization of salaries, redistribution of national wealth, diversification of agriculture, radical reduction of reliance on foreign assistance, creation of cooperatives and parastatals to guide the trajectory of economic growth, and state control of the import/export sector (Gershenbert, 1972). Rather than solving the crisis of legitimation, however, the declaration of socialism paved the road for Abote's ouster by Idi Amin. After the overthrow of Amin, the second Abote government dropped all pretences of socialism and adopted the second route, namely, deepening Uganda's integration into global capitalism.

Complicating Uganda's social ills and excessive dependence on primary commodity production is the country's equally excessive dependency on global finance capital. As bunker (1985) notes, until 1971, Ugandan peasant producers had some leverage with the government. Because the state was totally dependent on peasant cash crops for its sustenance in terms of directly taxing cash crop producers and earning foreign exchange by directly buying the crops from producers low and then selling the cash crops on the international market high. The latent fear was that, if the state overplayed its hand and squeezed the peasants to their bare bones in terms of onerous taxation and depressed prices for their crops, the peasants could potentially withdraw from commercial production into untaxable and internationally unmarketable subsistence food crops. The state's structural dependence on primary commodities thus gave Ugandan peasant producers some bargaining lever to obtain concessions relative to taxes and prices. However, the regime's access to global finance capital (even though under harsh conditions) appears to have nullified the peasants' potential power to withdraw from commercial production. The regime's dependence on grants and loans from global financial institutions and donor countries has eroded the exit option peasants used to have (Bunker, 1985f). Of course, the regime simply diversified the sources of its dependency without necessarily abandoning its dependency on peasant commodity production. Access to global finance capital allows the state not to concede to peasant demands. In fact, Uganda accelerated its dependence on global finance capital since 1980, reflected in the rapid rise of its debt service ratio from 17 percent in 1980 to over 50 percent in 1981 (Bunker, 1985).

Uganda's political economy is deeply grounded in several agricultural areas as described, all of which are vulnerable to the vagaries of weather and global price fluctuations. It is against the backdrop of the faltering primary commodity production that the Ugandan pirate state bourgeoisie hope to grow out of their external dependency by looking to the country's fishery mode of production. Although Uganda is a landlocked country, 20 percent of its total area is covered by lakes. Indeed, in terms of natural endowment, Uganda fares better than its immediate neighbors since 21.7 percent of its total area of 241,000 km^2 is considered suitable for cultivation compared with Kenya and Tanzania where the arable land is 12 percent and 4 percent, respectively. However, the integrity of this potential is undermined by the fact that over 80 percent of Uganda's workforce is still engaged in subsistence farming, a situation contextualized by extensive deforestation, soil erosion, overgrazing, water hyacinth infestation of the country's natural water systems, and increasing encroachment of its wetlands by landless people and capitalist developers.

On balance, Uganda remains a highly impoverished nation. One of the contributory factors to the country's level of poverty is its booming population, growing at 3.4 percent per annum with the average Ugandan woman having 7 children. (Only Yemen and Niger have birthrates higher than that of Uganda.) Moreover, 50 percent of Uganda's population is under age 14 and the median age is 15 years. In 2008, 7 million Ugandans were classified as chronically poor (BBC News Monitoring Africa, Aug. 23, 2008). In spite of the fact that Uganda is one of the most structurally adjusted nations in Africa, the country remains mired in the very conditions it sought to avoid by heavily borrowing from external sources. In May 2008, the World Bank, after praising Uganda for two decades as a model of sound macroeconomic policy, loaned the country $200 million to presumably fight extreme poverty in the country (BBC News Monitoring Africa, May 8, 2008).

Since the traditional means of primitive accumulation have reached their outer limits, the Ugandan elites have turned their attention to the country's water bodies for salvation. Historically, fish production in Uganda has been more crucial than in the other littoral states in terms of the sector's contribution to revenue generation, employment, livelihood maintenance, and protein intake. The importance of fish in the national diet is such that Ugandans obtain up to 60 percent of their animal protein from consuming fish (Rabi, 1996).

Uganda has a good number of large lakes, surrounded by a variety of smaller lakes, swamps, lagoons, and rivers where substantial subsistence and commercial fishing operations take place. Indeed, the country's natural water systems are so vast that they cover over 42,000 km^2 in area.

In 2002, over 150,000 Ugandans were directly engaged in the primary fishery sector as harvesters, and another 550,000 people were engaged in secondary and tertiary sectors as processers, traders, and providers of support services (FAO, 2005d). If dependents are included in the above figure, subsistence fisheries and commercial fishing operations provide livelihoods to more than 2 million Ugandans. Since the mid 20th century, fish catches from all Ugandan water systems have increased eight fold from 250,000 metric tons in 1950 to 2 million metric tons in 1999 (Ogutu-Ohwayo and Balirwa, 2006). Even though the 2 million metric tons figure is open to question or interpretation in the absence of reliable statistics, it nonetheless suggests hyperacceleration of demand for aquatic resources in the face of growing population pressure, terrestrial resource depletion, stark poverty, and government desperation for hard currency. According to official statistics, between 2001 and 2005, aggregate national fish production fluctuated between 200,000 and 250,000 metric tons per annum, with the Nile perch accounting for 42 percent, Nile tilapia for 35 percent, and *R. argentea* for 8 percent (European Union, August 2006).

Historically, all of Uganda's natural water systems have been biologically highly productive; however, Lake Victoria has, by far, been the jewel of the fishery mode of production in the past three decades. Uganda's lacustrine sovereignty over Lake Victoria covers 31,000 km^2—or 45 percent of the area.

Like the other riparian states, Uganda sees Lake Victoria as the epicenter of its fishery mode of production because of the rising popularity of Nile perch as an export commodity. The pirate state bourgeoisie have not been disappointed since the Nile perch quickly demonstrated their dominance in Uganda's fishery production, rising from less than 1,000 metric tons in 1980 to 41,000 metric tons in 1986 (Jansen, 2007). The productivity of the perch rose from five kg per ha in 1981 to 137 kg per ha in 1985, whereas the productivity of *haplochromis* dropped from 523 kg per ha to 108 kg per ha over the same period (Ogutu-Ohwayo, 1990). Reflecting the vertical expansion of fisheries, the number of boats with outboard engines also rose from 3,250 in 2002 to more than 5,000 in 2006, mostly chasing the perch and supplying the fish to the 15 fish processing factories; the number of dugouts and planked canoes increased from 8,000 in 1989 to 15,544 in 2000 (Beuving, 2010; Mkombo et al., 2007). Similarly, the number of Ugandan canoes operating on Lake Victoria steadily increased from 3,200 in 1972 to 15,452 in 2000, and then to more than 17,000 in 2008 (Ogutu-Ohwayo and Balirwa, 2006; *New Vision*, Jan. 5, 2010). In 2003, there were 550 landing sites on the Ugandan side of Lake Victoria serving Nile perch chasers (Grellier et al., 2004). The number of industrial fish-processing plants on the Ugandan side of Lake Victoria also rose from 2 in 1998 to 18 by 2007, heralding the international popularity of Nile perch fillet as well as its vital significance as a major hard currency earner for the pirate state bourgeoisie. Landings of Nile perch alone on the Ugandan side of Victoria reached 130,000 tons in 1992, sensationalizing the optimism of the elite (FAO, 2005d).

The strength of the Nile perch-based fishery mode of production in Uganda has been such that exports of fish and fishery products have become the second major nonagricultural exports after coffee in terms of hard currency generation. In fact, the national elites were so bullish about the biological productivity of Lake Victoria that they developed a long-term ambition to harvest fish worth $1 billion per annum from Lake Victoria alone by the end of the first decade of the 21st century (Yohannes, 2008). Initially, the flow of hard currency from fish export seemed to support the optimism of the elites, as export earnings rose from just $0.4 million in 1998 to over $145 million in 2008 (*New Vision*, Jan. 5, 2010).

Uganda's elites had equally been sanguine about the other natural water systems, especially Lakes Kyoga, Edward, George, and Albert, which are regarded as among Africa's most productive lakes. The Lake Kyoga complex, 2,700 km^2 in area and buffered by continuous growth of water lilies, vast seasonal and permanent swamps, and rich papyrus, occupies a saucer-like depression in central Uganda. The complex is known for hosting extensive and dense carpets of floating macrophytes on the edges of the open water, as well as submerged macrophytes throughout the entire lake—properties that make the complex exceptionally hospitable to diverse fish species (Hughes and Hughes, 1992).

The water body of the Kyoga complex opens off Victoria Nile, to the north of Lake Victoria, annually receiving 25.6 bcm of water from Victoria Nile, in addition to receiving six bcm from direct precipitation and 1.2 bcm from other inflowing rivers (Hughes and Hughes, 1992). The lake has a catchment area of 75,000 km^2. Because the maximum depth of the lake is in the neighborhood of 7 meters, Kyoga is polymictic, which means that its waters mix daily—something which explains its exceptional productivity. Light penetration and oxygen saturation are uniform throughout the lake that are exceptionally favorable to photosynthetic action, and, in turn, prevent thermocline formation and prolonged thermal stratification from occurring (Orach-Meza, 1990). Actually an assortment of minor lakes and swamps, the Kyoga complex used to host a very rich biodiversity (including up to 46 described fish species) before the complex was subjected to human biomanipulation intended to boost fish production by replenishing the native species with the introduction of exotic Nile perch, and such tilapine species as *O. niloticus, O. leucostictus,* and *T. zillii.*

In the early 1950s, there were few fishers and few fishing boats around Kyoga. However, the relocation of the Nile perch along with the tilapine species to the complex from Lakes Albert and Turkana in 1954 (the exotic species revolutionized Kyoga's fishery) made it the single most important source of fish in Uganda, overshadowing even Lake Victoria by a wide margin until the 1980s. By the early 1970s, the Nile perch and the exotic *Oreochromis niloticus* comprised 80 percent of the commercial catch in Kyoga. Thanks to the introduction of these exotic species, the lakes' productivity grew from 214 kg per ha per year in 1969 to 395 kg per ha per year in 1974. Consequently, the annual fish landed in Lake Kyoga in the 1970s stood at 146,000 tons as opposed to 74,000 tons from all other natural water systems, including Lake Victoria. Before the exotic species established full dominance, up to 15 species were commercially harvested from Kyoga, among which the most abundant were *Clarias spp., Bagrus spp., Lates spp., Protopterus spp.,* and *Oreochromis spp.* (Welcomme, 1979; Orach-Meza, 1990; Fryer and Iles, 1972; FAO, 2005d; Hughes and Hughes, 1992).

However, owing to extensive entropic degradation of the lake itself and its wider catchment area, the Lake Kyoga complex has undergone rapid siltation, eutrophication, and water hyacinth infestation. As a result, the loss of productivity has become a major concern to the predatory state. In view of the fact that the poverty rate among the people in the area is in the range of 40 to 60 percent, the poor have overexploited the forest in the lake's catchment area and the papyrus buffering the Kyoga complex. Additionally, the poor have overcrowded the floating islands of the complex, resulting in overfishing. Biological overfishing in the lake is so severe that fishers use undersized mesh to catch Nile perch in the range of 13 to 35 cm in length, while the allowable Nile perch caught must be 100 cm in length (World Resources Institute, 2009; *New Vision*, March 21, 2007; *The Monitor*, August 31, 2010). Furthermore, Lake Kyoga has begun to experience the effect of climate-driven change since its water levels have been dropping rapidly.

Concerned over the sharp reductions in fish harvests from a yearly average of 185,000 metric tons in the 1980s to 60,000 metric tons in 2008, the pirate state bourgeoisie prepared a development concept for Lake Kyoga basin and submitted it to international funders. As of this writing, however, nothing has materialized. At the same time, the pirate bourgeoisie have established 192 beach management units (BMUs) around the Kyoga complex to monitor, police, and manage licensed fishers and fishery output. However, the BMUs have so far failed to reverse resource depletion and degradation of the Kyoga complex (*New Vision*, March 21, 2007).

Lakes George and Edward, linked by the 36- km-long Kazinga channel, are another two important Rift Valley lakes found in western Uganda. Lake Edward, shared with the Democratic Republic of the Congo (DRC), is 2,200 km^2 in area, 65 km long and 38 km wide, with maximum depth of 117 meters and mean depth of 34 meters; Uganda's share of the lake is 610 km^2. The lake hosts 23 cichlid and 30 noncichlid fish species. Between 1957 and 1962, Uganda harvested 6,000 metric tons of fish per annum, of which the native Nile tilapia accounted for 60 percent of the total catch (Fryer and Iles, 1972; Hughes and Hughes, 1992). Traditionally, Lake Edward had been a rich source of catfish, tilapia, and lungfish. Between 1963 and 1988, its average annual catches were 5,500 tons (FAO, 2005d).

The polymictic shallow Lake George (which occasionally experiences diurnal stratification and bottom water deoxygenation) is 270 km^2 in area with an almost uniform depth of 3 meters. The lake hosts more than 32 fish species, including 17 species belonging to the genus *Haplochromis*. Based mainly on Nile tilapia, the fishery yield from Lake George was 4,700 metric tons per annum in the early 1970s (Fryer and Iles, 1972; Hughes and Hughes, 1992). In fact, in the 1950s and 1960s, the commercial operation regularly served Kampala and other towns with frozen tilapia. Between 1950 and 1990, the average annual catch from the lake was 3,000 tons (FAO, 2005d; Hughes and Hughes, 1992). In recent years, Lake George's fishery has seen sharp drops in catches. According to government statistics, a decade ago the average fisher used to harvest 20 kg per day as opposed to only 6.3 kg today (BBC News Monitoring Africa, March 16, 2010). In response to such sharp decline, the regime told the fisherfolk in March 2010 that the lake was about to be closed to fishing for 180 days supposedly to allow recovery of fish stocks. However, since fishing in Lake George employs 20,000 people, these fisherfolk faced a grim future. As one of the fishers said: "This lake [George] is our garden and market. Where shall we go since we have no land?" (BBC News Monitoring Africa, March 16, 2010).

Lake Albert, 5,270 km^2 in area, of which 54 percent lies in Uganda with the remainder lying in the DRC, ranks third in terms of its contribution to Uganda's fishery mode of production. The lake is 150 km long and 35 km wide on average, with maximum depth of 58 meters and mean depth of 25 meters. The lake's depth along with its water volume of 280 km^3 suggests that Albert is monomictic, meaning that its water mixes only once a year. This makes it sensitive to climate-driven change, pollution, and eutrophication. Based on three species, including the native Nile perch, annual commercial fish production in Albert fluctuated between 4,000 tons and 20,000 tons in the period 1955 to 1986 (Fryer and Iles, 1972; Welcomme, 1979; FAO Fisheries and Aquaculture, Uganda, Country Profile, 2004).

Despite early promises of continuous expansion of the fish harvest, Uganda's growth ideology was severely tested within a short decade as the fishery mode of production faltered. Overexploitation by both subsistence fishers and commercial accumulators began to have a visible impact on overall production on the Ugandan side of Lake Victoria. In 1990, Uganda hauled 130,000 metric tons of fish from Lake Victoria, which fell to 105,000 metric tons in 1995, and then stagnated at around 90,000 metric tons per year in the next five years. Likewise, the catches from Kyoga sharply dropped from over 150,000 metric tons per annum in the 1980s to less than 40,000 metric tons in 2009. Historically, on average, the 30 landing beaches on the Ugandan portion of Lake Albert used to handle 300 metric tons of fish per week; that number sharply dropped to 30 metric tons per week by 2009. In 2008, Uganda's 18 fish factories were operating at between 30 percent and 50 percent of capacity utilization for lack of enough fish to process (*New Vision*, Jan. 5, 2010). This limited growth called the elites' long-term ambition to annually export $1 billion worth of fish and fishery products into question. For example, Uganda's export of perch fillet to the European Union alone dropped from 23,800 metric tons in 2005 to 4,400 metric tons in 2008 (Josupeit, 2009).

Characteristically, the regime blamed the depletion of Lake Victoria fisheries on illegal fishing. As a prelude to the imposition of draconian measures on subsistence fishers, the regime produced spurious figures in which 60,000 tons of fish were claimed to have been lost annually to illegal fishing. According to official statistics, in 1996 and 1997 Uganda suffered a loss of $67 million in foreign exchange earnings owing to the so-called illegal fishing; in 2003, the claimed loss was a whopping $86 million, and $140 million was reportedly lost to "illegal" harvesting during 2008 and 2009 (*New Vision*, Jan. 5, 2010).

The official state fisheries institutes and, indeed, the whole media and the academic world, have continued to orchestrate the official version of the problem as caused by "illegal" fishing. Of course, there is a class interest behind this dubious orchestration. They may be from the lower and/or middle classes by origin—those who diligently echo the regime's characterization of events themselves are aspiring accumulators by function. Their continued employability and potential rise to higher status are closely tied to the prosperity of the national elites and the stability of the state. Moreover, the fishery agencies are almost totally dependent on the external funding to which the professional survival of those who staff the agencies is also tied. In effect, the professionals must serve as echo chambers for the global growth ideology.

In reality, though, the presentation of the crisis in the fishery sector exposes a fundamental contradiction between the necessity of survival of the broad masses of fishers and the logic of accumulation. The juxtaposition of illegal harvesting against what the regime defines as legal fishing is meant to obfuscate the true essence of this contradiction. In effect, the foreign exchange earnings from overexploitation of fish that directly flow into the coffers of the state is dubbed legal, while the fish which the toiling masses catch to make ends meet is characterized as illegal. Beneath this officially sanctioned public discourse lies the simmering contradiction between the crisis of accumulation and the crisis of legitimation, which can only be contained or temporarily postponed through the imposition of authoritarian coercion on a grand scale.

In any case, in order to control what the regime has designated as illegal fishing, and thereby protect the fish export sector, the regime proceeded to refine the instruments of violence, law, and bureaucracy as a containment strategy. These instruments consist of policing the lakes, arbitrary confiscation of fishing gear by state agents, and apprehension and imprisonment of so-called illegal fishers. In 2005, Uganda imported six modern speedboats to patrol its portion of Lake Victoria. As the supposed illegal fishing worsened, the Ugandan regime repeatedly threatened to close down 419 landing sites on Lake Victoria beaches unless the stern public warnings were heeded. In fact, the government followed through its threat by closing down seven fish-processing facilities in September 2008, after having accused them of receiving and processing immature fish; the regime blamed the loss of $60 million in foreign exchange earnings that year on illegal harvesting of immature fish by fishers, which the factories knowingly processed. A mature Nile perch is generally over 20 inches, but the regime reported that the fish factories were receiving 4-inch long Nile perch. The regimes' evidence was the sharp decline in standing Nile perch biomass from 1.9 million metric tons in 1999 to 370,000 metric tons by the beginning of 2009 (*New Vision*, April 6, 2009). Meanwhile, the regime confiscated and torched 10,000 undersized mesh nets from just one landing beach on Lake Victoria. As a follow up, 20 fishers were arrested and publicly paraded to shame them into total submission as a deterrent to other would-be illegal harvesters (*New Vision*, Jan. 5, 2010).

To help the Ugandan regime impose a form of police state on the country's water bodies, the African Development Bank also offered $1 million in 2009. This new fisheries development project had five legs: capacity building and infrastructure development, credit facility to fishers, research in fish farming, quality assurance, and management development. The external funding stipulated construction of a new fishery regime to be completed by December 2010. The overarching purpose of the program was to completely seal off Lake Victoria, in particular, from illegal immature fish harvesters. All fishing boats would have identifiable state plates, and all fishers would be screened before licensing in order to establish their legitimacy and competence. Landing sites, which hosted 5,000 fishers, were required to reduce their numbers to 2,000 fishers by December 2010. By the end of 2009, eight ice plants and 20 fish markets had already been built. To reassure international fish consumers of the quality of Ugandan fish, a quality-control laboratory was established in Entebbe. Moreover, eight modern patrol boats and 31 motorcycles were imported to help tighten the screws on would-be illegal harvesters through a robust monitoring and surveillance regime (*New Vision*, Jan. 5, 2010).

In addition to the imposition of vertical coercion from above, the Ugandan regime also began experimenting with horizontal coercion from below through the creation of beach management units under the guise of promoting comanagement. In theory, co-management calls for the participation of all stakeholders in running all aspects of fishing operations, including fish stock conservation, biodiversity protection, determination of number of fishers and types of fishing instruments, quantity and quality of harvest, price structures, and distribution of benefits.

As applied to Lake Victoria, however, the primary purpose of beach management units was to systematically foster intraclass struggle among lakeshore communities by creating a group of fishers with privileged access to the lake and deprive access to others. The operating assumption was that the privileged fishers would root out illegal fishers and illegal fishing gears. In the long run, however, vertical coercion in combination with horizontal coercion will not work because the whole exercise is predicated upon an unreasonable probability that the toiling masses of fishers will choose to obey the police state over fighting on to survive.

Short of a radical transformation of the fishery regime in favor of social justice for the broad masses of Ugandans, neither the crisis of legitimation nor the crisis of accumulation will be solved. Meanwhile, the country's natural water bodies will continue to undergo even deeper entropic degradation and exhaustion. This engenders a profound ecological crisis, which, in turn, contextualizes the ever-deepening contradiction between the necessity of survival and the logic of accumulation. The crucial point is, however, that rehabilitation, maintenance, and sustainable utilization of lake resources in the long run hinge on whether the character of the state is altered in fundamental ways. This ultimately requires the contestation, negation and replacement of the prevailing social order.

As constituted at present, the Ugandan state represents the interests of a distinct class of crackpot democrats and kleptocrats, who have temporarily thrived on the construction of subterfuges, deceptions, and manipulations of public discourse. It should be recalled that Yoweri Museveni's National Resistance Movement (NRM) stormed to power on the tidal wave of popular support by dangling a ten-point plank, ostensibly signifying a social contract between the NRM and the Ugandan masses. These points included (among others) pledges to promote democracy and human rights governance, solidify national unity, rebuild the economy, restore social services, resettle the internally displaced Ugandans, and eliminate corruption (Flanary and Watt, 1999; Tangri and Mwenda, 2008; Robinson, 2007).

To demonstrate his intentions, Museveni created a commission in 1989 to review the structure and effectiveness of government services. Subsequently, Museveni approved the 255 recommendations that the commission crafted to make government efficient and free of corruption. Decentralization of public services and privatization of the economy were marketed as the antidote to public corruption. The Office of the Inspector General was set up to keep an eye on potentially corrupt officials (Flanary and Watt, 1999; Tangri and Mwenda, 2008; Robinson, 2007). In official parlance, privatization of the economy, liberalization of trade, reduction of the public workforce, and eradication of corruption received priority. By pretending to take important steps toward achieving these priorities, Museveni temporarily accumulated legitimacy with his own people and credibility with external donors.

In response to the orchestration of simulated reform measures, the international financial institutions and donor countries actually rewarded Museveni handsomely by funneling over $5.4 billion between 1987 and 1996; and between 1997 and 2007 the average annual flow of foreign assistance to Uganda was $800 million (Tangri and Mwenda, 2008). These generous external financial flows were ostensibly meant to give momentum to the reform effort.

The ideology conjured up to mask the class character of the transition was dubbed "nonparty democracy" as if it were a new concoction. Religion, regionalism, ethnicity, and kinship were defined as vices by virtue of the assumption that they would give rise to a destructive multiparty system. Nothing was new in this formulation, however. The same reasoning was used by the old guards of African liberation in support of the one-party state.

In any event, in practice, decentralization of public services meant decentralization of corruption based on familial, kinship, and ethnic connections, and patron-client relationships. For example, Uganda's state workforce ballooned from 150,000 in 1995 to 218,000 in 2004, resulting from the expansion of political patronage (Robinson, 2007). Indeed, Museveni continued to count on kinship and ethnicity to solidify his grip on power through the selective induction into the political bureaucracy of persons from his ethnic power base in western Uganda. Selection, recruitment, appointment, promotion, and transfer of personnel have been predicated on political considerations; political connection and loyalty have been more important than competence and accountability in Museveni's Uganda. Likewise, privatization of state-owned firms meant transferring them to members of the ruling elites and their hangers-on, using wives, relatives, and friends as cover.

The 1998 survey showed the prevalence of corruption in Uganda in the forms of manipulation of revenues and expenditures; exploitation of the public procurement system; petty and grand theft; embezzlement and misappropriation of public funds and assets; payment for ghost workers; bribery and extortion; false declarations of import-export transactions; and payment for services or goods not delivered. Even a World Bank mission was compelled to acknowledge the prevalence of corruption in Museveni's Uganda during its 1998 visit to the country (Flanary and Watt, 1999; Tangri and Mwenda, 2008; Robinson, 2007). According to the World Bank, Uganda was losing $300 million annually since 2005 due to corruption in public procurement alone. Transparency International reported that half of Uganda's annual budget of $950 million is pilfered by members of the national elite. Oddly, half of the country's budget is categorized as coming from donor countries, which suggests that foreign donors subsidize Uganda's social corruption in order to enable the regime to perform its legitimation functions on behalf of global accumulators (Mugerwa, 2008; Rotberg, 2007).

However, corruption is not unique to Uganda's pirate state bourgeoisie. Indeed, corruption is integral to accumulation even in advanced capitalist states; the character of capitalist corruption remains the same wherever it occurs, only its form changes. In the Ugandan case (as in the case of all postcolonial states, for that matter), de facto ownership of the state by the pirate bourgeoisie is the functional equivalent of ownership of the means of production. As some keen observers have correctly pointed out (Shaw, 1976; Burdette, 1984), the postcolonial African elites acquired pseudobourgeois status, not by owning the means of production, but rather by controlling and perpetuating state power and by maintaining their parasitic association with global capital.

In this parasitic association, the African elites are demonstrably more responsive to the demands and exigencies of global capital than to the needs and demands of their own people. Domestically, the pirate bourgeoisie use the state as an instrument of accumulation by steering public resources to themselves and their relatives, ethnic members, and political sycophants. The tax collection system, privatization of parastatals, public procurement systems, and keeping nonexistent employees and soldiers on the public payroll are used as methods of steering public resources to the new class of kleptocrats. In 2003, for example, ghost soldiers on the army's payroll in Uganda were costing the government $30 million a year. Moreover, a minister in Museveni's cabinet engineered a number of legal cases against the government so that his law firm and his partners could collect money from the government. After winning the cases against the government, the firm was awarded $120 million between 1999 and 2003, distributed proportionately among the connivers (Tangri and Mwenda, 2008). In the absence of this condition, the pirate state bourgeoisie cease to be a ruling class and the political economy becomes anything but dependent capitalism. Internationally, the state plays a mediating role between the pirate state bourgeoisie and global capitalism for which the local accumulators harvest the benefits.

To be sure, as the observations in the preceding paragraphs suggest, the crisis in the fishery mode of production is intimately linked to the crisis of legitimacy, contextualized by the crisis in the other components of the food ecology. After all, the so-called illegal fishers are, for the most part, people who migrated to lakeshores on account of depletion or degradation of rural resources, which began with the introduction of colonial capitalism into the country. This trajectory of rural to urban or lakeshore migration is rooted in the alienation of land. In fact, Museveni made the resolution of the land issue a priority when he accepted the World Bank recommendation to privatize the land tenure system. In 1993, he instituted a freehold land system supposedly to promote private property rights and agrarian efficiency. Under the prevailing Ugandan conditions, the effect of this land policy was that those who had money were able to buy land while those without money were left landless or were driven to marginal areas or to lakeshores. As of the mid-1990s, 16 percent of households in Uganda had no access to land, and 52 percent of the households were cultivating, on average, no more than 1hectare of land. Moreover, the gendered character of landhold inequity was glaring since female ownership of land was limited to just 7 percent at a time when one-third of Uganda's households were headed by women (Nyannugasira, 1996; World Bank, 1993). Because the regime miserably failed to solve the agrarian crisis, the landless Ugandans drifted either to lakeshores to survive by fishing or to forest reserves in search of land to farm. Just as the number of fishers has increased over the years, the number of forest encroachers has also increased. Between 2006 and 2008, for example, the number of forest encroachers in Ugandan nature reserves increased from 180,000 to 300,000.

As of this writing, resolution of Uganda's agrarian crisis does not appear to be in prospect. When Museveni, under the stewardship of the World Bank, officially launched the market-oriented macroeconomic policy in 1987, it was predicted that Uganda was going to be transformed from being dependent on subsistence agriculture and on net food importation to being a modern agroprocesser and food exporter, not only reducing poverty but also eradicating underdevelopment and poverty once and for all. It is noteworthy that, according to the 1988-89 survey, 50 percent of Ugandans were in absolute poverty and 45 percent of children under five years old were undersized (World Bank, 1993). In 1992, 89 percent of Uganda's 17 million people were still tied to rainfed farming, growing sorghum, maize, cassava, sweet potato, and plantain (World Bank, 1993).

In the 18 years since then, the economic status and food security of Uganda's masses have hardly improved. The GDP per capita in the country in 2009 was $420 compared to $170 in 1992; if adjusted for inflation, the former figure does actually indicate stagnation in welfare. Even though the Paris Club and the international financial institutions reportedly gave Uganda $2 billion in debt relief since the year 2000 under the so-called Highly Indebted Poor Countries Initiative to purportedly broaden the livelihood options for people in extreme poverty, the pendulum hardly moved in a positive direction. In 2009, 82 percent of Uganda's labor force was still dependent on the same rain-fed crops for subsistence, with per capita output decreasing every year due to input deficit, fragmentation of landholdings, and meteorological variability. In 2005, 55 percent of Uganda's 26 million people were in absolute poverty, which fell to 37.7 percent in 2009. In absolute terms, however, this means that more Ugandans were in extreme poverty in 2009 than there had been in 1992. Furthermore, even though the number of undersized children under five years old at 38 percent in 2009 was slightly lower than the 1992 percentage, the absolute number of children in that category actually increased significantly in relation to the population (UN OCHA, November 30, 2009).

The poverty indices are reflected in the fact that the crises in water, sanitation, and health remain acute. Bacterial diarrhea, hepatitis A, cholera, typhoid fever, bilharzias, malaria, tuberculosis, and HIV-AIDS remain uncontrolled in Uganda. Extreme poverty in the country is exacerbated by natural conditions, especially cycles of drought, flood, and crop infestation. In 2009, for example, Ugandan farmers harvested only half of the anticipated food crops, leaving more than 1 million people in the north of the country totally dependent on international food aid. In fact, in November 2009 the UN launched an appeal for $197 million in international humanitarian assistance to support 2 million Ugandans during 2010 (UN OCHA, Nov. 30, 2009).

At the time when the UN launched its appeal, a new strain of cassava disease was spreading like wildfire in Uganda. Cassava is generally considered an insurance against starvation during years of bad harvest. The tuber is drought-resistant, less labor intensive, and does not require storage since it can stay in the ground until needed. Over 70 percent of Ugandans consume cassava; in fact, cassava is the mainstay of people in the eastern and west Nile regions of the country. Ugandan subsistence farmers used to produce 6 million metric tons of cassava per annum from about 200,000 ha, worth $6 million, until the early 1990s when the cassava mosaic disease (CMD) decimated the tuber completely. However, with a resort to CMD-resistant varieties, the tuber had recovered by 2005 with 500,000 ha placed under cassava cultivation, even though overall output stagnated around 5.5 million metric tons per annum. Unfortunately, in 2009, a new strain of viral infection known as cassava brown streak disease began attacking the tuber in 30 out of 38 cassava-growing Ugandan districts and even spread to mainland Tanzania and western Kenya (*The East African*, July 13, 2009; UN IRIN, May 20, 2010). Spread principally by white flies, the new form of the viral infection attacks all cassava varieties at every altitude as opposed to CMD, which restricts its attacks to lower altitudes, eroding the coping capacities of Ugandans. It is in the face of such vulnerability of the general food ecology to widespread degradation that more and more Ugandans are today flocking to Lake Victoria and other satellite lakes in search of livelihood, thereby worsening the crisis of overharvesting and lacustrine degradation.

IN PLACE OF CONCLUSION

Aquatic resource sustainability is a function of the difference between exactions on and insertions into the resource environment. Demand for virgin resources determines the magnitude of exactions on the environment; at the same time, the demand for a sink into which high entropic wastes are inserted determines the capacity of the environment to process and absorb the wastes. Thus, the hydrobiological state of Lake Victoria, including its evolution, growth, maintenance, and decline, has moved in tandem with demand conditions.

Demand for stock resources in Lake Victoria has been a function of the interplay of deliberate state policy and population pressure, coupled with rising international demand for fish and fishery products. In addition, natural rainfall variability in conjunction with income differentials between terrestrial and aquatic systems has made Lake Victoria a magnet for migrant settlers. The increasing commercialization of the fishery mode of production has served to complicate the issues of sustainability and livelihood maintenance.

Even though the problem of overfishing in Lake Victoria dates back to the introduction of colonial capitalism into the Great Rift Valley region well over a century ago, postcolonial regimes, by embracing the same growth ideology, have taken lacustrine resource overexploitation to a new height.

In the late 1960s, for example, international fish experts asserted that the deep portion of Lake Victoria on the Tanzanian side abounded with haplochromine cichlids, the harvesting of which could yield 100,000 tons annually; Uganda, too, could potentially harvest another 90,000 tons of benthic haplochromine cichlids per year from its portion of Lake Victoria (Welcomme, 1979). This orientation served to inform the Dutch-Tanzanian fisheries agreement of the late 1960s. Blinded by the growth ideology to the consequences of industrializing the entire lake, the two governments developed an appreciation of the fish fauna of Lake Victoria as integral to Tanzania's economic modernization. The Netherlands dispatched fish experts and technicians along with an experimental trawler to help Tanzania develop a fishing industry. Accordingly, a fishing trawler was deployed in 1974 to the Mwanza Gulf of Lake Victoria under Dutch supervision. The immediate plan of action was to build a fishmeal factory to process 60 metric tons of haplochromine cichlid fish a day for domestic and international markets.

The haphazard nature of the plan later prompted Goldschmidt (1994: 4) to raise rhetorical questions: "60 metric tons of cichlids per day? How many days will it take before the fish supply in Mwanza Gulf was depleted? Had anyone taken that into consideration? Had the size of the cichlid population been assessed? Was anything known about the resilience of the ecosystem? How many metric tons of cichlids could be caught each year without permanently damaging the ecosystem?"

Of course, neither the Dutch government nor Tanzanian authorities knew anything about the long impacts of trawl fishing on the processes of the aquatic ecology and the species therein. Indeed, the instance cited above illustrates how the search for quick acquisition and accumulation of wealth has taken a heavy toll on Lake Victoria.

Since the advent of capitalism in East Africa in the late 19th century, the fishery mode of production of Lake Victoria has been marked by two phases. The first phase extended from 1900 to 1980 in which subsistence and artisanal fishery were dominant. Then, construction of the railway networks from Mombasa to Lake Victoria served to popularize fish and fish products domestically and regionally. Concomitantly, to meet the growing demand for fish, artisanal fishers increasingly resorted to new methods of harvesting. For example, between 1917 and 1920, 1.7 million pieces of the popular *Oreochromis esculentus* were shipped by rail annually from Kisumu to Nairobi—a figure which sharply increased to 5 million pieces by the early 1940s (Gibbon, 1997).

The growth of markets for fish and fishery products led not only to the intensification of effort, but also to the importation of modern instruments of production that replaced the traditional nets made from natural fibers. Of the imported gill nets made from artificial fibers, linen and nylon nets were particularly most efficient because they had a long life since they were not susceptible to the elements and decay.

In combination, population pressure, undersized mesh nets, expansion of markets and communication, introduction of outboard engines, and improvement in fishing gear eventually led to depletion of important commercial fish species, especially the status of *esculentus* became very disconcerting to the colonial authorities. For example, between 1949 and 1952, the number of licenses issued by the authorities increased from 35 to 60,000 (Ogutu-Ohwayo, 1990; Fryer and Iles, 1972; Crul, 1995). In addition to being known for its low fecundity, *esculentus* inhabits and feeds in inshore areas, making it easily accessible to artisanal fishing. In keeping with the logic of "picking first the lowest hanging fruit," fishers harvested bigger fish using large nets. When the big ones were fished out, they began to use smaller mesh sizes to catch smaller fish. In the early years, people were able to catch up to 100 fish per single 5-inch net, which progressively fell to 7 or 5 fish per net by the 1920s (Worthington and Worthington, 1933; Graham, 1929; Fryer and Iles, 1972; Thado, 1998; Kudhongania and Chltamwebwa, 1995).

A vicious cycle involved the progressive reductions in mesh sizes. Temporary increases of harvest were achieved from a combination of ever-increasing effort and increasingly smaller mesh sizes. However, every time the threshold was reduced, more and more fish were caught before they could reproduce to keep the fish population stable. In fact, the number of fishers, the number of gill nets, and the extent of effort increased with every passage of time. By 1953, the number of gill nets set by some 30,000 fishers in Lake Victoria each night amounted to more than 2,000 km in length, which could have covered a distance from London to Casablanca. Indeed, the period 1952 to 1971 saw even more marked rises in nylon net use as Ugandan fishers alone were believed to have been setting a huge number of nylon gill nets each night with a cumulative length of 13,000 km (Fryer and Iles, 1972). However, eventually a point was reached when effort and yield were in distressful disequilibrium. In the 1950s, for example, fishing effort increased by 100 percent while yield increased by only 10 percent (Kudhongania and Chltamwebwa, 1995).

For three decades, the colonial administration saw the answer to fish depletion in Lake Victoria in regulating mesh sizes, prohibiting all mesh sizes below 5 inches (Fryer and Iles, 1972). Regulation did not work, however, and the continuing decline of fish stocks in Lake Victoria resulting from the increases in both the number of fishers and the number of gill nets, especially nets of smaller mesh sizes, and the inability of the colonial authorities to control and regulate the harvest sector eventually necessitated a change in direction.

The colonial administration was particularly concerned over the prospect of commercial fishing in Lake Victoria as the *Oreochromis esculentus* and *Oreochromis variabilis* (two target species most favored by fishers) were on the verge of total collapse. Having run out of both patience and any effective means to regulate fishing effort and mesh sizes, the authorities moved to biomanipulate the species composition of the lake with a view to replenishing the dwindling fish stocks by introducing exotic species without evaluating the long-term consequences of the action and in total disregard for the warnings from some fish ecologists.

Between 1954 and 1962, they relocated four exotic Nile tilapiines together with the Nile perch from Lakes Turkana and Albert to Lakes Kyoga and Victoria. The biomanipulators were, of course, pleasantly surprised when the commercial fish harvest from Lake Kyoga rose from 45,000 tons in 1956 to 170,000 tons in 1977, 80 percent of which was accounted for by the introduced *O. niloticus* and the Nile perch (Ogutu-Ohwayo, 1990). Meanwhile, the Nile perch was quietly flourishing in Lake Victoria until it visibly displaced the native species by the early 1980s. By its remarkable dominance in the lake, the Nile perch ended the artisanal and subsistence-based fishery regime and inaugurated the second phase in the modern evolution of the fishery mode of production, transforming Lake Victoria as the most prized lake in the region and, perhaps, the most prolific lake in the world.

The traditional fishery mode of production relied on more than 300 cichlids and 14 genera of noncichlid species, such as *Propterus, Bagrus, Barbus, Synodontis, Mormyrus, Labeo,* and *Clarias* (Balirwa, 1998). In the 1960s and 1970s, the annually reported catches of 100,000 metric tons of fish from Lake Victoria consisted of primarily *Haplochromis* (30-40 percent), tilapines (15-20 percent), such small pelagic species as *Rastrineobola argentea* (10-20 percent), catfish (10 percent), lungfish (5-10 percent), and another five species of fish comprised the balance (FAO, 2005d).

The second phase began in the 1980s, however, when thanks to the Nile perch, the species composition of the catches irreversibly changed. The annual fish harvest in the lake steadily rose from 100,000 metric tons in the late 1970s to 500,000 metric tons in 1989, with the Nile perch now accounting for 63 percent, *R. argentea* for 19 percent, and *O. niloticus* for 9 percent of the catch. By the late 1980s, the annual fish harvest from Lake Victoria represented 27 percent of the total inland fisheries of Africa. The annual market value of this quantity of harvest has been $270 million, shared by the three riparian states and their fishing communities. The harvest bonanza, driven by the rapid proliferation and dominance of the Nile perch, was such that the value of fish production increased by 500 percent on the Kenyan side of the lake while on the Tanzanian side it rose by whopping 750 percent. In 1990, Tanzania exported 35,000 metric tons of Nile perch fillet, almost a fifth of the perch its fishers hauled from Lake Victoria that year. In 1991, Kenya also exported 10,000 metric tons of Nile perch fillet, valued at $20 million (Crul, 1995; Greboval and Mannini, 1992; Ogutu-Ohwayo and Balirwa, 2006; Njiru et al., 2008). The veritable consequence of this remarkable shift to a fishery mode of production, dominated by the introduced exotic species, was the fact that native target species had become extinct or sharply declined in abundance and commercial importance; an estimated 300 haplochromine cichlids have gone for good either as a result of predation by the Nile perch, competition with the exotic tilapiines, and/or in response to a set of multiple stressors that have triggered ecomorphological, hydrobiological, and physiological changes in the lake itself (Hecky et al., 2010; Njiru et al., 2005).

By the mid-1980s, Lake Victoria's commercial fishery had for all practical purposes been transformed from a multispecies fishery regime to one based on three species: the Nile perch, the exotic *Oreochromis niloticus*, and the native *Rastrineobola argentea* . The affiliated outcome of the Nile perch-dominated fishery regime has been the globalization of the lake in which foreign finance capital displaced local ownership of the means of production, exemplified by the fact that 35 fish-processing facilities have found hospitable domiciles around the lake; some of the processing factories have capacities to process up to 50,000 metric tons of fish per annum.

The factories have primarily been targeting the Nile perch. This top predator is particularly attractive to large-scale operators because the perch grows up to 2 meters in length and a fully mature perch weighs 200 kg. In addition to being pliable to a variety of gill nets and industrial fishing techniques, the boneless white fillet of the perch is highly attractive to exporters (Greboval and Mannini, 1992; Njiru et al., 2005). The Nile perch dominance of the fishery mode of production also had profound social ramifications as well as adverse ecological impacts. Following the popularity of the perch and the globalization of fish processing operations, people from the Lake Victoria basin began to migrate to the lake in large numbers. As a result, the number of fishers and their dependents increased from 480,000 to over 3 million over a period of a decade (Njiru et al., 2008).

It is this magnitude of exactions on Lake Victoria from two directions—from fish- processing factories and from increased human crowding around the lake—that makes it most vulnerable to accelerated entropic production and consequent dissipation of lacustrine resources. The first principal factor accelerating entropic degradation of Lake Victoria is the sheer number of people living around the lake and its basin. In 1930, the number of people living in the lake's basin was just 4 million. That number has grown to over 40 million today and is still growing at 3.8 percent a year with 3.5 million of these people actually living in the immediate vicinity of the lake (Bootsma and Hecky, 1993; Hecky et al., 2010). People are taking too much low-entropy resources from the lake as well as from its wider catchment area and putting too much high-entropy wastes into the lake, resulting from a combination of the rapid modernization of fishing gear and methods, poor farming practices, deforestation, and wetland destruction. For example, a single modern boat with outboard engines can drag up to 40 nets over a 3 km area, which makes the boat highly productive in terms of accelerating the harvest rates but also destructive both economically and biologically. These active gill nets destroy passively set nets by small fishers and they collect too much of the fish. In fact, the efficiency of bigger gill nets combined with the use of undersized mesh nets have resulted in the depletion of bigger fish and in the overexploitation of juvenile fish. In the early 1980s, the average size of a perch caught by fishers was 50 kg, which fell to 10 kg by the mid-1990s. Ninety-five percent of Nile perch caught since the beginning of the 2000s are said to be below first maturity. In 2000, it was reported that 36 percent of the 290,000 gill nets operating in Lake Victoria were below the required minimum mesh size of 127 millimeters. In contrast, in 1990, only 9 percent of the 75,300 gill nets employed in the lake were below the required mesh size. In 2002, of the 408,237 gill nets in action, only 15 percent had mesh sizes of 7 inches or above. As a result, 98 percent of Nile perch caught was juvenile including perch as small as 2 kg (Njiru et al., 2008; Mkombo et al., 2007).

Moreover, the supporting facilities, such as net making, boat building, production of packing materials, and so on, are causing significant damage to the riparian environment. For example, the amount of riparian material needed for boat building increased from 15,236 cubic meters in 1998 to 60,000 cubic meters in 2002. In addition, 68,000 cubic meters of wood is consumed annually to process fish (Njiru et al., 2008). In the pre-Nile perch era, subsistence fishers relied on grass to smoke the small haplochromine cichlid fish without causing damage to the riparian ecology. Smoking or deep-frying the Nile perch, however, requires intensive use of firewood, reflected in the fact that the number of smoking kilns at Wichlum Beach in Kenya rose sharply from just 10 in 1984 to 50 in 1991 (Njiru et al., 2008).

The expediency to generate hard currency by the littoral states is another major factor exacerbating the entropic degradation of the lake. During the first five years of the new century, the three littoral states of the area had been exporting over 100,000 metric tons of fish from Lake Victoria alone, worth over $300 million. This is, in large part, due to increased effort since the number of fishers and the number and types of fishing vessels and gear have sharply increased. Between 2000 and 2006, the number of fishing vessels with outboard engines increased from 4,108 to 12,765, the total number of fishing craft rose from 42,493 to 69,160, the number of fishers increased from 129,305 to 196,426, and the number of gill nets increased from 650,000 to 1,222,308. At the same time, the number of hooks tripled from 3 million to 9 million (LVFO, Resource Monitoring, March 7, 2009; LVFO, Strategic Vision 1999-2015).

These increases are reflected in the volume of fish production. The 2005 gross fish production in Lake Victoria was 800,000 metric tons with beach value of $340 million, mostly composed of immature species. The contributions of the *R. argentea*, the Nile perch, haplochromines, and tilapia were 44 percent, 32 percent, 13 percent, and 10 percent, respectively. In terms of monetary value, however, the Nile perch still accounted for 71 percent of the total catch, demonstrating its hegemonic role in the fishing regime. The fish production in 2006 was 1,061,107 metric tons with beach value of $600 million, predominantly comprising immature species. While the share of the *R. argentea* rose to 54 percent of the total catch and that of the haplochromines remained constant at 13 percent, those of the Nile perch and the tilapia fell to 24 percent and 7 percent, respectively (LVFO, March 7, 2009).

The increases in the number of fishers as well as in the quantity and quality of fishing crafts have four important implications for the spatial distribution, abundance and commercial composition of fish species.

First, the introduction of the Nile perch has radically transformed the evolutionary trajectory of Lake Victoria. The 1969-71 bottom trawl survey revealed that haplochromine cichlids constituted 81 percent of the lake's biomass. A decade later, however, the predator perch represented 80 percent of the total lacustrine biomass. Furthermore, *O. esculentus, O. variabilis, Labeo victorianus, O. leucostictus,* and *Bagrus docmac* have long been depleted. As a result, the lake has been transformed from a multispecies fishery to one dominated principally by three commercial species: the Nile perch, the exotic *Oreochromis niloticus* and the pelagic indigenous cyprinid *Rastrineobola argentea.*

Second, since the fishery sector in Lake Victoria has been dominated by artisanal and traditional fishers, inshore fishery has been subjected to heavy exploitation. For example, during 2000-01 the fish stocks in inshore waters were four times greater than those found in offshore waters. In August 2006, however, the offshore fish stocks were found to be higher than inshore stocks. Acoustic surveys show that in the 1999-01 period the mean standing stock of the Nile perch was 1.29 million metric tons, accounting for 59 percent of the total mean stock, which fell to 820,000 metric tons by August 2006, representing 39 percent of the total mean standing stock. On the other hand, the mean standing stock for the *R. argentea* rose from 480,000 metric tons (22.4 percent of total) to 830,000 metric tons (38 percent during the same period). The stocks of *O. niloticus* and haplochromines stood at 370,000 metric tons (17.2 percent) and 470,000 metric tons (23 percent) in August 2006, respectively. In fact, trawl surveys present a different picture. Between 1997-2000 the stock of the Nile perch was estimated at 623,000 metric tons in the entire lake with a mean density of 11 km^2, which then declined to 571,000 metric tons with a mean density of 6.8 km^2 during June/July 2005-

06. According to estimates from the Jinja-based Fisheries Research Institute, the total stock of the Nile perch further declined to 221,175 metric tons by 2007 (LVFO, March 7, 2009; FAO, 2008b).

Third, with the depletion of inshore fisheries, fishers have had to upgrade their fishing crafts in order to travel further offshore in search of fishing grounds. The social implication of this upgrading has been that the majority of fishers have been pushed out while those with capital could purchase motorized vessels, thereby completing the proletarianization of labor and the de facto commercial enclosure of the lake.

Fourth, the depletion of fish species, intimately connected to changes in demand conditions, especially for Nile perch, has served to accentuate competition, not only among fishers, but also between the littoral states of Lake Victoria. All three states have pinned their expectations for higher foreign hard currency earnings on the productivity of the Nile perch. In consequence, the three states have been attempting to impose a police state on their respective lacustrine jurisdictions to deny the poor fishers access to the same resource.

Demand conditions in Lake Victoria are not limited to extraction of fishery resources; there are pressures on the lake's water resources from other competing interests, namely, agriculture, aquaculture, industry, and nature reserves. Agricultural interests are poised to make more demands on Victoria's water resources. To improve their food security situation, the littoral states have intensified their efforts to modernize the agricultural sector. The model is still Western: scientific stimulation of agrarian production through the rapid expansion of irrigation works. To fully modernize, the process will require a great deal of water and chemical fertilizer. In addition to providing a continuous supply of water to run hydroelectric generators, the lake must supply water continuously for irrigation. The three littoral states have blueprints on how to irrigate more than 1 million ha in order to boost food production by drawing water from Lake Victoria. To realize this ambition, they will need up to 10 bcm of water annually from the lake (Waterbury, 2002; Yohannes, 2008). Given the high rainfall variability and recurring drought frequently experienced in the region, the riparian states are naturally predisposed to look to Lake Victoria as the provider of last-resort-resources in the hope of improving their food security. Kenya already has a 2,400-ha rice irrigation scheme and a 900-ha sugarcane plantation on its side of Lake Victoria. Similarly, in addition to existing growth schemes, Tanzania has identified 61,000 ha of land as suitable for various irrigation projects on its side of the lake. The area is already the principal cotton, coffee, and livestock producing region. On the Ugandan side, there are well established sugarcane plantations, horticulture and floriculture farms, widely using wetland and lake water resources (Awange and Ong'ang'a, 2006).

There is also renewed ambition in all littoral states to make aquaculture integral to accumulation under the familiar guise of national food security. Colonial authorities had toyed with introductions of fish farms in the 1940s in all three littoral states. In 1948, the colonial authorities set up two fish farms in Kenya. However, mesmerized by the productivity of Lake Victoria, Kenya's pirate state bourgeoisie almost completely neglected the aquaculture subsector. In 2003, for example, 7,790 fish farmers produced a mere 1,012 metric tons comprising the African catfish, tilapiine species, common carp, largemouth bass, and rainbow trout (FAO, 2005b). Now, however, Kenyans have begun to give high priority to promoting commercial aquaculture to grow the sector by 1,000 percent in a matter of few years.

In 1949, British authorities also introduced fish farms into Tanzania (then called Tanganyika), and by 1968, 8,000 fish ponds were in operation, characterized by low productivity. As in Kenya, the Tanzanian elite gave low priority to freshwater fish farms. In 2005, there were 14,100 small fish farmers tending to 14,000 freshwater ponds, who produced 1,522 metric tons (FAO, 2005b).

Relatively speaking, aquaculture is far more developed in Uganda than in its neighbors. In 2005, there were 20,000 subsistence and commercial fishponds with an average area of 500 square meters, on average producing 1,500 kg per ha per year for subsistence farmers, and 2,500 kg per ha per year for commercial operators. There were 500 large-scale commercial hatchery operators whose ponds ranged between 5,000 and 50,000 square meters in surface area. Now internationalization of Uganda's commercial hatchery operations has increased since the beginning of the new century. As a result, foreign finances have been flowing into the subsector—either in the form of direct investment or joint venture with local businessmen. As a result, freshwater aquaculture production picked up from just 285 metric tons per annum in the 1980s to 15,000 metric tons in 2005 (FAO, 2005d).

Uganda's pirate state bourgeoisie hope that aquaculture production can make up for the decline in Nile perch production in Lake Victoria. Uganda has already set up an aquaculture fishery unit within the Ministry of Agriculture, Animal Husbandry and Fisheries, headed by a principal fishery officer with his own senior fishery officers and support staff. The ministry has identified 31 districts, mostly along the Lakes Victoria-Kyoga-Edward-George axis, as suitable for commercial hatchery operations (FAO, 2005d).

Drivers of the renewed interest among the East African states in aquaculture involve sets of domestic considerations and external enticements. In terms of internal considerations, since Lake Victoria and other satellite lakes are overfished and terrestrial sources of food and revenue are degraded, the pirate state bourgeoisie and private capital accumulators see great potential in commercial aquaculture production. Likewise, international agencies, such as the FAO and the World Bank, see great promise in the promotion and development of aquaculture as Africa's answer to the challenges of stagnation of inland-capture fisheries, poverty, structural unemployment, and malnutrition. Indeed, the FAO and its partners have long expressed disappointment at sub-Saharan Africa's dismal performance in aquaculture production despite its enormous potential. By some estimates, 31 percent of the continent's land surface area is considered suitable for small-scale fish farms and 13 percent of the land area is suitable for commercial aquaculture, the utilization of which could supposedly improve the nutrition, income, and employment of the continent's poor people (Aguilar-Manjarrez and Nath, 1998). In 1999, for example, sub-Saharan Africa produced 117,000 tons from aquaculture operations, representing a mere 0.4 percent of total global aquaculture production (Dugan, 2003).

Some writers (Changadeya et al., 2003) attribute Africa's poor performance not to lack of suitable sites and water resources, but rather to poor genetic diversity in cultured species, slow growth of cultured species, poor stock selection and management, lack of scale economy, and low government priority. Consequently, these writers and consultants have been actively promoting the idea of private commercial aquaculture in order to supposedly close the food deficit in sub-Saharan Africa.

More important, however, the growing prominence of fish and fishery production from capitalist aquaculture has begun to impress the pirate state bourgeoisie throughout the continent. Having grown from less than 1 million metric tons per annum in the early 1950s to 51.6 million metric tons in 2006, with a value of $78.8 billion, global aquaculture has become the fastest growing subsector and is projected to overtake capture fisheries in the near future. In 2006, the aquaculture subsector already contributed 47 percent of global fish food supply. In 2006, more than half of total global aquaculture production was from freshwater environment (FAO, 2008b).

In light of this growing importance of aquaculture, the private and state actors appear to be excited about the prospects of bonanza from commercial aquaculture by converting the wetlands in the Lake Victoria basin into dense clusters of fishponds. China and Vietnam now supply the model template for the promotion of commercial aquaculture in East Africa. Whereas in 2006, China produced only 17.5 million metric tons of captured fish, it produced 34 million metric tons from aquaculture, accounting for 67 percent and 49 percent of global aquaculture production in terms of quantity and value, respectively. That year, China exported $9 billion of fish and fishery products, becoming the leading exporter in the world in that sector (FAO, 2008b).

However, this new drive for aquaculture growth has socioeconomic, biological, and environmental consequences for Lake Victoria.

First, since the driver of commercial aquaculture production is profit and revenue generation for both private investors and the pirate state bourgeoisie, the potential outcome of this renewed interest in the subsector is likely to result in freezing out subsistence fishers. Given the fact that the quantity of fish exported is determined by the global value chain on the basis of quality, commercial aquaculture will, by definition, favor heavy capitalization of producing and processing fish in large quantity for export.

The enduring lesson we glean from the Nile perch industry experience in Lake Victoria is that generation of exportable fish requires unit scale in terms of pond size, species composition, chilling, refrigerating, and freezing the fish and cold storage to keep the processed fish until shipped out. In effect, only those individuals or firms with capital can afford to set up large-scale fish farms and fish-processing facilities to the quality and standard specifications of the global distribution channels and global retail chains. Without access to capital and credit facilities, technical expertise, managerial skills, and information about markets, the indigenous communities would be pushed out from the wetlands on which they have hitherto depended for biomass production and subsistence farming. The final outcome is likely to be the de facto commercial enclosure of the wetlands and floodplains in the Lake Victoria basin. International agencies and researchers are already forwarding a series of recommendations favoring state disengagement from the aquaculture subsector, production and marketing of high-value fish through selective breeding and genetic manipulation of species, increasing production intensity and efficiency, supply of high-quality fish seed and fish feed, and investment in technologies as requisite conditions of large-scale commercial aquaculture. The recommendations underpinning the strategy of selective breeding, hybridization, and chromosome manipulation of fish species are to grow fish bigger, fatter, and faster in large ponds under conditions of mechanization (Dugan, 2003; Changadeya et al., 2003).

Second, there is a growing temptation to boost aquaculture production through genetic manipulation of native species or introduction of exotic species in order to enhance performance of the subsector and thereby insure commercial profitability and ostensibly food security (FAO, 2008b). However, biological engineering in aquaculture production of existing species or introducing alien species can have unintended consequences for wild species if the genetically altered or introduced species escape from the ponds. Molecular manipulation of native fish species or culturing exotic species in fishponds can result in two possibilities. The gene flow from cultured species into their wild relatives may erode the genetic diversity of the species involved. Species that are genetically engineered to perform well in the hatchery may not adapt well to conditions of wild life. So if they escape from their hatchery into the wild, hybridization and introgression may occur, which may spell genetic contamination or extinction of the species altogether (Rhymer and Simberloff, 1996). Alternatively, the cultured exotic species may escape to the wild environment and displace indigenous species altogether by preying on them or by altering the habit and food requirements. For example, In Zambia and Zimbabwe, *O. niloticus* have been introduced into aquaculture production systems because the species are considered to grow faster and bigger than indigenous species. In due time, however, some of the introduced species have escaped and have contaminated the gene pool of indigenous wild species because of the resulting unplanned hybridization. Similarly, some *O. niloticus* have escaped from their designated hatchery in the Limpopo valley in South Africa, leading to hybridization with *O. mossambicus.*

Proponents of species introduction try to reassure us by suggesting that poor management, rather than the reality inherent in the nature of the capitalist drive for faster accumulation, is blamed for the escape of the species (Changadeya et al., 2003). The reality is that there are many variables beyond the control of managers that may generate unexpected consequences. After all, aquaculture is known to be the leading gateway for the spread of aquatic invasive species globally and consequently for biological pollution of indigenous species. For example, Fish farmers in the United States made a series of introductions of Asian bighead carp, silverback carp, and black carp in the 1960s and 1970s to filter the water by removing algae and suspended matter out of catfish ponds, as well as to control yellow grub populations and trematode parasites in the catfish ponds. However, all three species of carp have escaped from their designated places when the ponds overflowed following heavy floods in the 1990s. Now the carp have overrun the Mississippi River basin and have spread to many water systems in many states (Ferver, 2001; Naylor et al., 2001; Chick and Pegg, 2001). The carp are very fecund and their populations can undergo exponential explosion; individual carp can grow big quickly. Between 1988 and 1992, for example, commercial fishers managed to collect 600 kg of bighead carp and silverback carp per annum from the Mississippi and Illinois Rivers, which jumped to 5,000 kg in 1994. Since 1997, however, the harvesters have been annually hauling 50,000 kg of bighead and silverback carp from the same fishing grounds (Chick and Pegg, 2001).

All Asian carp are known for their exceptional fecundity, big size, hardiness, and ravenous appetite. The average black carp, which can grow up to five feet in length and weigh up to 33 pounds, consumes up to four pounds of freshwater mollusks a day. Since 67 of the 300 mollusk species in the United States are either vulnerable to extinction or are already extinct and another 69 species are under Federal protection, predation of freshwater mollusks in the magnitude indicated above by carp can result in a biological crush of mollusk populations. A female black carp lays between 129,000 and 1.18 million eggs annually, and a female bighead carp lays up to 1 million eggs in a season. The ecological devastations from carp populations in terms of threatening to radically alter the faunal composition and abundance of native species became so widespread that President Obama appointed a "carp czar" in September 2009 to run an $80-million program to prevent the carp populations from moving up to Lake Michigan. The fear is that, if the carp populations manage to make up to the Great Lakes region, the $7-billion commercial and sport fishing industry, employing more than 800,000 people, may crash.

Today, all five species of carp are widely distributed and well established in North America, posing grave risks in shallow lakes and wetlands, destroying the food sources, habitats, and nursery grounds of native species. As the experience with the introduced carp populations in North America suggests, even under optimal managerial conditions, gene contamination can occur in and around Lake Victoria due to natural fluctuations in rainfall and floods. As we saw earlier, the common carp (*Cyprinus carpio*) have colonized Lake Naivasha after they have been carried by heavy flooding during the 1997-98 rainy period in the lake's catchment. Apparently, the flood dislodged the carp from fishponds and carried them to the River Malewa, which returned the favor by transporting the carp to Lake Naivasha where they have firmly established themselves at the expense of other fish species. Similar events can produce similar effects in Lake Victoria. It is noteworthy to recall the great floods of 1961-64 when Lake Victoria received 173 bcm of additional water, raising its water levels by 2.5 meters (Collins, 2002; Georgakakos et al., 2005). In any event, escapees can irreversibly pollute Lake Victoria's aquatic species. The resulting hybridization may potentially lead to introgression or stuntedness, eroding the genetic diversity, stability and resilience of aquatic populations.

Third, a recent study on 10 Ugandan commercial aquaculture production schemes in terms of the water quality, site selection, fish seed, pond fertilizer, fish-rearing facilities, and post-harvest practices reveals widespread exposure to hazardous conditions. In nine of the ten ponds sampled, animal manure, such as chicken droppings, is used to fertilize the ponds; five ponds are on sites where municipal sewage flows; four ponds receive waste from domestic sources; three ponds receive agricultural runoff; and owners of all ponds make their own fish feeds without any oversight (Bagumire et al., 2009). In addition, another study on 38 Ugandan commercial ponds found the presence of mercury, lead, cadmium, and an assortment of organochlorine pesticide residues, even though detections of these traces are not as yet above the thresholds for safe consumption (Bagumire et al., 2008).

In so far as industrialization is concerned, the riparian states, individually or collectively, are committed to making the lake the pivot of rapid regional economic growth. In August 2008, the East African Community unveiled a $5 billion Lake Victoria Development Plan to transform the entire lake basin into a modern economic zone. The centerpiece of the plan is to build a ring highway around Lake Victoria and a series of feeder roads leading to the shores of the lake. These projects in turn are expected to spur growth of industry, agriculture, fisheries, and tourism and to speed up the transportation of people and goods in and around the lake. Even in the pre-Nile perch era, movements of passengers and goods were already active around Lake Victoria. In 1972, for example, the East African Railway and Harbors Corporation transported 500,000 passengers and 500,000 metric tons of goods around the lake (Crul, 1995). In response to the prospects of huge gains from the plan, international, regional, and local investors are expected to build tourist facilities, lodges, hotels, cruise ships, and other strategic industries all around the lake (*Africa News*, 2008).

Anticipating implementation of this master plan, Zain's subsidiary CETEL and Ericsson have recently committed to expand cellphone services by upgrading the existing infrastructure and by adding 21 new radio sites. Upon completion, 90 percent of the fishing zones of the lake will have full cellphone coverage. Even more enticing, from the perspective of the cellphone operators, is the expectation that 3.5 million people who live in the immediate vicinity of Lake Victoria are potential cellphone service users (Africa News, March 17, 2008). This suggests that the radical transformation of the urban ecology and the concomitant necessities required to support this transformation pose far-reaching threats to Lake Victoria. More water and more energy will be needed to support the present and future industrial establishments and urban dwellers. And the littoral states are naturally bound to look to the lake for both water and energy.

Thus, demand-driven conditions, involving multi-uses and multi-users, are bound to increase the stressors on supply conditions. For example, Uganda's controversial decision to increase its hydropower generation capacity at the two dams on the outlet of Lake Victoria by increasing its water share from Lake Victoria has adversely affected water levels in the lake. Historically, Uganda relied on the Nalubaale Dam (formerly the Owen Falls Dam) built on the outlet of Lake Victoria in the 1950s with 180-mw installed capacity, now expanded by another 90-mw. In keeping with the capitalist logic of destructive creation, in 1954 British engineers removed the natural rocks that had acted as a weir at the mouth of Lake Victoria where the lake empties its surplus water into Victoria Nile. Then they replaced the natural weir with a manmade dam by which they effectively transformed Lake Victoria into a supergiant hydroelectric dam. According to the formula known as the "agreed curve," the amount of water passing the turbines at Owens Falls was determined to be between 300 and 700 cubic meters per second. In order to maintain the hydrological stability of Lake Victoria and the integrity of the agreed curve, Uganda was placed under an obligation not to tamper with the arrangement. This way, the manmade weir, would mimic the natural weir in which case the lake level would not be affected by the outflow of water from the lake.

However, the engineers were behind the curve, insofar as their understanding of the capitalist logic of destructive creation is concerned, because capitalist growth is not meant to come to a standstill so that nature could take its own course. In their defense, the engineers could not have anticipated what would come to pass half a century later. It is in this sense that Shackles (1972) insightfully observes that reason is a prisoner of time in which humans (however reasonable) have little say about the yet-unformed future.

In any case, as Uganda's urban demand for more hydro-energy shot up in the 1990s, the World Bank and the African Development Bank raised $248 million to build a second hydropower plant just eight kilometers from Lake Victoria, which came on stream in 2002 with installed capacity of 250 mw (Pearce, 2006; Georgakakos et al., 2005). This violated the agreed curve formula, which was supposed to preserve the natural integrity of the lake. Now the two hydroelectric dams began releasing, on average, 1,250 cubic meters of water per second, which is 55 percent more than allowed under the agreed curve formula. Between the opening of the second dam in 2002 and the end of 2005 alone, Lake Victoria lost 75 cubic km of water, amounting to a loss of 3 percent of the lake's water volume; lake level dropped by six feet, 55 percent of which is attributed to overreleases by Uganda's hydropower stations while 45 percent is due to drought (Pearce, 2006; Georgakakos et al., 2005). Consequently, towns around the lake began to run low on drinking water, fishing boats got mired in mud, and ferries were stranded far away from their jetties.

The consequence of the sharp decline in water levels damaged aquatic macrophytes and riparian vegetation fringing the lake as well as the aquatic species that use the wetlands and riparian vegetation for food, habitation, refuge from the predator Nile perch, breeding, and nursery. A study by Kiwango and Wolanski (2008) attributes the collapse of fisheries in Lake Victoria during the period 2000 to 2006 to a 20 to 50 percent increase in water uptake by the two Ugandan hydroelectricity dams. In response to decreased water level, the papyrus wetlands fringing the lake collapsed. As a result, the fishery of tilapia fell by 80 percent and the food insecurity of fishery-dependent communities sharply rose. Such development is predictable since plant species composition, distribution, and vegetated margins of Lake Victoria are integral to the fish ecology of many species, which rely on these resources for shelter from predation, as well as for habitat, feeding, spawning, nesting, and nursery (Toowoomba, 2000).

Moreover, the integrity of the lacustrine ecology has been severely undermined since the papyrus wetlands have been critical in absorbing up to 50 percent of nitrogen and 25 percent of phosphorus before being inserted into the Lake Victoria through influent rivers and streams (Kiwango and Wolanski, 2008). The degradation or collapse of the papyrus fringing the lake also has a cascading effect on the overall health of the lake. The decay of the submerged papyrus in combination with more insertions of nitrogen and phosphorus into the lake in the absence of buffering papyrus, eutrophication of Lake Victoria has grown significantly.

As though this colossal fiasco was not enough to demand halting further action, Uganda has today embarked on the Ayago and Merchison hydropower projects, which will have 840 mw installed capacity when they come on stream in the very near future. In fact, by 2045, Uganda will have six hydropower stations on Lake Victoria with combined installed capacity of 3,553 Mw (Georgakakos et al., 2005). Uganda's riparian neighbors are not going to idly watch this hydro development. They, too, will rush to expand the use of their share of the lake's water for purposes of both irrigation and hydroenergy generation, perhaps totalizing the entropic disintegration of the lake.

Nature reserves are another source of demand for lake waters. The tourism sector (the principal hard currency earner in East Africa) is intimately connected to nature reserves. The continuous flow of hard currency from tourism hinges on whether the reserves are maintained or not. A recent study by Gereta et al. (2009) suggests that the future of nature reserves in the region could be precarious for lack of influent rivers. The study points to the case of the Serengeti reserve in the face of potential water deficit. The Serengeti nature reserves in Tanzania and the Maasai reserves in Kenya largely depend on the Mara River, which rises in the Mau forest complex. However, the perennial flow of the Mara River has been called into question in the face of growing deforestation and water diversion for irrigation and hydropower generation. In light of these development projects and recurring drought, the flow of the Mara River has been decreasing substantially in recent years, making maintenance of the Serengeti and Maasai nature reserves exceedingly difficult. What makes this ecoregion particularly unique is the fact that more than 1 million wildebeests and 200,000 zebras make an annual pilgrimage between the Tanzanian and Kenyan sides of the ecoregion, essentially following the Mara River. As the aforementioned study suggests, if the 1972-type drought occurs in the future, up to 80 percent of the wildebeests may die off. To stem such a scenario, it may be necessary to construct artificial wetlands along the banks of the Mara River in the Serengeti, and to expand the western side of the reserves by 5 km for purposes of connecting the nature reserves to Lake Victoria's perennial waters, thereby exerting more pressure on the lake.

Before concluding this chapter, the three types of crises mediating the social relations of production and the natural evolutionary processes merit some highlighting. The populations of the three littoral countries are growing, not only in number, but also in destitution. With economies in tatters and urban employment nonexistent, people are increasingly relying on terrestrial and aquatic sources of food to survive. This situation has created a crisis of legitimation in which state regulatory and control mechanisms are ignored, defied, and contested. If the biodiversity and health of Lake Victoria are to be restored, maintained, and enhanced, alternative means of livelihood for the toiling masses must be found. However, since finding alternative sources of livelihood is not in prospect as of this writing, the legitimation crisis is likely to remain unsettling for both the state and nature.

Then there is the crisis of accumulation. The elites in all three countries still view Lake Victoria as a cash cow, a strategic source of hard currency. Since export earnings from such cash crops as tea, coffee, cotton, and cashew nuts are increasingly hard to come by in the context of the highly volatile global price structures, the elites still hope that export of the Nile perch would close the gap. Thus, in order to contain the crisis of accumulation, the elites are predisposed to place high priority on the fishery mode of production to strengthen the export sector. This means that the lake is neither going to be given ample time to heal itself, nor will there be adequate time for the international effort to restock the lake with fish species to work. The danger is that the national elites of the littoral states still harbor the illusion that rehabilitation, maintenance, and enhancement of the Nile perch-based fishery could be had through semimilitarization of the lake. However, the prevailing conditions of deprivation, frustration, and desperation of the toiling fishers suggest that an outright confrontation between security forces and poor fishers is likely to be the defining feature of the fishery regime.

Finally, there is the ecological crisis, which stems from the fundamental contradiction between the growth ideology and the conservation imperative. Both aquatic and terrestrial ecologies have been targeted for decades to provide all of the throughput to sustain national growth. The consequent partial industrialization and increased urbanization have been transforming the low-entropy materials into high-entropy wastes, in the process engendering multiple stressors consisted in pollution, sedimentation, and eutrophication of Lake Victoria and other Rift Valley lakes. In the context of the sharpening legitimation and accumulation crises, the ecological crisis will continue to receive peripheral attention—if any at all. Thus, conservation of lacustrine biodiversity becomes a fanciful rhetoric.

In the neighborhood where I grew up, people used to say that when two bulls fight over a meadow, it is the grass that gets trampled upon and damaged; when the fighting is over, both bulls end up with nothing. In the context of the struggle between the masses toiling to survive and the national elite fighting to maintain and aggrandize their wealth, the ensuing ecological crisis will determine the fate of them all.

3 | Lake Victoria: The Supply Conditions (Part I)

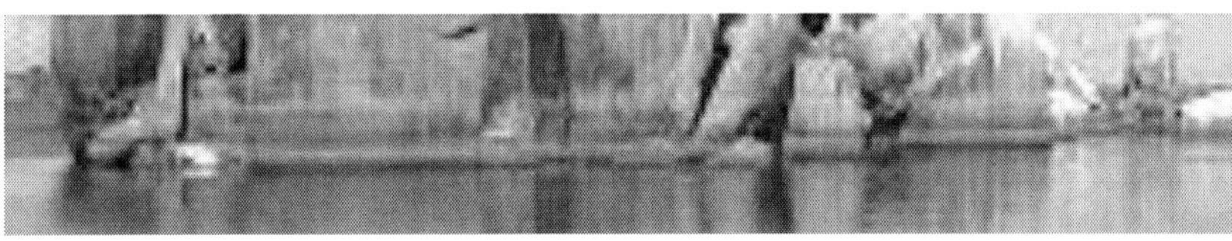

INTRODUCTION

From a thermodynamic point of view, Lake Victoria is a tiny, yet important hydroecological piece of the earth's living system. Thus, what takes place in any of the various components of the earth's living system may affect the natural state of the lake—more so than what happens inside its own ecology. The tautological statement here is that the crisis in Victoria's fishery mode of production could be a microcosmic expression of a general crisis in the earth's living system, driven in no small measure by the character of our contemporary capitalist civilization. Indeed, scientific views are crystallizing around the idea of how humans have become the dominant force in the evolution of the earth's living system by adding the anthroposphere (made up of socioeconomic forces) to such natural systems as the atmosphere, geosphere, hydrosphere, biosphere, and cryosphere as the determinant factors. According to paleoclimatic evidence, humans were minor players in the development of the earth's living system over the past 10,000 years (the Holocene epoch). However, the anthroposphere now appears to be interfering with the trajectory of the earth's natural development.

Some writers have suggested that we may be in transition from the Holocene period—in which the earth fluctuated within a narrow range of natural parameters, without losing its resilience, stability, and coherence—to an anthropocene period, where humans have become the dominant players with unlimited capacity to throw the other natural spheres off their trajectories (Rockstrom et al., 2009; Hanson et al., 2008; E. Wilson, 2002; Sachs, 2008). In this anthropocene period, humans have become a powerful geophysical force, taking too much out of nature beyond its regenerative capacity, in terms of low-entropy throughput and inputting high-entropy wastes beyond the waste absorptive capacity of nature. Moving in tandem, the necessity of survival of the poverty-stricken global population and the capitalist logic of destructive creation are undermining the complex biological processes that make the planet habitable. The concomitant potential to change the whole trajectory of how the natural components of the earth's system function may push complex systems of plants, animals and whole biological processes to a tipping point (Millennium Ecosystem Assessment, 2005).

The observation noted above suggests that our understanding of the present and future status of Lake Victoria cannot be isolated from our effort to comprehend the effect of anthropospheric processes on the operation of other components of the earth's system as a whole. This is especially true since many triggers of Victoria's state transition may originate in the various components of the other natural spheres. If it appears that I am using Lake Victoria as a proxy to understand the impact of changes in the other natural spheres on the biolimnology of the lake, then I am on the right track. What may appear as a digression to the reader is actually a strenuous attempt on my part to advance a dialectical understanding of the particular nexus between Lake Victoria and the earth's living systems. I hope to advance an understanding of the inexorable relationship between nature and society by overcoming what E. O. Wilson (2002) calls the problem of ontological reductionism, scientism, and simplism—not to mention abject intellectual poverty and timidity.

In this chapter, I will analytically move back and forth between the general and the particular, in the hope of capturing all the causes and consequences of Lake Victoria's state transition. Contemporary researchers are still divided over 3 rival hypotheses regarding the present biolimnological state of Lake Victoria and its impact on supply conditions. As we shall see in this chapter and the next, climate-driven change, cultural eutrophication, and human biomanipulation of the lake's species composition are offered as competing hypotheses to explain the state transition of Lake Victoria. The reductionist qualities implied by these rival hypotheses aside, the fact that even natural scientists are unable to discriminate among the causal variables driving the state transition of Victoria demonstrates the limitations of existing quantification modeling processes. Specifically, they are unable to precisely demarcate the boundaries between naturally generated phenomena and human-induced change, on the one hand, and they are unable to capture the complex joint operations of the laws of thermodynamics and evolutionary biology, on the other. The difficulties involved in the present state of our knowledge point to the urgency for a radical epistemological shift in our understanding of the complex link between nature and modern capitalist societies, not only at the local level but also at the global level. This chapter is about how human-aggravated climate change, human-driven terrestrial resource depletion or degradation, and the metabolic transformation of the urban ecology have contributed to a human-induced state transition of Lake Victoria. All of these factors have far-reaching implications for people who depend on the lake for livelihood, energy, drinking water, and irrigation.

As natural scientists (E. Wilson, 2002; Rockstrom et al., 2009; Hanson et al., 2008; Barnosky et al., 2011; Millennium Ecosystem Assessment, 2005) observe, the current general crisis in human existence stems from the fact that man has become a geophysical force. Mankind possesses the seemingly limitless ability to radically change his environment, but only at his own peril. Researchers have identified anthropogenic Habitat modification, Introduction of invasive species, OverPopulation, Pollution, and Over-harvesting of natural resources (encapsulated in the acronym HIPPO) as the major culprit of ecosystem change. HIPPO refers to the 5 major anthro-pospheric determinants undercutting the evolutionary trajectory of biosocial development and, by implication, all supply conditions. In addition to HIPPO, the influence of climate change is a 6th culprit compounding the problem.

From an evolutionary biological point of view, the problem begins when too many people take too much low-entropy throughput out of nature in order to meet their basic needs and satiate their ever-expanding wants. When overharvesting of low-entropy throughput exceeds the regenerative capacity of nature, then the synergistic downward spiral of resource depletion is set in motion. The introduction of alien species into local environments, intended to boost natural production, simply defeats the purpose as it consequently undermines local sustainability.

On the other end of the problem, the export of high-entropy wastes to the environment overwhelms the absorptive capacity of the environment when it exceeds nature's ability to assimilate or process the wastes into harmless substances. The degree to which humans have been overtaxing the regenerative capacity of nature and overwhelming its waste absorption capacity is measured by an ecological footprint. This footprint consists of the average amount of resources appropriated by each person in the world in the form of natural resources (land, water, energy, and timber), transportation, commerce, and waste absorption. In 1978, it was said that the human population on earth had already exceeded the carrying capacity of the planet. Since then, the overshoot of the living human population has overwhelmed the planet's biocapacity. If we use ecological footprints as a yardstick to evaluate the seriousness of the problem, then 4 additional planets would be needed in order for the average human being to enjoy life in the same way it is currently being enjoyed by the average American. Indeed, a closer examination of the living planet index portends an ominous ramification of resource depletion and species extinction for human civilization. Between 1970 and 2000, the abundance of terrestrial vertebrate species and marine species had already fallen by 30 percent, while that of freshwater species had declined by 50 percent (Millennium Ecosystem Assessment, 2005; Rockstrom et al., 2009; E. Wilson, 2002; Sachs, 2008).

In the developing world, the current average ecological footprint available to each person is 2.5 acres, in contrast to 24 acres per person in the United States. If each person in the developing world (accounting for 5.6 billion of the 7.1 billion in the global population) seeks to reach the living standard of the average American, then the negative consequences could be beyond comprehension (E. Wilson, 2002).

It is in the context of the above general discussion that the biolimnological state of Lake Victoria must be situated. Indeed, microlevel manifestations of the great global thermodynamic conundrum abound in the Lake Victoria basin. The human population in the basin has grown from 5 million in the 1950s, to more than 40 million today, and continues to grow at 3.5 percent per annum (Cohen et al., 1996; Balirwa, 2007). At this annual growth rate, the basin's population will double every 20 years. Since 70 percent of the people in the Lake Victoria basin already live in abject poverty, the necessity of survival will continue to inform their collective decisions on how to eke a living out of nature. This entails liquidation of natural capital stocks in the basin. The consequent catchment degradation, associated with extensive deforestation and wetland destruction, will inevitably expose Lake Victoria and other satellite natural water systems to continuous insertions of undesirable and harmful nutrients and sediments, thereby accelerating their sedimentation and eutrophication.

After all, fish population dynamics, trophic structure, and productivity in Lake Victoria are controlled in large measure by factors and processes external to the fishery mode of production. Natural variations in climate, precipitation and flood patterns, nutrient supply, terrestrial habitat modification, and species introduction can have unpredictable consequences for the hydrobiology, physical properties, and trophic profile of the lake. For instance, the differential heat budget—between the tropics and the polar regions, as well as between oceans and land—associated with global warming is expected to have an indeterminate impact on the seasonality, frequency, and intensity of rainfall patterns; the indeterminate impact will also affect rates of evaporation and evapotranspiration and, by implication, aquatic productivity (FAO, 2009; Allison et al., 2009). To be sure, it is in the context of the 6 determinants of biosocial evolution (which biologists have identified) that the supply conditions in Lake Victoria must be examined.

Today, all of the determinants of nature-society relationships about which biologists write are having direct bearings on the limnology, biodiversity, and productivity of Lake Victoria. As a subset of the general environment, the lake naturally receives energy and matter from the environment, frequently through the mediation of both natural processes and human actions. Both the quantity and quality of this exchange between Lake Victoria and the general environment matters a great deal, with respect to ecological sustainability, spatial distribution, abundance and resilience of species. Such important natural phenomena as extreme weather events (including storms, floods, and droughts), significantly affect the hydrobiology, geochemistry, and physiology of Lake Victoria. As we shall see later, global warming is now having significant influences on the hydrologic cycle, altering the seasonality, frequency, and intensity of precipitation in Lake Victoria's basin. Thus, it stands to reason that in reaction to events external to the lacustrine ecology of Lake Victoria, the availability and quality of fish food to lakeshore communities may be substantially diminished. Consequently, this shortage will increase the vulnerability of fishery-dependent communities to the erosion of livelihoods, unemployment, hunger and poor health.

Although natural processes produce many unanticipated surprises and uncertainties that affect supply conditions, the issues that should concern us most here are the effects that demand conditions have on supply conditions, such as the removal of riparian vegetation, wetland destruction, deforestation, capitalist agriculture, industrialization, and the slumization of urban centers. After all, we cannot control what nature does to itself and to us, other than to develop coping and adaptive mechanisms. However, we can manage our decisions and actions in ways that avoid or minimize the effects of our actions on climatological, biological, and ecological conditions. As a group of UN scientists recently pointed out, the greatest challenge facing humanity today is "how to avoid the unmanageable and how to manage the unavoidable" (*UN Chronicle* 2007).

The expression of this view suggests 3 pointers: First, in studying Lake Victoria, we need to take the global ecology as our unit of analysis. This does not mean that we have to study every aspect of the global ecology in order to fully understand the hydrodynamics and hydrobiology of the lake. Rather, we will have a dialectical understanding of how the interconnectedness of all things in the global environment affects the biolimnology of Victoria. As indicated earlier, the global ecology is divided into seemingly separate and yet interdependent compartments. The connecting and unifying force is the biogeochemical cycle. Living things, matter, and chemical properties circulate not only from one ecosystem to another within each sphere, but also between the spheres. These factors influence the structures, characteristics, processes, and functions of the ecosystem they come into contact with. Of course, the greater the spatial proximity among ecosystem compartments, the greater the impact of the biogeochemical cycle on biophysical and hydrobiological processes. In this sense, Lake Victoria cannot be studied or understood in isolation from the various ecosystem components with which it comes into contact; climate, forests, wetlands, rivers, streams, humidity, temperature, and precipitation are all integral to the lacustrine ecology. Forests and wetlands, for example, are vital players in multiple stress mitigation or reduction that may affect the lake if inserted into it. It is helpful to constantly remind ourselves of the simple fact that Lake Victoria is not only a natural water reservoir, but also a nutrient sink for its entire watershed. Lake Victoria constantly receives loads of dissolved and undissolved particles of phosphorus, nitrogen, pesticides, and other organic and inorganic substances that generate multiple stressors, including biogenic and physical triggers for state transition of the lake.

Second, scientific understanding of the structures, characteristics, processes, and functions of Lake Victoria's hydroecology is central to the capacity of fishery-dependent communities to put mechanisms of adaptation and mitigation in place in order to avert (or at least reduce) the vulnerability of people to natural disasters. Of course, this pointer begs the question of whether the development of mitigation and adaptation options is feasible in the context of the neocolonial capitalist logic of destructive creation and accumulation entrenched in Lake Victoria's riparian countries.

Third, understanding of Victoria's lacustrine ecology alone is insufficient to insure the long-term welfare of lakeshore communities. Long-term welfare also requires understanding and contesting the prevailing neocolonial mode of production and proposing an alternative social system, which is simultaneously pro-human welfare and anti-entropic degradation. Giving a fuller expression to these pointers is the overarching purpose of this chapter and the following chapter. As the paragraphs in this section have already suggested, in order to fully assess the impacts of mutually inclusive forces on the fishery mode of production, I will situate the master narrative in macrolevel analysis by highlighting supply conditions on Lake Victoria as effected by global warming, deforestation, wetland destruction, species introduction, pollution, and eutrophication.

THE IMPACT OF CLIMATE CHANGE ON SUPPLY CONDITIONS

Climate-driven change is probably the greatest threat to the long-term ecosystem health of Lake Victoria for reasons elaborated in this section. Climate change is the difference between incoming heat energy and outgoing heat energy, in the forms of ultraviolet radiation and infrared radiation, respectively, as determined by the relative concentration of greenhouse gases in the atmosphere. Balanced concentrations of greenhouse gases in the atmosphere are essential to animal and plant life on earth. Life without them is unimaginable. The atmosphere is made up of essential life-supporting greenhouse gases, which are responsible for regulating the amount of ultraviolet radiation entering the earth and the infrared radiation reflected from the earth to the atmosphere. When solar radiation hits the earth, a great amount of it is absorbed by primary producers (generally green plants) to process their own food by combining sunlight, water, and carbon. However, most of the heat energy is reflected back to space in the form of infrared radiation when earthly objects give off energy resulting from molecular collisions within the material objects themselves. The problem occurs when plants are destroyed and their abilities to capture solar energy for photosynthetic actions and carbon storage are foregone.

Even worse, the actions and processes involved in the destruction, burning, and decomposition of plants result in emissions of CO^2, nitrous oxide, methane, and other gases into the atmosphere. When humans burn fossil fuels, they worsen the problem by further thickening the concentration of greenhouse gases in the atmosphere. For example, in 2007, global burning of fossil fuels emitted 29 billion metric tonnes of CO^2 and deforestation contributed another 7 billion metric tonnes of CO^2 (Sachs, 2008). The increase in the concentration of greenhouse gases reduces the transparency of the gases in the atmosphere. Instead of allowing the outgoing infrared radiation to dissipate and scatter in space, they absorb more of the outgoing heat energy in the form of CO^2, methane, nitrous oxide, and water vapor, resulting in overall global warming. Thus, climate change is directly related to the extent to which the transmission and absorption of heat energy by greenhouse gases is regulated. The greatest danger of climate change can be a radical alteration of the hydrologic cycle. This determines the global distribution of water, which, in turn, determines the productivity of all ecosystems. Since evaporation and rainfall regimes are a function of temperature differentials, some regions can become wetter and other regions drier; rainfall episodes can become extremely more erratic and shorter, adversely affecting food growing seasons and, by extension, generating food insecurity (World Bank, Sept. 2009).

Here, one crucial point warrants highlighting. Nonequilibrium thermodynamic biotic systems, driven by solar energy, are subject to constant dissipation of energy in keeping with the laws of thermodynamics. Human extraction of low-entropy throughputs from the environment, and transformation of those throughputs by economic institutions, may disrupt the natural evolutionary process or may accelerate the dissipation matrices. The result can be erosion of the atmosphere's ability to cleanse itself through the natural oxidation process (Alexander et al., 1997). The erosion of this critical capacity for self-cleansing and self-renewal generally results from human alteration of the biochemical cycles. After all, life is supported by how nitrogen, phosphorus, hydrogen, oxygen, carbon, and sulfur are cycled and combined in the right proportions, since these 6 elements make up 95 percent of all living things on earth (Alexander et al., 1997). Thus, human appropriation or synthetic mobilization of these elements to produce a vast variety of goods can affect the biochemical cycles in ways that that would impact climate and hydrology. Through reactive nitrogen mobilization to make fertilizer, for example, we change the nitrogen cycle by generating nitrous oxide, which contributes to the thickening of atmospheric greenhouse gases and results in climate change.

It is in the context sketched above that climate change becomes the biggest threat to the limnology, biodiversity, and productivity of Lake Victoria. Current projections suggest that temperatures in the tropics will rise 3 to 5 degrees Celsius higher in the future than observed in the last century (Hecky et al., 2006; Bootsma and Hecky, 1993; Georgakakos et al., 2005; Alexander et al., 1997). The conundrum inheres in the fact that the causes of climate change are multiple and global in scope. This is due to the biogeochemistry of the sources and circulation of the causes of climate change, which find palpable expression through the gaseous composition or deterioration of the atmosphere. Regardless of the origins and who is directly responsible for the emissions, all greenhouse gases end up in the atmosphere with decidedly global implications. Lake Victoria is particularly sensitive to climate change of any magnitude for 3 fundamental reasons:

First, thermodynamically speaking, temperature, volume, pressure, wind velocity and direction are the major determinants of how a given ecosystem functions (Jorgensen and Szirezhez, 2004; Crul, 1995). Insofar as Lake Victoria is concerned, the tropical location of the lake magnifies the impact of climate change on land, air, and water temperatures and, by extension, on the hydrodynamics, biogeochemistry, biodiversity, and productivity of the lake. The intrinsic relationship between energy through-flow and the productivity of the lake (although complex) is very straightforward. This is because the amplitudes of energy fluxes determine the annual heat budget and hydrodynamics of the lake. In turn, these determine density stratification, vertical temperature structure, wind velocity, and internal distribution of nutrients on which the trophic profile of the lake and primary production depend (Coulter and Spigel, 1991; Bootsma and Hecky, 1993; Lehman et al., 1998).

Second, increases in regional temperatures are apt to increase the evaporation rates of lake surface waters, as well as the rates of evapotranspiration from watershed vegetation. In turn, this affects the water budget of the lake, especially if the increased evaporation and evapotranspiration rates are not counteracted by equal and corresponding rates of rainfall. For example, during the period from 1960 to 1982, the hydrologic equilibrium condition in Lake Victoria was maintained because the mean annual lake evaporation of 90 bcm was counteracted by an equal amount of lake precipitation; likewise, the mean annual lake outflow of 42.9 bcm to the Nile River was matched by an equal amount of watershed outflow to Lake Victoria (Georgakakos et al., 2005). Thus, it stands to reason that, given its large surface area and shallow water base, Lake Victoria is bound to lose significant amounts of water to evaporation and evapotranspiration in response to climate change—of any magnitude—unless this loss is counteracted by precipitation.

There is an added danger, in that Lake Victoria's vulnerability to high evaporation and evapotranspiration rates (due to its tropical location) is apt to reinforce thermal stability, low oxygen saturation potential, and high oxygen demand. Thus, maintenance of a stable balance—between the losses from evaporation and the gains from precipitation—is particularly critical to the hydroecological balance of the lake. This is critical since Lake Victoria's heterogeneous catchment areas are known to be prone to patterns of cyclical droughts, which have frequently resulted in lower rainfall and decreased river inputs. For example, 45 percent of the 2.5 meter drop in Lake Victoria's water level between 2000 and 2005 was attributed to regional drought; excess human appropriation of lake water was responsible for the other 55 percent of the drop. Unfortunately, all hydrologic models now indicate that future drought conditions in Victoria's basin are going to be more frequent and prolonged (Georgakakos et al., 2005).

Third, as noted earlier, lake productivity is measured by the extent to which exchange of nutrients and oxygen takes place between the epilimnion (the relatively warm and mixed upper layer of lake waters) and the hypolimnion (the noncirculating and cold bottom layer of lake waters). Under mixed conditions in the absence of lake stratification, nutrients released by bacterial action are transported upward from the hypolimnion, where wind currents evenly distribute these nutrients to primary producers in the epilimnion (Cox, 1993). However, when a rise in regional and global air temperatures affects the heat budget of the lake, stratification of the lake ensues. This phenomenon is expressed in terms of the epilimnion tending to warm faster than the hypolimnion, thereby slowing or blocking the transport of nutrients between them, and simultaneously impeding the conveyance of oxygen from the surface waters to deep waters of the lake as the metalimnion (separating the upper and lower portions of the lake) strengthens. In the absence of vertical mixing, the resulting decrease in internal nutrient circulation short-circuits primary productivity. In turn, this limits fish production, since internal nutrient circulation is a major source of phosphorus and silica that are integral to aquatic plant nutrients.

Moreover, since the internal movement of water regulates the distribution of nutrients and oxygen—on which the distribution and abundance of lacustrine organisms depend—prolonged stratification results in oxygen depletion and progressive expansion of deoxygenated regions, thereby reducing habitable space for fish. While rises in air and water temperature reduce the conveyance of oxygen from surface to bottom waters, they simultaneously increase decomposition of dead organic matter in the bottom regions of the lake. Decomposers utilize scarce oxygen to carry out their functions, further compounding the problem of oxygen depletion. Until the 1960s, Lake Victoria was known to have been oxygenated to its deepest portions. However, by the early 1990s, oxygen concentration in the lake's bottom floor had decreased by more than half during periods of thermal stratification (Cohen et al., 1996; OReilly et al., 2003; Berburg et al., 2003; Lehman et al., 1998; Bootsma and Hecky, 1993; Hecky et al., 2006; Tierney et al., 2010).

It is in the context of the preceding discussion that several authors (Lehman and Franstator, 1994; Lehman et al., 1998; O'Reilly et al., 2003) have forwarded climate-driven change hypothesis to explain the present warming of Lake Victoria. After all, there have been discerning trends in temperature anomalies in Lake Victoria basin for quite some time. Even the temperatures in the hypolimnial regions of the lake—that used to be homothermal throughout the year—have recently seen a significant rise in temperature. As a result, Lake Victoria has become much warmer than it was in the 1960s. Moreover, as recently as the 1960s, thermocline formation and thermal stratification were not observed occurring in Victoria because its water column was mixed year round; if occasional thermal stratification occurred in deep waters, it was ephemeral with insignificant long-term effects on primary production and the spatial distribution of species (Hecky et al., 2010). Today, Victoria is thermally well stratified and the thermocline remains well entrenched at depth of 30-40 meters during much of the year. The effect of this phenomenon on primary production and ecosystem health is unmistakable. The conclusion that emerges from the climate-driven change hypothesis is that the unprecedented anthropogenic regional and global energy input—and the consequent rise in temperatures—may be responsible for the present limnological condition of Lake Victoria, as well as for the recent poor levels of lake productivity.

Even more troubling is that the projected trends in climate change do not bode well for the lake, as well as for the millions of people who depend on it for food, water, and energy. In fact, when the U.S.-based Water Resources and Energy Management International submitted its investigation findings of Lake Victoria to the Ugandan government in 2005, it drove home the unpleasant future scenario regarding the lake's vulnerability to climate change. Basing its investigation on a combination of the global climate circulation model, biocorrection, and downscaling procedures, the hydrologic model for Victoria's watershed, the river basin model for the lake, and the downstream river system, the firm comprehensively drew a grim picture of Lake Victoria. The findings depicted far-reaching environmental and hydroecological ramifications.

According to the findings, the average air temperature over Lake Victoria and its watershed is predicted to increase by 4-5 degrees Celsius by the end of this century. Since lake water balance is a function of lake precipitation, lake evaporation, watershed outflow, and watershed evapotranspiration, a change in climate of this magnitude will produce drier hydrologic conditions, lower lake levels, and lower watershed outflow. These factors will indeed have inordinate impacts on the limnology, biodiversity, and productivity of the lake. Indeed, observable data and historical patterns indicate that Lake Victoria is expected to experience 20 bcm of water deficit per year, as the net basin supply decreases by 50 percent per annum (Georgakakos et al., 2005). Since lake precipitation equals lake evaporation, and watershed outflow corresponds to lake outflow, Lake Victoria's vulnerability to climate change looms even larger if lower lake precipitation is followed by higher lake evaporation, lower watershed precipitation, and higher watershed evapotranspiration. Unfortunately, this is predicted to be the case.

From the fishery point of view, climate change as predicted by the models discussed above will have 3 unwelcome outcomes:

Fish resources are crucially important to the livelihoods and diets of the poor people in the littoral countries of Lake Victoria, as they provide inexpensive sources of protein, essential macronutrients, minerals, vitamins, and fatty acids. Therefore, a collapse of the fishery mode of production is likely to lead to worse malnutrition, hardship, and protein deficiency. After all, the population in Kenya, Tanzania and Uganda obtain more than a quarter of their protein from consuming fish. Moreover, the food status of the poor can further be compounded by the severe impacts climate change can have on agriculture and forest products. Large proportions of the people in the region depend on these sources to supplement their diets. Following the breakdown in agriculture, people are likely to flock to lakeshores to eke out a living as fishers, compounding the problem of overfishing and depletion of aquatic resources.

Next, climate change-driven fishery collapse is likely to drive tens of thousands of fishers, processors, traders, and other ancillary service providers out of business, exacerbating the already grim labor market conditions in the littoral countries.

Finally, since fish export is integral to hard currency generation for the 3 Lake Victoria riparian states, the collapse of the fishery mode of production is very likely to further erode the financial base of legitimacy. Since tourism and other climate-sensitive industries are bound to be affected, the situation can indeed worsen. Even under ideal governance conditions (assuming the countries have a relatively good class of rulers), the lack of institutional and financial capacity to absorb sudden shocks and recover from climate change-driven hazards will become extremely difficult (Allison et al., 2009). The effect can indeed be magnified by the apparent absence of a system alternative to the prevailing mode of production. The present system, embedded in the logic of destructive creation, driving capital accumulation, and the necessity of survival on the part of the masses, is likely to make the search for mitigation and adaptation options an exceedingly difficult undertaking.

On another level, climate-driven change may influence the patterns of the hydrologic cycle and environmental change in Victoria's basin in ways that can spell disaster for fish species by producing natural fluctuations and extreme weather events. For example, during the wet years of the early 1960s, Lake Victoria's water levels rose by 2.5 meters, its water storage capacity increased by 170 bcm, and its surface area expanded by more than 6,000 km^2—leading to extensive destruction of littoral vegetation by wave action (Georgakakos et al., 2005; Kudhongania and Chltamwebwa, 1995; Collins, 2002). Furthermore, if rain flushes out deoxygenated water from the bottom of a lake or adjoining swamps after a prolonged drought, the anoxic water can decimate a large number of species. In July 1953, an upwelling of hypoxic water in Lake Victoria was observed to have killed a vast number of *Haplochromis* and *Tilapiine* cichlids. A similar event, which occurred in Lake George in September 1957, was believed to have killed 250,000 tilapia, along with a vast number of haplochromine cichlids (Fryer and Iles, 1972). On the other end of the spectrum, prolonged drought can result in sharp reductions in river and subterranean water flows. In turn, a lack of these water flows will affect lake levels, trophic structure, and productivity. For example, during the prolonged Sahelian drought, fish catches in the Niger basin fell from 90,000 tons in 1970 to 45,000 tons in 1985. Similarly, the fish harvest in the Chad basin dropped from 220,000 tons in 1974 to 35,000 tons in 1985 (Welcomme, 2003).

Climate change directly affects not only lake productivity but also other aspects of the terrestrial ecology in the catchment areas of Lake Victoria. For example, Mount Kenya (one of Kenya's five water towers) has lost seven of its 18 ice caps, and the remaining are rapidly fragmenting and thinning. Likewise, the glaciers on Zwenzori Mountain, which supply fresh water to more than 1 million local Ugandan and Congolese people, have shrunk from 650 ha in 1905 to 108 ha by 2005; if the present trends persist, Zwenzori Mountain may have no glaciers left by 2025 (UN IRIN, Nov. 10, 2006).

Thus, as climate changes, it is not only the fishery mode of production that is adversely affected, but the entire food ecology and cash crop economy is equally affected since agriculture throughout Africa is rain-fed. More than 460 million of Africa's 850 million people were subsistence farmers in 2004, and 85-90 percent of the continent's agriculture is rain-fed, thus making agriculture extremely vulnerable to climate change of any magnitude (Barnett and Whiteside, 2006). According to the 4th report of the Intergovernmental Panel on Climate Change (IPCC, 2007), most of the marginal regions of agriculture are likely to be taken out of production for lack of rainfall. As a result, crop yields in sub-Saharan African countries may fall by as much as half by 2020 and as much as 90 percent by 2100. In addition, by 2050 between 350 million and 600 million Africans are projected to experience severe water stress (Boko et al., 2007). In a continent already beset by colossal food insecurity, such reductions of food could have serious environmental, social and political implications.

For example, if the average temperature rises by 2 degrees Celsius, Uganda's robusta coffee crop may disappear entirely from central Uganda. Uganda is currently the second largest coffee producer in Africa, after Ethiopia. However, 90 percent of Uganda's coffee is robusta, which grows well within 20-25 degrees Celsius (Cocks, 2007).

Moreover, in addition to the prevalence of poor governance and misdistribution of national resources, the growing food insecurity in the Rift Valley countries is in part attributable to climate instability. For instance, between 1973 and 2009, there were no less than 9 major famines of biblical proportions in eastern and northeastern Africa. These famines exposed millions of people to hunger and starvation while, at the same time, deepened the dependency of the regions on external food aid. Even as of this writing, the Greater Horn of Africa was witnessing its worst drought in 60 years, making more than 12 million people vulnerable to famine of enormous proportions. As of June 2011, there were 3.5 million people who needed $525 million in external emergency food aid in Kenya alone (Reuters, June 28, 2011).

Lake Victoria's riparian societies are impacted not only by climate change, impairing their immediate food ecology, but also by the circular nature of global food supply disruptions or by the sudden rise in global price structure due to climate change. For example, between early 2010 and early 2011, 44 million more people worldwide were driven into poverty as the price index of 55 food commodities rose to record levels—global prices for maize rose by 74 percent, for wheat by 69 percent, for palm oil by 55 percent, for soybeans by 36 percent, and for beef by 30 percent (BBC News, Aug. 14, 2011). In part, these changes in global food prices are a result of weather events. When Australian wheat was downgraded, due to spoilage by heavy rains in 2010, the demand for American wheat soared and led to a 100 percent price increase in American wheat over the previous year.

In light of the reasons discussed in this section, climate changes appear to have enormous impacts on the limnology, biodiversity, and productivity of Lake Victoria. The disturbing fact thus far is that global emissions of greenhouse gases that worsen climate change have continued to grow. Unless emissions are drastically reduced, it appears probable that the consequences of climate change may assume catastrophic dimensions in the future, since global temperatures may rise between 3 and 6 degrees Celsius in the coming decades. This may push global surface temperatures beyond the tipping point (Hanson et al., 2008; Bales and Duke, 2008; Lemonick, 2008; World Bank, Sept. 2009; Millennium Ecosystem Assessment, 2005). Recent studies prepared by both natural and social scientists seem to corroborate the trends toward increasing warming. These studies envision 3 scenarios, depending on how soon the problem of global climate change is addressed.

According to the first expected scenario, average global temperatures may rise by 1.3 degrees Celsius by 2040. By then, the number of people living in countries experiencing chronic water shortages would have risen to 4 billion from 500 million in 2000. Consequently, mass dislocations and migrations, competition for rapidly dwindling resources, chronic water shortages, spread of diseases, and severely weakened adaptive capacities are likely to trigger internal and regional conflicts with serious hydrological, social, economic, and political implications—especially in Africa where 2-thirds of the continent's land mass consists of desert and dry land (Campbell et al., 2007; Oxfam, 2011; World Bank, Sept. 2009). Projections based on direct observations of climate patterns and computer simulations suggest that by midcentury, there is a 90 percent probability that the lowest temperatures in the tropics and subtropics during food growing seasons will exceed the hottest temperatures recorded between 1900 and 2006. This includes the whole continent of Africa.

As a result, it is projected that the production of staple crops in these regions may fall by 20 to 40 percent between 2040 and 2060 alone (Battisti and Naylor, 2009; World Bank, Sept. 2009). Since 3 billion people live in the tropics and subtropics (expected to double by century's end), the catastrophic consequences of global warming are inconceivable. It is useful to note that in 2003, the hottest recorded summer temperature in France was 3.6 degrees Celsius above the long-term climate patterns. The net result of the heat fluctuation was that 50,000 Europeans died from heat waves. Furthermore, the production of French corn, fruit and wheat plummeted by 30, 25 and 21 percent, respectively. In fact, in September 2009, the U.K.-based Hadley Center projected that the kind of heat wave of 2003 is 4 times more likely to occur in the future. Moreover, the global average temperature rise of 4 degrees Celsius can happen sooner that previously projected—as soon as the 2050s. Even though the global temperature rise is expected to be 4 degrees Celsius, different regions are likely to experience different temperature gradients—perhaps up to a 10 degree Celsius rise in parts of Africa (Battisti and Naylor, 2009).

Under the second scenario of severe climate change, an average rise in global temperatures of 2.6 degrees Celsius by 2040 may produce even graver nonlinear environmental, epidemiological, economic, social, and political events and conditions; each are likely to overwhelm societies and governments worldwide (Campbell et al., 2007; Battisti and Naylor, 2009). If this second scenario of climate change materializes, the competition among Lake Victoria's riparian societies for the lake's freshwater is bound to be severe and potentially bloody. Indeed, the littoral states of Lake Victoria may not survive as we know them today.

Under the third catastrophic scenario, global temperatures may rise 5.6 degrees Celsius by 2100, becoming the norm and representing the tipping point. Such an eventuality may not only radically change human relationships within and between societies, but may also irretrievably alter man's relationship with nature (Campbell et al., 2007).

Whichever scenario occurs, climate-driven change is likely to have far-reaching implications for the biolimnological state of Lake Victoria and the societies that depend on it for survival. There are 2 fundamental conditions that are likely to make the effects of global warming on the limnology, biodiversity, and productivity of Lake Victoria loom even larger in the future.

First, the lake's littoral states are poised to continue imitating the fossil fuel-based mode of capitalist production, thereby contributing significantly to the adverse changes in the microclimate of the region through dirty industrialization, deforestation, overgrazing, and wetland destruction. Moreover, the majority of the people in the Lake Victoria basin depend on charcoal as their primary energy source, which is a contributor to deforestation, nutrient loading and warming. After all, over 90 percent of primary energy consumption in Victoria's riparian countries is from firewood.

Second, the major causes and effects of climate change are global in scope, and there is not much the littoral states and societies of Lake Victoria can do alone without concerted global actions to reduce the causes and mitigate the effects of global warming. As far as we could tell, wealthy capitalist countries failed to reduce their own emissions and assist poorer nations in developing alternative sources of energy. Furthermore, the so-called emerging market countries have also decidedly chosen to adopt the 19th-century mode of dirty industrialization, exemplified by China. As the World Bank report (Sept. 2009) outlined, global emissions of greenhouse gases are so massive that it would take an energy sector investment of $100 to 700 billion per year to stem the drive toward the tipping point. In fact, other observers suggest that it would take an outlay of $45 trillion to cut global fossil emissions in half by 2050 (Bales and Duke, 2008). In light of the above 2 points, global trends in emissions do not bode well for the biolimnological future of Lake Victoria., since the lake is an integral hydroecological piece of the biosphere.

FOREST ECOLOGY AND SUPPLY CONDITIONS

The interdependence of forests, wetlands, and bodies of freshwater has social and biophysical dimensions. The social dimension relates to the fact that the current principal struggles in Victoria's riparian countries are over those three elements. For 80 percent of the population in the region, these 3 sectors of the ecology are central to their food ecology and coping capacities. For the pirate capitalists, they still remain the main source of primitive capital accumulation. Therefore, any biological deterioration in either of these sectors—resulting from the intensification of competition—reverberates in other sectors. In biophysical terms, forests and wetlands—by retaining soils and sediments, as well as agrochemicals and other trace elements—perform services that are crucial to the healthy state of bodies of freshwater. In the event that degradation or liquidation of these natural capital stocks occur, natural water systems are bound to suffer from widespread sedimentation, eutrophication, and consequent impoverishment of aquatic food sources. Various studies have shown that clearing of forests within the watershed ecology, habitat fragmentation, wetlands conversion to other purposes, and modification of biotic conditions in the tropics, are strongly associated with increased soil erosion, agrochemical runoff, and greater transport of dissolved materials and silt.

As a result, between 10 and 100 aquatic vertebrates and between 100 and 1,000 aquatic invertebrates in tropical lakes succumb to extirpation each year (Cox, 1993). The state of sturgeon in Lake Michigan illustrates how human-induced ecosystem disturbances can adversely affect the biodiversity, productivity, and limnology of lakes. In 1879, fishers harvested over 1.78 million kg of sturgeon from Lake Michigan. However, by the turn of the century, the same number of fishers barely managed to harvest 50,000 kg of sturgeon. The hydrobiology, chemistry, and physical properties of the lake were substantially modified by rapid watershed modification, industrialization of agriculture, drainage of wetlands, and urban sprawl (Cox, 1993).

The preceding paragraph suggests that Lake Victoria is inseparable from the wetlands that fringe the lake and the forests that comprise its wider catchment area, since the lake depends for its biophysical integrity on the ecosystem services of riparian vegetation and forests. In this section, I shall highlight the organic relationship between the lacustrine environment and the forest ecology, deferring discussion on the role of wetlands to the next section. I will begin with a general discussion on the forest ecology and then particularize the discussion to Victoria's riparian conditions.

Much has been written about the importance of forests not only as sources of food, fiber, and construction material, but also as critical performers of nonconsumption services, such as macro- and micro- climate regulation, rainfall generation, oxygen production, carbon sequestration, nutrient cycling, soil formation, and sediment retention. As a matter of fact, over 150 years ago, the trailblazers of modern environmentalism projected the potential implications of colonial vandalization of tropical forests by the British and Dutch East India companies for global climate, as well as consequent desiccation of terrestrial vegetation and species extinction (Pepper, 1996). It is important to review some of the vital ecosystem services that forests provide, in order to appreciate how prescient the environmental trail blazers were to recognize the intrinsic relationship that exists between forests and the flow of ecosystem services, in the forms of energy, materials, climate regulation, carbon sequestration, rainfall generation, biodiversity maintenance, and information.

To begin with, forests are rainfall generators, carbon dioxide absorbers, and the planet's thermostat. A close look at how water is formed suggests the existence of an intrinsic nexus between water, forest and climate. Significant proportions of the precipitation that takes place on land result from evapotranspiration with forests, riparian vegetation, and wetlands acting as natural pumping stations. Without forests, it would be inconceivable for cloud formation, rainfall generation, and freshwater purification and retention to take place in most global ecoregions. For example, it is estimated that the Amazon forests pump out 20 billion gallons from their stored water of 3 trillion metric tonnes into the atmosphere each day. This is said to be equivalent to the energy of 80,000 coal-fired supergiant power stations performing the same job every day. It is no coincidence that the Amazon basin receives half of its rain from its own hydrologic cycle. In the Congo basin, 75 to 95 percent of rainfall comes from recycled moisture within the basin itself (Mitchell et al., 2008; Meyers, 1997; Daly and Farley, 2003; Nasi et al., 2002).

The efficiency of forests as generators of water and regulators of the hydrologic cycle are functions of the spatial distributions, structures, characteristics, densities, and contiguities of forests, which determine the frequencies and intensities of precipitation. In Southeast Asia, on the average, undisturbed contiguous forests intercept 35 percent of rainfall, while logged forests capture less than 20 percent, and rubber plantation trees intercept even less than twelve percent of the falling rain (Meyers, 1997). Integral to forests' ability to generate rainfall is their capacity to recapture much of the rainfall, retain it, induce infiltration, reduce runoff, and in turn, translate these variables into more evapotranspiration and more subsequent rainfall. The complex evolutionary interdependence of the structural components of nature is such that even the Amazon and Congo basin rainforests are interconnected through biogeochemical cycles and other biohemispheric processes, each with profound implications for climatic conditions and precipitation regimes in other regions. The 2 tropical regions are connected by the natural back and forth oscillation of atmospheric movements across the Atlantic Ocean. Consequently, heavy rainfall and floods in the Congo basin coincide with droughts in the Amazon basin and vice-versa.

Moreover, these precipitation patterns affect the climate and hydrology of other regions. For example, recent observations of these weather patterns indicate the annual deforestation of the Congo basin by 1.5 million ha has resulted in rainfall decrease in the Great Lakes region of the United States by between 5 and 15 percent; a similar impact was discerned in the Ukraine and some parts of Russia (Mitchell et al., 2008). From this observation, one could infer that the 200 million ha of the world's tropical rainforests lost between 1950 and 1990 and the 427 million ha of additional tropical forests that had undergone significant degradation might have had significant bearings for local, regional, and global climate and rainfall patterns. It is helpful to keep in mind that, of the 500 billion metric tonnes of carbon stored in terrestrial vegetation, 86 percent or 430 billion metric tonnes is securely sequestered in the tropics and subtropics. Therefore, when 100-year-old tropical trees are burned for energy or to make way for farms, they release their heavy loads of CO^2 in no time. However, it will take 100 years to fully recapture the carbon by growing their replacement. In the meantime, the CO^2 will still be in the atmosphere aggravating climate change (ETC Group, 2010; Haan et al., 1997; Meyers, 1997).

Therefore, when we speak of how forest loss affects Lake Victoria, we are talking about how the water balance of the lake and its basin is intimately related to the state of the total forest ecology. Since lake precipitation, lake evaporation, watershed precipitation, and watershed evapotranspiration are key determinants in Lake Victoria's hydrological balance, reduction or fragmentation of forests within its own basin and the Congo basin can disturb the processes of cloud formation and rainfall generation. This can have serious implications for the lake's water balance. During the period from 1960 to 1982, the mean annual lake precipitation was 90 bcm and the mean annual lake evaporation was 90 bcm, balancing each other out. During the same time frame, the mean annual lake outflow of 42.9 bcm to the Nile River was matched by 42.9 bcm of watershed outflow to the lake per year (Georgakakos et al., 2005). It is in the context of such hydrological reality that the role of the forest ecology in Victoria's water balance must be understood. Forest fragmentation and destruction of riparian vegetation in the basin have already substantially eroded the necessary conditions for evapotranspiration and subsequent rainfall. For example, as we shall see below, fragmentation and destruction of the Mau Forest Complex in southwestern Kenya are having serious consequences for water levels in Lake Victoria.

The destruction is also having serious consequences for the tea-growing regions of the country. Most of the rivers that flow into Lake Victoria rise in the Mau Forest Complex. In addition, most of Kenya's commercial tea plantations are concentrated near the Mau Forest. However, due to much reduced cloud formation and rainfall generation in the nearby forest, tea transnational James Finley recently reported that yields from its tea plantations near the Mau Forest Complex have progressively declined. As a result, the company closed down 4 of its tea processing operations in Kenya. It is not cash crops alone that are affected by the decline in rainfall. Rather, the entire food ecology is impacted.

Climate regulation is another critical ecosystem service provided by forests. By virtue of their monopoly possession of chlorophyll, which is the only substance on earth capable of capturing solar energy for photosynthesis, plants are the critical players in the regulation of all temperatures. Forest canopies are believed to absorb up to 90 percent of sunlight that strikes them, which is essential to drive the evapotranspiration process (Daly and Farley, 2003). Additionally, running on solar energy, forests are dominant players in terrestrial net primary production on which billions of people and wildlife depend. Deforestation and forest fragmentation also mean undercutting the ability of plants to capture solar energy and use it to process life-supporting services, such as creating an amenable climate. Reductions in forest cover mean increasing the earth's albedo and, in effect, increasing warming. As stated earlier, temperature levels are determined by the exchange of outgoing and incoming solar energy and by the gaseous composition of the atmosphere. The greenhouse gases (GHG) in the atmosphere allow ultraviolet radiation to enter the earth, some of which is captured by green plants. In exchange, the earth reflects infrared light back to the atmosphere. If the infrared radiation is absorbed by greenhouse gases and water vapor in the atmosphere, the earth's temperatures are likely to change since the atmosphere's ability to allow the exchange between incoming and outgoing energy is determined by the level of concentration of GHG. It is in this context that forests act as natural thermostats. Forests insulate the earth's surface from searing wind and heat, thereby reducing extreme weather events. Forests also generate cooling effects by the cover they provide. Stated differently, since forests regulate the water cycling process via the evapotranspiration regime, deforestation affects the patterns of air circulation, moisture, and humidity, thereby changing the earth's albedo and aerodynamic drags (Nasi et al., 2002).

The production functions are probably the most recognized services of forests. Today, 1.4 billion poor people in the world still depend on forest resources to survive. Forests provide food, fiber, fodder, timber, and biomass fuel. In 1992, it was estimated that 2.9 million Kenyans, who lived near natural forests, were harvesting $94 million worth of forest products in terms of food, timber, polewood, energy, and water (UNEP, 2002). Furthermore, as of 2002, there were 60,000 Kenyan woodcarvers, each hiring 5 individuals to harvest the wood and to sand and polish the carvings. As this subsistence woodcarving sector expanded, the export value of Kenya's carvings steadily rose from $60,000 per year in the 1950s to $20 million per year in 2001 (UNEP, 2002).

As in Kenya, 87 percent of Tanzanians who live in rural areas heavily depend on forest resources for a variety of means of livelihoods. Over 90 percent of the primary energy used in the country is derived from forests (Milledge et al., 2007). Moreover, since forests are natural homes and sources of food for countless wildlife, millions of people throughout Victoria's basin depend on these wildlife populations for nutrition. For example, poor communities in Tanzania, driven by the necessity of survival, poach over 40,000 wild animals each year from the Serengeti game reserves for bush meat (World Bank, 2008).

In order to produce the essential forest food sources, global forests are generally dependent on the services of over 100,000 vertebrate and invertebrate pollinators and tens of thousands of animal seed dispersers, all of which live in and are sustained by the forests (Nasi et al., 2002; Costanza et al., 1997). Given this natural condition, deforestation and forest fragmentation veritably result in the elimination or substantial reduction of these useful pollinators and seed dispersers. In turn, this results in substantial reduction in food supply for hundreds of millions of people in the tropics and subtropics. It is helpful to keep in mind that the annual services provided by pollinators to global agriculture alone is estimated at $517 billion (Costanza et al., 1997). A partial focus on how honeybee colonies contribute to the general food ecology, forest productivity, and biodiversity maintenance, should help to clarify the observation already made about the importance of pollinators. Honey is a multipurpose product widely used in foods, cosmetics, pharmaceuticals, candle making, and lubricants. By providing pollination services, honeybee colonies are thereby essential to the productivity of forests, biodiversity maintenance, flowering trees and food plants, and income of indigenous communities throughout East Africa.

Tanzania's 35 million ha of forests are known to host 9.2 million honeybee colonies, capable of producing 138,000 metric tonnes of honey and 9,000 metric tonnes of beeswax per annum, worth $156.4 million at 2003 dollars. At present, only 3.5 percent of this potential is harnessed. About 2 million Tanzanians derive significant portions of their income from the production and sale of honey and beeswax. In the first 3 decades after independence, Tanzania annually exported 368 metric tonnes of beeswax and 467 metric tonnes of honey (Mwakatombe and Mlingwa, 2005). Uganda also has the potential to produce 500,000 metric tonnes of honey per annum, even though present annual production is 5,000 metric tonnes. Between 1 and 2 million Ugandans are believed to be actively engaged in honey production (*New Vision*, January 27, 2009). A large number of rural Kenyans are also active in honey production. Annually, Kenya has the potential to produce over 100,000 metric tonnes of honey and more than 10,000 metric tonnes of beeswax (*The Nation*, April 17, 2010).

All told, insofar as the production functions of forests are concerned, the annual dollar amounts of goods and services furnished by forests is over $4.7 trillion, of which tropical forests account for $1.9 trillion (Costanza et al., 1997). Of course, this approximation pertains only to one year's worth of benefits. If forests are lost forever, this would amount to the liquidation of forests as natural capital assets, denying us of the continuous flow of materials, energy, food, fiber, water, and information. It is in this context that one should understand the importance of the forest ecology to Lake Victoria and the 40 million people who live in its basin. These people derive substantial proportions of their livelihoods from the forests. Land cover change in the basin means that not only do these millions of people lose critical sources of food and material, but the lake itself also suffers from the entropic degradation of forests, manifested in terms of soil erosion and consequent lake sedimentation.

There is also a critical nexus between forests and soil formation and fertility. Through disposal of their waste and dead organic matter, forests continually provide the necessary means for soil formation and soil renewal. By holding soil in place, forests also prevent or contain soil erosion. Thus, deforestation means soil erosion, since between 34 and 75 billion tons of soil are annually lost worldwide. Over 84 percent of soil erosion and degradation is caused by water and wind in the absence of the protective shield provided by forests and vegetation cover. The direct and indirect costs of soil erosion as measured by water storage, conveyance, treatment, replacing water and nutrients on cropland, and human health worldwide is $400 billion per year (Daily, 1997).

At the risk of being redundant, it can be said that forests are most critical in the provision of protection services. By providing canopy and ground cover, forests prevent soils from turning into solid mud that prevents water infiltration and increases runoff. The canopy layers of tropical forests capture up to 90 percent of rainfall before it touches the ground and then gently release the water without causing erosion. For instance, the Mount Kenya forests annually save Kenya $55 million in terms of protecting the catchment from erosion by water and wind (Daly and Farley, 2003; E. Wilson, 2002; Nasi et al., 2002; Costanza et al., 1997; UNEP, 2002). Through the services they provide toward soil formation, stability, and sediment retention, forests protect the watershed ecology against water or wind-driven erosion. This stands to reason that deforestation in Victoria's basin will exacerbate soil erosion, thereby causing the transportation of silt and nutrients by rivers to the lake to exacerbate the sedimentation and eutrophication of the lake.

From the biolimnological point of view of Lake Victoria, tropical forests are particularly important because of the lake's location in the tropics, where weather patterns are substantially affected by land-cover change. For example, changes in the Congo basin tropical forests—which straddle 6 countries and account for 20 percent of world tropical forests—have direct bearings on the state of Lake Victoria and its basin. After all, the Congo basin forests are home to half of the continent's plant and animal species. Yet, the Congo basin forests are under assault from corporate loggers, illegal harvesters, commercial plantation operators, and bioenergy marauders. Today, the 33 countries in the tropics have placed their forest resources under commercial vandalization, as they annually generate $20 billion from the sale of tropical timber. The effect of such commercial plunder is the loss of 13 million ha of tropical forests every year, including 5.5 million ha of rainforests (E. Wilson, 2002; Mitchell et al., 2008; UNEP, 2002). In particular, tropical rainforests are highly vulnerable to rapid degradation because of their nutrient-poor soils and extremely low resilience potential. It is a small wonder that 15 of the 25 hotspots in the world are in tropical rainforests. If the present levels of deforestation and the destruction of other ecosystems continue unabated, half of all plant and animal species may become extinct by 2050, since up to 27,000 species are estimated to become extinct each year in tropical forests as a result of human-induced rainforest loss (E. Wilson, 2002; Goldschmidt, 1994; Sachs, 2008; Barnosky et al., 2011). Nothing more illustrates this observation than the state of the African black rhinoceros, which numbered more than 1 million in 1900, but was reduced to 15,000 by 1985 and to 4,800 by 2000 (E. Wilson, 2002).

The biological protection service that forests provide to agriculture is also enormous. It is believed that up to 99 percent of the potentially harmful pests to agriculture are controlled by biological agents, such as bacteria, fungi, spiders, flies, birds, and a galaxy of bugs. These agents thereby obviate the need for the use of financially and environmentally costly agrochemicals. For example, in the United States alone, the presence of biological control agents in forests saves farmers $54 billion per year in cost avoidance (Nasi et al., 2002). As Naylor and Ehrlich (1997) have pointed out, the costs of increasing reliance on synthetic pesticides and herbicides outweigh the benefits from their application to control pests and weeds. After all, 90 percent of the potential pests are actually providers of natural pest control services. Thus, the application of 2.5 million metric tonnes of pesticides worldwide each year is eroding the natural means of controlling pests. The human costs entail the fact that 25 million people worldwide suffer from pesticide poisoning each year, resulting in 220,000 deaths. Over 90 percent of the pesticide poisoning and 90 percent of the deaths occur in countries of the global south. In ecological terms, applications of pesticides cause an estimated $25 billion worth of external damages by liquidating natural pest control services. In fact, if all the costs of health hazards to humans, replacement of natural pest-control services, research on pest-resistant chemicals, and the environmental damages caused by pesticides are included, the global annual costs of dependence on synthetic pesticides rise to between $954 billion and $1 trillion (Naylor and Ehrlich, 1997).

Paradoxically, despite heavy uses of biocides and synthetic chemicals, more than 100 weeds have become resistant to herbicides and another 150 plant pathogens have become resistant to chemicals. In the midst of these mammoth losses, the chief beneficiaries of this dependence on chemicals are agrochemical and pharmaceutical companies, since sales of pesticides have steadily increased over the decades from $2.7 billion in 1970 to over $21 billion in the mid-1990s (Naylor and Ehrlich, 1997). The relevance of the preceding observation is that deforestation of the Lake Victoria basin means not only loss of crucial biological pest control agents but also dependency on imported biocides that are costly in all financial, environmental, and human terms.

Forests are also natural waste-treatment plants. They are critical in detoxifying and recirculating wastes, recovering nutrients from wastes by transforming them into useful substances, and removing other nutrients or inducing the breakdown of excess toxic nutrients and compounds. In this sense, they are natural waste treatment facilities (Costanza et al., 1997). To illustrate this point, consider New York City's experience with its poor water quality resulting from the deforestation in upstate New York. This forest historically was critical to the city's reservoir. The great conundrum facing the city was whether to build a modern water-treatment plant, which would cost $6 to $8 billion, or to restore the watershed area to its natural condition. After weighing the costs and benefits of the 2 choices, the City decided to spend $1.5 billion on the land in upstate New York, and restored the watershed to its natural condition. By protecting the natural catchment area, the City now gets a safe, abundant, and cheap drinking water supply for its citizens. While nature does the purification process for free, residents also enjoy the aesthetic and recreational benefits for free (Daly and Farley, 2003).

Another pivotal role that forests play is in carbon storage and sequestration. Day in and day out, forests continuously trade massive amounts of life-supporting oxygen for corresponding amounts of carbon dioxide. The global forests, covering 42 million km^2, store carbon in the order of 120 to 400 metric tonnes per ha. A typical hectare of primary tropical forests stores 280 metric tonnes of carbon. All in all, forests hold 1,200 billion metric tonnes of carbon compared to the 750 billion metric tonnes concentrated in the atmosphere. Moreover, forests are responsible for 65 percent of plant growth and carbon fixation on terrestrial ecosystems (Meyers, 1997; Nasi et al., 2002; Danielsen et al., 2008). For reference, old-growth tropical forests alone remove 4.4 billion metric tonnes of CO^2 from the atmosphere annually, which is equal to 15 percent of annual global GHG emissions (Mitchell et al., 2008). The carbon removed from the biosphere is securely stored in the biomass unless the forests are cut down and burned or decay. So, when we clear-cut forests, not only do we lose the vital ecosystem services, but we also worsen the GHG emissions since the carbon stored in the biomass is released during burning or decay. Thus, deforestation or forest fragmentation means loss of critical capacity to convert solar energy fluxes to biomass and plant growth, to enhance biotic carbon sequestration, and to boost oxygen production.

Today, the forest ecology's capacity to regulate the global carbon cycle is increasingly being weakened by intensive commercial clearing of forests and peatlands. The conversion of peats to agrofuel plantations in particular presents numerous dangers. Peats are accumulations of partially decomposed vegetation matter. Owing to the compression and concentration of organic matter, peats hold massive quantities of carbon. Even though peats constitute a mere 3 percent (covering 4 million km^2) of the earth's landmass, they hold 528 billion metric tonnes of carbon or roughly 75 percent of the CO^2 currently in the atmosphere (Smolker et al., 2007; Mitchell et al., 2008). This fact makes peatlands the most dangerous emitters of GHG when anthropogenic actions disturb them. What is troubling, from the point of view of carbon cycling, is that commercial logging, proliferation of oil palm and rubber plantations, and illegal forest harvesting throughout the tropics are proceeding unchecked. For example, half of the peatland and forest cover in Indonesia and Malaysia is already under commercial concession. This is spurred by the fact that Southeast Asian countries together earn $10.4 billion per year from commercial logging concessions, with informal logging adding billions more. In Indonesia alone, informal logging generates $4 billion per year. In addition to the 6 million ha commercially cleared for agrofuel plantations, loggers in Indonesia have cleared more than 18 million ha for timber extraction in less than 2 decades. If the prediction comes to pass that the 150 million ha of forests and peatlands in Indonesia, Malaysia, and Papua New Guinea are converted to agrofuel plantations and food production by 2050, the ecological impact of such devastation could be incomprehensible (Mitchell et al., 2008; Smolker et al., 2007).

The Amazon forest ecology's capacity to regulate the carbon cycle is also being eroded by anthropogenic actions. The 2.3 million km^2 of forest that Brazil plans to convert to biofuel feedstock production (mostly oil palm) stores 42.5 billion tons of carbon in above-ground biomass alone; if Brazil succeeds in converting the forest on this massive area to bioenergy plantations, the amount of carbon released would be more than the 2006 total global emissions of 36 billion tons (Butler, 2008; Sachs, 2008). The trends in Brazil are not encouraging, as the country is poised to make good on its promise to become the premier model of agrofuel production. After all, between 2006 and 2007, the rate of deforestation in several Brazilian states was already more than 84 percent (Smolker et al., 2007). The current rate of deforestation in Brazil releases 150 million metric tonnes of greenhouse gases and accounts for 70 percent of that country's GHG emissions; this suggests that half of Amazonia might be completely deforested in 20 years, since the annual average rate of deforestation in legal Amazonia between 2000 and 2005 was already 2.2 million ha (Mitchell et al., 2008; ETC Group, 2010).

The present competition among agrofuel barons over the Congo basin forest resources presages similar concerns. The Congo basin forests (covering 2 million km^2) are believed to hold between 24 and 39 billion metric tonnes of carbon and annually release 237 million metric tonnes of CO^2 because of anthropogenic actions (Mitchell et al., 2008). Deforestation and degradation of the Congo basin mean not only the loss of carbon sink, but also the loss of critical ecosystem services, such as rainfall generation, microclimate regulation, erosion prevention, and provision of livelihoods to 50 million people in the basin. As of this writing, 30 percent or 600,000 km^2 of the Congo basin forests (the size of France) are under commercial concessions by multinational logging firms, which have been ravaging such endangered tree species as the afromosia and other luxuriant trees regarded as priorities for conservation (Mitchell et al., 2008; Green Peace Canada, 2009). The dominance of global logging corporations in the Congo Basin is such that they simply bribe national and local agents with beer, sugar, soap, and salt as access fees. Even after ransacking the forest resources, they evade taxes. For example, in 2008, the German-Swiss Danzer logging firm avoided 8 million Euros in taxes by simply bribing some hangers-on. In the estimation of Greenpeace Canada (2009), 8 million Euros was enough to vaccinate 700,000 Congolese children. If the present rate of forest resource exploitation continues in the basin, the DRC (which contains 63 percent of the remaining Congo basin forests) is bound to lose 40 percent of its forest cover by 2050, while releasing 34.4 billion metric tonnes of CO^2 into the atmosphere (Simms, 2009).

To conclude the general discussion on the benefits of forests, the following points merit some emphasis:

First, it is against the backdrop of the ecosystem services and benefits from forests discussed in this section that the impacts of the loss of forest resources in the Lake Victoria basin must be assessed. It is not only the particular extent of deforestation that matters, but also the state of global forests as a whole. So long as the operation of the biogeochemical cycles remains global, it ultimately may not make a difference where deforestation is occurring. The destruction of the Congo basin rainforests have as much impact on the biolimnological state of Lake Victoria as the deforestation taking place within the lake's own ecoregion.

Second, nature is constantly at work to achieve sustainable coherence by balancing natural extinction occurrences with speciation processes. However, human beings are undermining the trajectory of the evolutionary processes by taking too much low-entropy throughput out of nature and putting too much high-entropy waste into nature. This produces extreme synergistic ecological stressors, driving unknown numbers of plant and animal species to extinction. We are often reminded that there have been 5 extraordinary events in the last 540 million years, which probably wiped out between 50 and 95 percent of life forms on earth, the most recent great event being responsible for the total extirpation of the dinosaurs. Full restoration of biological diversity to the previous state (if at all) would probably take millions of years. According to scientists, the 6th great event of mass extinction may already be under way since human-induced climate change, the spread of pathogens, resource overexploitation, habitat fragmentation, the introduction of invasive species, pollution, and expanding human biomass are creating multiple triggers for explosive extinction. Scores of species are being wiped out of existence each day, since anthropospheric processes have now increased the rates of global species extinction by 1,000 percent over natural extinction rates. If the present rates of extinction continue, we may probably lose nearly half of animal and plant species on earth by 2100 (Millennium Ecosystem Assessment, 2005; Wilson, 2002; Barnosky et al., 2011). It is in the context of this empirical observation that the state of aquatic species in Lake Victoria ought to be understood and evaluated.

Third, forest depletion or fragmentation erodes the ecoregion's capacity to generate rainfall, which is critical to Lake Victoria's hydrological balance. Since the lake depends on direct precipitation for 80 percent of its water budget, while river inflows supply the remaining 20 percent of the balance, deforestation is very likely to influence the ecoregion's capacity to generate the necessary conditions for cloud formation and rainfall production.

Fourth, since forests act as natural air conditioners, their absence or fragmentation may drastically affect the microclimate of the region, which means more warming. In Tanzania alone, deforestation is said to release 100 million metric tonnes of greenhouse gases annually, which is twice the total annual emissions in Norway (Obulussa, 2008). So, if the 5 countries that comprise the Lake Victoria basin continue to deplete or fragment their forest resources, the climatic impact of deforestation on the state of the lake is unmistakable.

Fifth, forest resources are integral to the survival strategy of millions of people in the basin, in terms of providing nonwood food and bush meat, as well as biological agents that are essential to pollinating their traditional crops and controlling pests. In light of this reality, deforestation would mean forcing people to migrate to Lake Victoria and other satellite water systems in search of livelihoods as fishers, thereby worsening aquatic resource depletion.

Sixth, deforestation means widespread erosion of the basin, exacerbating the siltation, sedimentation, and eutrophication of Lake Victoria.

Finally, since the state of terrestrial resources is functionally inseparable from the conditions obtained in the aquatic ecology of Lake Victoria, any effort undertaken to address the crisis in the fishery mode of production cannot succeed unless similar efforts are simultaneously undertaken to address the crisis of terrestrial resource depletion and degradation. More important, the integrity of Victoria's forest ecology cannot be maintained so long as the neoclassical conception of natural resources as mere raw material to be harvested is changed. The pirate state bourgeoisie and the new global biomasters see the region's standing forests in terms of their commercial value or profit potential. In their view, these forest resources have no value until capital and labor is applied to them. The contributions they make to carbon sequestration, climate stability, hydrology, and the shelter and food they provide to wildlife and the indigenous poor people do not enter the bookkeeping of the new biomasters and their local hangers-on. Having outlined the benefits and costs of forest resource extraction, in the remainder of this section I shall turn to the specific drivers responsible for deforestation and forest fragmentation in the Lake Victoria basin.

Insofar as deforestation or forest fragmentation in the Lake Victoria basin are concerned, the main drivers are the exponential growth of human and livestock biomass, which are extremely dependent on the appropriation of primary plant production, commercial logging, and expansion of agrocommercial plantations. Put differently, the primordial necessity of human survival associated with population pressure, and the objectification of nature driven by the logic of capitalist accumulation, contextualize forest resource extraction and degradation in the basin. When their world was empty, the ability of humans in the basin to harvest natural capital stocks was constrained by the primitive character of their instruments of production and by the need-based mode of social existence. More important, in an empty world, people had the luxury to migrate to new places and return to their original biosocial habitat after the environment had ample time to renew itself. However, things have now changed in many radical ways, and the people in the basin face a full world locked in profound contradictions and struggles. Such struggles express themselves in aggressive competition from greedy wealth accumulators, shrinking living space, resource depletion, habitat destruction, proliferation of invasive species, and pollution of natural water systems. Moreover, the geographical immobility of communities now puts greater pressure on the immediate environment for all life-supporting services. Inevitably, conversion of more and more forested areas to cropland, overharvesting of nontimber forest products to supplement food needs, overextraction of fuelwood for energy, and other associated activities have become inimical to the forest ecology throughout the Lake Victoria basin.

As stated before, the Lake Victoria basin has one of the densest populations in the world, which predominantly survive on liquidating the basin's natural capital stocks. Nothing illustrates this point more than the present state of the Kakamega Rainforest in Western Kenya, just 40 kilometers northeast of Lake Victoria.

Until a century ago, West and East Africa were connected contiguously by the Guinean-Congolian lush tropical rainforest belt. On the eve of colonialism, Kenya was home to over 240,000 ha of rainforest, which has been reduced to less than 23,000 ha. With the advent of colonialism, the country's rainforest fell under the combined assault of uncontrolled commercial logging, clear cutting for cultivation, and over extraction of nontimber forest products. The rapacity with which the rainforest was subjected to logging was so alarming that in 1933, the colonial administration gazetted the remnants of the rainforest at Kakamega for state protection. Even though the measure slowed the rate of deforestation, fragmentation and degradation of the rainforest continued unabated under growing pressure from the communities surrounding the forest. In 1999, the population density (with a typical household having 8 members) around the Kakamega Rainforest reached 548 persons per square kilometer, compared to the national average of 54 persons per square kilometer. Furthermore, the poverty rate among the populace in the Kakamega area was 70 percent (Orimo, 2008, Muller et al., 2009).

The increasing population led to increasing demands for arable land to grow food, pasture to graze herds, fuel wood to cook meals, timber to build homes, and overharvesting of non-timber forest products to supplement survival. As a result, the Kakamega Rainforest has lost half of its forest cover in the last 25 years. The Kakamega case is a microcosm of the general environmental condition in the Lake Victoria basin. In its original state, 40 percent of the basin was under forest cover. Most of the forest resources are now severely fragmented or liquidated for multiple purposes. Since 75 percent of the 40 million people who live in the basin are engaged in subsistence farming, overpopulation, landlessness, and poverty have forced the people to increasingly rely on forest clearance for cultivation and forest resources for fuelwood, construction material, boat making, and fish smoking. Furthermore, deforestation and ecosystem degradation in Victoria's basin have become inseparable from subsistence activity, since poor rural communities are placed under objective material compulsion to operate within the fundamentals of the market. In their struggle to survive, they log trees and present them to middlemen that haul the logs to companies. They extensively engage in charcoal production to eke out a living by meeting the energy needs of urban dwellers that are not connected to the electricity grid. After all, only less than 10 percent of the total population in the countries covered in this study are connected to electricity. This means that over 90 percent of the population derive their energy for cooking and heating from fuelwood and charcoal (Awange and Ong'ang'a, 2006).

Human overpopulation in the Lake Victoria basin is compounded by the increasing importance of the livestock mode of production, which has added another dimension to the crisis of deforestation, soil degradation, and desertification of the basin. In 2007, the 5 countries that share the Lake Victoria basin had a combined livestock population of over 178 million, including cattle, goats, sheep, camels, pigs, and chickens. The same year, the total meat production in these 5 countries was roughly 1.2 million metric tonnes (www.countrystat.org, 2007).

Throughout East Africa, livestock are not only a major source of livelihood for millions of people, but also a source of exchange value, and a measure of wealth and social status. Therefore, the tendency to increase personal livestock population naturally comes to the millions of herders. Furthermore, as in all developing countries, there is a growing demand in the Lake Victoria cobasin countries for animal protein. This growing demand, associated with increased population growth and the shift in urban dwellers' preference for more animal protein, has already signaled a transformation of the livestock mode of production in favor of the industrial system. More important, the pirate state bourgeoisie in the countries covered in this volume have added the livestock sector to their portfolio of foreign exchange earning undertakings, such as coffee, fish, tea, tobacco, and so on. Each requires greater industrialization of livestock production. This shift from grazing and mixed farming systems—traditionally based on the symbiosis and synergy of cropping and animal raising—to industrial livestock production will have important impacts on the forest ecology, soils, and natural water systems throughout the basin. In light of the growing depletion of natural forests and grazing pastures, the default option for the riparian states has become promoting industrial livestock production around Lake Victoria in close proximity to water resources and urban consumers. This is in line with the global trend towards the industrialization of meat production.

In the 1990s, a number of authors had already begun speaking of the coming livestock revolution. A team of researchers, led by Christopher Delgado of the International Food Policy Research Institute, entitled their 1999 research product "Livestock to 2020: The Next Food Revolution," implying the prospective replication in the global livestock sector of the experience of the "green revolution" in cereal production. However, as we shall see below, the optimism was overdrawn. This is because a livestock revolution, driven by the logic of destructive creation, will violate all sustainability thresholds with far-reaching implications for terrestrial, freshwater, and near-shore marine ecosystem health, as well as for global climate. As a large group of researchers led by Johan Rockstrom (2009) pointed out, of the 9 planetary boundaries considered essential for safe operating spaces for sustainable human well-being, 3 boundaries (climate, biodiversity, and nitrogen cycle) are already violated. Furthermore, the livestock sector has, in no small measure, contributed to the transgression of the boundaries because livestock production depends on appropriation of huge amounts of primary resources and massive applications of reactive nitrogen to produce animal feed.

As the 2006 FAO report shows, 56 billion land animals are slaughtered globally each year to feed the human population, which is projected to double by 2050; this is commensurate with the rise in human population and urbanization (Steinfeld et al., 2006; Koneswaran and Nierenberg, 2008; FAO, 2009b). To close the protein consumption gaps between the global north and the global south, such developing countries as Victoria's riparian states are projected to be in the driver's seat of the next livestock revolution, as their combined urban populations are projected to jump from 30 percent in 1980 to 50 percent by 2020. Between the beginning of the 1970s and the early 1990s, developing countries had already increased their annual meat consumption by 70 million tons and their milk consumption by 105 million tons. This is compared with increases in meat consumption of 26 million tons and milk consumption of 50 million tons in developed countries. Indeed, by 2020, developing countries will account for 188 of the 303 million tons of projected global meat demand and 391 of the 655 million tons of the projected global milk demand (Delgado et al., 1999).

However, counting on FAO data, Pelletier and Tyedmers (2010) have taken a dim view of the potential impacts of the livestock revolution on all sustainability regimes. This results from the exponential appropriation of plant biomass, the mobilization of synthetic nitrogen, and the consequent emission of greenhouse gasses. If projections hold, the livestock revolution can potentially overshoot the critical sustainability thresholds of what the authors call "the safe operating spaces" in the domains of net primary production, climate change, and industrial nitrogen mobilization. Pelletier and Tydmers estimate that, in 2000, the livestock sector occupied 52 percent of the safe sustainability threshold for greenhouse gases and 72 percent for biomass appropriation, while the mobilization of industrially fixed nitrogen already overshot the safe operating threshold by 117 percent. By 2050, the "safe operating space" for greenhouse gas emissions taken over by the livestock sector is expected to increase by 70 percent. Furthermore, biomass appropriation is expected to increase by 88 percent and synthetic nitrogen mobilization is expected to increase by 294 percent. From the biolimnological point of view of Lake Victoria, the rise in the regional and global livestock populations, and the methods by which they are raised, will have sixfold implications for the state of climate, animal, and plant composition, the integrity of natural water systems, and the livelihoods of the poor. These 6 entropic consequences of plant biomass appropriation by livestock, together with associated services, should contextualize the discussion on how land cover changes and ecosystem degradation—both at the basin-wide and global levels—have contributed to the state transition of Lake Victoria.

First, at the level of low-entropy resources, livestock are direct competitors with humans for land and water resources as consumers of plant biomass, animal feed, water, and energy. In fact, in the semiarid and arid regions of Victoria's littoral countries, the majority of the land, plant biomass, and water have been taken over by the livestock population. In the last half century, livestock population growth in the Lake Victoria basin has paralleled human population growth. This rate of growth has put enormous pressures on forests, pastures and water resources. Since these regions of East Africa are vulnerable to extreme meteorological variability, pastoralists and agropastoralists practice extensive grazing. This means moving their livestock from place to place, following rainfall patterns. With a livestock population limited to the carrying capacity of the ecosystem, extensive grazing systems could probably be used sustainably. However, in addition to periodic drought, the exponential increases in animal herds have deprived the ecosystem of the necessary time to regenerate. Moreover, high levels of human density in traditional food-growing regions have led the poor to migrate to the semiarid and arid regions; here, they have come to compete with traditional pastoralists and agropastoralists for plant biomass, land and water resources (Kategile and Mubi, 1992; Niboye, 2006).

Government policy to develop the livestock sector, through the promotion of ranches based on intensive grazing systems, has also reinforced the appropriation of plant biomass, land, and water resources. The pirate state bourgeoisie, counting on advice from expatriate advisers, have always seen modernization of the livestock sector as integral to what they call long-term national development.

Uganda established several hundred commercial cattle ranches throughout the 1960s and early 1970s. It all began when the World Bank offered financial loans to establish 5 state-owned beef cattle ranches, each with 38,000 to 45,000 acres, intended to galvanize expansion of the 50 existing private cattle ranches. Even though the turmoil in that country disrupted full-scale industrialization of the livestock sector, Museveni's regime pursued a vigorous policy to restock the commercial cattle ranches with $10.5 million in aid from Kuwait. The long-term livestock policy was to increase the cattle ranch population by ten million by 2000 (Byrnes, 1990). Later, under the National Livestock Productivity Improvement Project funded by the African Development Bank, the regime accelerated the restocking program. Museveni himself set the example in the drive for livestock accumulation, establishing several personal ranches. The largest was the Kisozi ranch, with 6,000 head of cattle on 5 km^2 of fenced pasture land, watered by the Katanga River. By 2005, Museveni was collecting USh105 million per annum from the sale of 700 bulls and cows (*New Vision*, Nov. 9, 2005). Uganda's livestock herders responded to Museveni's example and other government initiatives, by increasing their livestock possessions. As a result, Uganda's cattle population rose from 5.9 million in 2000 to 11.4 million in 2008, accompanied by 15.9 million goats and sheep, 39.4 million poultry, 3.2 million pigs, and 32,870 camels (*New Vision*, Oct. 2, 2009).

Upon independence, Tanzania also rapidly moved to commercial beef cattle ranching operations under its land nationalization and sedentarization program. The understanding of the young pirate state bourgeoisie was predicated on the notion of the pastoral and agropastoral mode of production as a manifestation of backwardness. Therefore, with a loan from the World Bank, the government created the state-owned National Ranching Corporation (NARCO), which commercially enclosed 630,000 ha to begin cattle ranching operations on a national scale. However, the livestock revolution that Tanzania's pirate state bourgeoisie hoped for did not materialize. This forced the regime to give the sector a fresh impetus in 2004 by trying to imitate the Botswana model.

In 2004, the regime dispatched a team of researchers and observers to Botswana to investigate if that country's ranching system could be replicated in Tanzania. The question that roamed in the minds of Tanzania's elite was how was it that Tanzania barely managed to earn $7 million from meat export—despite having 18.5 million head of cattle, 16.7 million goats and sheep, and 30 million chickens—while Botswana could generate $600 million per year and revenue from the export of 20,000 metric tonnes of beef to the European Union, even though the country had only 3 million head of cattle (*The East African*, March 4, 2004). In the minds of the elite, the answer laid in modernizing the livestock sector by creating fenced ranches for exclusively commercial livestock production. This could be achieved either through the active promotion of public-private partnerships or further privatization of commercial livestock production. Since pastoralists and agropastoralists owned 94 percent of cattle in Tanzania, the regime sought to modernize the sector in favor of large fenced ranches. According to the 2004 master plan, by 2007 NARCO was expected to increase its cattle herd from 33,000 to 90,000 head of cattle on 124 ranches. The overarching aim of the policy was to make the livestock sector attractive to foreign investors (*The East African*, March 4, 2004).

Although NARCO was as little successful as before, despite unhampered access to commercially fenced pasture and water, traditional herders continued to increase their livestock possessions. This was reflected in the growth of national milk production from 1.4 billion liters in 2004 to 1.65 billion liters in 2010. The number of milk-processing plants rose from 22 to 46 over the same period. Meat production increased from 378,500 tons in 2005 to 449,673 tons in 2010 (*Tanzania Daily News*, Oct. 14, 2010). From the perspective of the elite, these increases in animal products were not enough.

After all, the commercial dairy subsector continued to stagnate around 700,000 head of cattle, compared with 3.5 million head of cattle in Kenya's dairy farm subsector. Thus, the regime initiated a new policy to accelerate the modernization of the livestock sector through the extension of subsidies and subleasing state ranches to private investors. As the president outlined, in addition to raising hybrid cattle that would produce more milk and more meat, cultivating high-quality animal feed and storing the feed in granaries for distribution to livestock keepers would the first steps toward modernizing commercial livestock production. By 2010, with this outline in mind, NARCO had subleased 134 commercial ranching blocks to private investors on 389,134 ha of pasture land. By the beginning of 2010, 48,301 head of cattle, and 7,182 goats and sheep were being raised on these privatized ranches (*Tanzania Daily News*, Oct. 14, 2010). Of course, commercial enclosure of ranches meant keeping traditional herders out from grazing pasture and water.

In comparison to its neighbors, Kenya has a more developed commercial livestock subsector. By 1986, Kenya could count 2.5 million dairy cattle, 10 million multipurpose head of cattle, 19 million sheep and goats, 18 million poultry, 900,000 camels, and 120,000 pigs (Kategile and Mubi, 1992). From an ecological point of view, livestock overpopulation in Kenya is particularly problematic, since 83 percent of the country is semiarid and arid. By 1986, these regions were supporting 35 percent of the country's total human population, 67 percent of its cattle population, 86 percent of its sheep and goats, all of its camels, and the majority of the nation's wildlife population (Kategile and Mubi, 1992). Despite human and livestock overpopulation, Kenya's pirate state bourgeoisie continue to push livestock keepers to produce more milk and more meat. In 2007, for example, the National Ministry of Livestock and Fisheries issued guidelines on how livestock keepers could overcome the problems of land resource depletion, poor animal nutrition, and low genetic productivity of their herds, by shifting to intensive production system. The case of Kisii District in the Lake Victoria basin was used to illustrate the matter. In 2007, the human population density in the district was 600 persons per km^2. Yet, the people in the district had to share their ever-shrinking land and water resources with 146,000 head of indigenous Zebu cattle, 159,000 head of dairy cattle, 134,000 sheep and goats, and 1 million chickens (*East African Standard*, July 30, 2007).

However, a shift to an intensive livestock production system could hardly satisfy the wants of the pirate state bourgeoisie, since in industrial or mixed systems, green pasture is not always available. This means that ranchers will have to rely on commercially produced feed concentrates. As of writing, for example, there were 80 provender millers in Kenya, with a total installed capacity of 700,000 metric tonnes per annum. They use ingredients such as pulses, tubers, soya, wheat, barley, maize, oats, millet, sunflower, animal products, oil cakes, and fishmeal. This means that food that was traditionally reserved for human consumption would now be diverted to concentrate animal feed. This not only worsens the food situation, but also makes animal feed expensive to smallholder herders. For example, even though Kenya's animal feed millers had an installed capacity of 700,000 metric tonnes per annum, they produced only 400,000 metric tonnes in 2010; as a result, prices for animal feed shot up and made access to manufactured animal feed difficult for small livestock producers (*The Nation*, Feb. 26, 2011).

It is in light of the above considerations that the livestock sector has become a major driver of deforestation in Victoria's basin, as more and more forested areas are converted into growing pasture or livestock feed. It is only logical that increases in livestock production necessarily lead to increases in deforestation, environmental degradation, greenhouse gas emissions, and water shortages. These conditions are bound to worsen as the pirate state bourgeoisie in Victoria's riparian countries continue to imitate the global patterns of livestock production, entailing conversion of the basin's forested areas and wetlands for commercial grazing pasture and concentrate feed production. Of course, this is not out of the ordinary since the global livestock sector currently uses 3.9 billion ha of land for grazing and feed crops. In 2005, 1,543 million metric tonnes of farm-grown cereals, pulses, brans, oil cake, oil crops, and roots and tubers were fed to livestock; this is projected to double by 2050, commensurate with the doubling of global livestock population (FAO, 2009b). In fact, by 2050, the amounts of cereals and coarse grains directly appropriated to feed the additional global livestock population can be enough to feed extra 4 billion people (Tudge, 2004).

Therefore, the livestock revolution (now being imitated in East Africa) is bound to exert enormous pressure on Lake Victoria and on other freshwater resources throughout the basin. Water withdrawals are intensively used for drinking animals, and also for growing and processing animal feed, cleaning and cooling farm animal facilities, and for processing livestock products. As Colin Tudge (2004) has pointed out, today's average farmer uses 500 liters of water to grow 1 kilo of potato, 900 liters to grow 1 kilo of wheat, and 2,000 liters to grow 1 kilo of rice. In contrast, the farmer uses 3,500 liters of water to raise a kilo of chicken and 100,000 liters of water to obtain a kilo of beef. This suggests that the drive toward commercial livestock production, in the countries covered in this volume, is bound to accelerate freshwater depletion, deforestation, desertification, soil erosion, and lake sedimentation.

Second, many writers blame the state transition of Lake Victoria on the insertions into the lake of primarily reactive nitrogen, phosphorus, and other synthetic agents. If this is true, which I believe it is, then basin-wide promotion of the livestock revolution will make any effort to contain and reverse the eutrophication of Lake Victoria extremely difficult, if not impossible. The livestock revolution implies a transition from the extensive grazing system to industrial livestock concentration and production in close proximity to Lake Victoria. This transition is based on reactive nitrogen mobilization to grow animal feed, and petroleum-derived fuels needed to drive tractors, to transport vehicles, to process the crops into concentrates, to transport the animal products to markets, and so on. The high primary production of the wetlands buffering Lake Victoria, lake water availability, and proximity to markets and transport networks, make the areas surrounding Lake Victoria particularly attractive to the captains of the livestock revolution. As a result, gaseous emissions from livestock production are bound to be pushed above the sustainable scale in that domain. After all, the current global livestock sector accounts for 18 percent of the total greenhouse emissions, which is greater than the emissions from the transport sector. This will certainly increase in the future in proportion to the rise in the number of livestock population. Indeed, by 2050, global meat, milk, and egg production is projected to add 39 percent to greenhouse gas emissions above the 2000 levels; furthermore, appropriation of plant primary biomass by the livestock sector will generate an additional 21 percent, while the mobilization of synthetic nitrogen will add another 36 percent (Pelletier and Tydmers, 2010). It thus stands to reason that confinement of livestock farm operations in delimited geographic areas around Lake Victoria, as well as the increasing dependency on concentrate animal feeding operations, will simply serve to accelerate the demand for more biomass and the mobilization of reactive nitrogen needed to boost production. Since industrially fixed nitrogen determines the productivity of cropland dedicated to animal feed, more of it has to be fixed. After all, the global amounts of anthropogenically mobilized nitrogen are already more than the total amounts provided by all natural terrestrial systems. What the livestock revolution does is increase reactive nitrogen mobilization, most of it eventually ending up in water bodies. By some estimates, livestock manure by itself accounts for 2-thirds of anthropogenic nitrous oxide emission, which is 300 times more potent in trapping heat than carbon dioxide (Smolker et al., 2007; Pelletier and Tyedmers, 2010).

It is important to bear in mind that emissions result not only from animal manure, urine, and other animal wastes, but also from clearing forests to create pasture, the application of nitrogen and phosphorus to boost animal feed production, the intensive use of fossil energy to grow and transport animal feed, the transportation of processed animal products to markets, and from the amount of energy expended on heating and cooling animal operation facilities. Already, the world annually produces 100 million metric tonnes of petroleum-derived reactive nitrogen to grow livestock feed. All told, livestock populations are responsible for emissions of 2.4 billion metric tonnes of carbon dioxide per annum, from combinations of livestock-driven deforestation, soil cultivation, land deterioration, and desertification. It is a small wonder that the current livestock sector accounts for 65 percent of anthropogenic nitrous oxide and for 35 to 40 percent of methane emissions (Steinfeld et al., 2006; Koneswaran and Nierenberg, 2008; FAO, 2009b; Pelletier and Tyedmers, 2010). In light of the preceding considerations, radical alteration of the nitrogen cycle may be significantly disrupted, resulting from the intensification of industrial livestock production around Lake Victoria. Since nitrogen passes through several cycles from one form found in the atmosphere to ammonia, nitrate, nitrite, and other forms, before it returns to its original state through denitrification, its impact on the biolimnology of Lake Victoria is likely to be compounded by the spread of the livestock revolution in the lake's basin.

Third, promotion of the livestock revolution around Lake Victoria (which appears to have grabbed the attention of the pirate state bourgeoisie) is likely to increase the generation of untold amounts of nutrient-rich solid and liquid wastes. These wastes will inevitably enter Lake Victoria and other natural waterways, worsening the eutrophication and pollution of the lake and other freshwater bodies throughout the basin. This unsavory observation is empirically supported by four cautionary lessons drawn from the experiences of the advanced capitalist countries.

The first lesson relates to the fact that the transition of the political economy of livestock production, from one based on traditional grazing and mixed farms to industrial livestock production, does accelerate the pollution and contamination of land, water, and air. Concentration of livestock operations in geographically confined areas inevitably leads to production and accumulation of huge amounts of manure, excretes, and other forms of waste over a small landmass with little capacity to assimilate. From the accumulation of this waste, nitrogen, phosphorus, and other pollutants enter the general aquatic and terrestrial environments. In the Netherlands, for example, of the 208,000 tons of ammonia generated in 1993, 181,000 tons came from animal manure in (Haan et al., 1997).

The second important lesson pertains to the fact that, if animal waste production exceeds the waste absorptive and processing capacity of terrestrial and aquatic systems, the continuous flows of animal waste can lead to dangerous oversaturation of nutrients; this will lead to eutrophication and acidification of lakes, as well as to surface and ground water contamination, soil pollution, biodiversity loss, and aquatic resource degradation (FAO, 2009b).

The third important lesson to be learned from the experiences of early movers in the livestock revolution is that the quantity of high-entropy waste is directly related to the scale of low-entropy transformation. In the United States, for example, 800,000 corporate farm operators annually generate $60 billion in revenue. However, the environmental cost is equally hefty, because livestock operations generate 133 million tons of manure per year, which is 13 times larger than the solid human sanitary waste production (Burkholder et al., 2007). The manure is overloaded with such macronutrients as nitrogen, phosphorus, growth hormones, antibiotics, microbes, copper, zinc, and other elements; all of which enter natural water systems through direct dumping, overflow of lagoons, leakage from storage facilities, and deposition by wind and rainfall (Burkholder et al., 2007).

The fourth lesson to keep in mind is that during the digestion process of their protein intake, livestock release nitrogenous gases to the environment through their urine and manure. For example, it is estimated that cattle excrete 80 to 90 percent of the nitrogen and phosphate intake from their feed, while pigs excrete 65 per cent, and chickens excrete 70 percent. When nitrogen evaporates in the form of ammonia, especially if accompanied by phosphate leaching from manure, the process accelerates the eutrophication and acidification of aquatic ecosystems (Haan et al., 1997; Steinfeld et al., 2006; FAO, 2009b).

Fourth, the intensification of the capitalist mode of livestock production around the entire Lake Victoria basin is bound to totalize degradation of the terrestrial ecology and worsen the sedimentation of the lake. Historically, the relationship between nature's biocapacity and traditional livestock keeping in Victoria's riparian countries has been marked by symbiosis and synergy. Such animal waste products as manure are known to improve soil fertility and vegetation biodiversity. By feeding on excess biomass, farm animals help to control plant overgrowth and they also carry plant seeds in their manure and/or hoofs for dispersal. Throughout East Africa, such symbiosis has historically been critical to the livelihoods of pastoral and agropastoral communities. These communities largely depend on grazing systems, primarily occupying the ecologically delicate semiarid and arid areas. What is unique about grazing systems is that they approximate closed systems, in that what is generated within the systems is internally absorbed without exporting the waste to other systems (FAO, 2009b).

However, scale had always been integral to the biocapacity of grazing lands. As mentioned earlier, population pressure in the rural regions of Victoria's littoral countries, as well as increasing demand for animal products in growing urban centers, have resulted in livestock population growth. Furthermore, the factors have resulted in the expansion of human settlements and cultivation, leading to the shrinkage of grazing land and ecosystem degradation. The traditional flexibility of alternating between wet and dry season grazing has diminished; livestock population now heavily graze year-round in geographically delimited areas around Lake Victoria, resulting in overgrazing, deforestation, soil erosion, soil compaction, and loss of water storage capacity (Niboye, 2006). Actually, this phenomenon is not unique to East Africa. Several studies have pointed out that, by 1990, 680 million ha of grazing range lands had become degraded worldwide. According to another study, 73 percent of the world's 4.5 billion ha of range lands had undergone substantial modification by 1990; these modifications ranged from moderate to severe degradation, and were attended by significant deforestation, contributing to biodiversity loss and habitat alteration (Haan et al., 1997; Steinfeld et al., 2006; FAO, 2009b).

Similar phenomena are being observed in the category of mixed farm systems, which are minor players in the livestock mode of production in Victoria's riparian countries. Kenya may claim a partial exception because it has inherited somewhat large mixed farm systems from the colonial era. Today, Kenya has 3.5 million dairy cattle in addition to some 13 million multipurpose head of cattle. Unfortunately, with the growing popularity of the livestock revolution, these mixed farm systems may rapidly lose ground to industrial livestock farming. This will be consistent with the phenomenon manifested worldwide, since the annual global supply of milk from mixed farms declined from 92 percent in the 1990s to 33 percent by 2007, and the contribution of sheep and goat meat fell from 70 percent to 53 percent over the same period (Haan et al., 1997; FAO, 2009b). Within limits, there is symbiosis in mixed farms, between livestock contribution to soil fertility—through the supply of organic fertilizer—and recycling of crop residues. The danger with livestock overpopulation in the mixed farm systems is less crop residues available for animal feed and soil cover, leading to high rates of soil erosion.

Fifth, the transition to industrial livestock production has the potential to completely erode the food ecology of traditional communities in Victoria's riparian countries, with the concomitant effect of inducing migration of dispossessed livestock herders to the shores of Lake Victoria. The livestock sector in Victoria's riparian countries, accounting for up to 40 percent of agricultural output, is critical to the livelihoods of the majority of rural people. In fact, livestock are the sole source of livelihood for pastoralists, while they are the dominant source of livelihoods for agropastoralists. From the point of view of these traditional communities, the livestock revolution presages total erosion of their livelihoods and their displacement. Moreover, with the growing prominence of industrial livestock systems, more and more arable land is bound to be diverted away from growing food to growing animal feed. It is worth noting that, currently, more than 50 percent of global corn and 80 percent of soybeans is already devoted to feeding the global livestock population (Koneswaran and Nierenberg, 2008).

Sixth, as we shall see in the next chapter, microorganic pollution is a growing health problem in and around Lake Victoria, which is associated with the continuous insertions of human and animal excrements into the lake. Congregations of livestock and concentrated animal feeding operations around the lake are very likely to worsen the health risk. As suggested earlier, water availability from the lake, high primary production from its surrounding wetlands, and proximity to consumers are the factors that make the congregation of livestock operations in the area very much in prospect. If the lake's water resources and adjacent wetlands provide the necessary means of livestock production, the 40 urban centers and burgeoning towns around Lake Victoria supply ready-made markets for animal products. It thus stands to reason that the colocation of animal farms, in close proximity to the sprawling urban centers around Lake Victoria, pose serious public health risks associated with microbial and chemical contamination of air, water, and soil. The continual insertions into the lake of manure, urine, ammonia, hydrogen sulfide, inhalable particulate matter, malodorous vapors, and veterinary drugs from industrial animal farms and concentrate animal feeding operations will worsen the health hazards if the livestock revolution is fully realized in the area.

There are important lessons to be learned from the advanced countries that have consummated the livestock revolution. Several studies on the health impact of industrial animal farms indicate a strong relationship between concentrate animal feeding operations and human health. Health impacts observed include mucus membrane irritation, acute respiratory and digestive illnesses, organic dust toxic syndrome, asthma-like distress, bronchitis, sleep disturbances, and mood swings. The increasing use of antimicrobial growth promoters and a regiment of antibiotics are also health risk (Delgado et al., 1999; Steinfeld et al., 2006; Heederik et al., 2007; Donham et al., 2007; FAO, 2009b). Moreover, contrary to formal assertions that the livestock revolution lends itself to poverty alleviation by producing high-quality and abundant animal products, industrial livestock production can actually worsen the misdistribution of animal products.

The poor will continue to suffer from malnutrition-driven issues, stunted growth, and rheumatic heart diseases from lack of sufficient animal protein. Conversely, the rising commercial class, the pirate state bourgeoisie, and their hangers-on will get fatter and sicker from having too much of the good thing. These groups will develop costly overweight-related diseases such as hypertension, diabetes, and cardiovascular maladies. In 2000, there were 937 million obese people, 62 percent of them in the global south; and there will be 1.2 billion obese adults by 2030 (ETC Group, 2010). It is useful to mention that there were 970 million people suffering from hypertension in 2004, 640 million of which lived in the global south. In fact, the number of people suffering from hypertension is projected to rise to 1.56 billion by 2025, with the majority of them living in the global south (Kearney et al., 2005; Mackay and Mensch, 2004). Additionally, there were over 300 million diabetic people in 2010, costing the global economy to the tune of $376 billion per annum. By 2030, the number of people living with diabetes is actually projected to reach 438 million, absorbing almost $500 billion per annum in expected global healthcare expenditures (International Diabetes Federation, 2010).

Another lesson to be learned is that lifestyle changes, often resulting from dietary conditions, are major contributors to cardiovascular ailments. Indeed, more than 17.5 million people worldwide die each year from cardiovascular diseases, representing 30 percent of all deaths. According to the World Health Organization, income losses due to noninfectious chronic diseases could be as high as $558 billion in China and $237 billion in India per year by 2015 (Bayer, 2010). Coping with epidemic noninfectious diseases is costly, both financially and in human terms. In 2001, for example, the United States spent a staggering $181 billion to treat heart patients alone (Tudge, 2004).

There are convoluted triangular relationships between producers and consumers of livestock products, on the one hand, and between industrial livestock producers and pharmaceutical corporations, on the other. Commercial livestock producers depend on drug companies for the supply of nontherapeutic drugs and growth hormones to grow their farm animals bigger, fatter, and faster. Producers further depend on the drug companies for the series of regiments of antimicrobial drugs and antibiotics to keep their animals alive until they are slaughtered. Heavy consumers of livestock products also depend on pharmaceuticals to fight back their lifestyle-driven hypertension, diabetes, and cardiovascular diseases. In this sense, both commercial livestock producers and livestock products consumers are central to the reproduction of the pharmaceutical industrial complex. This is the brave new world into which the East African societies have been born, midwifed by globalization. As described above, to grow their animals, and to fight back a variety of tropical diseases and newly generated infections, commercial livestock operators in Victoria's basin are increasingly dependent on foreign pharmaceutical companies for regiments of growth hormones, antibiotics, and vaccines. The Lake Victoria basin is already known for regular outbreaks of rinderpest, foot and mouth diseases, and peste des petits ruminants (PPR), which decimate the livestock population from time to time. In 2004, for example, cattle lung disease killed 87,197 head of cattle in Tanzania's 12 districts surrounding Lake Victoria (*Business Time*, March 12,

2004). In Kenya, pastoralists in the Turkana District also lost over 100,000 goats and sheep to an outbreak of PPR in late 2006 (UN IRIN, July 24, 2007).

In sum, the orchestration of the livestock revolution, as integral to national economic development and poverty eradication, is predicated upon illusion or deception. Modernization of the livestock sector by itself is not the problem. The problem is that traditional herders will be fenced out into extinction from commercially enclosed industrial livestock and concentrated animal feeding operations. Pastoralists and agropastoralists may soon die out as a species, with nowhere to go but the shores of Lake Victoria and the burgeoning urban centers, swelling the reserve army of slum dwellers.

Up to this point, I have highlighted the direct and indirect impacts of human and livestock overpopulation on the natural forest and water resources in Lake Victoria's basin. However, human biomass and livestock crowding are not the only drivers of deforestation, rangeland degradation, soil erosion, and microorganic pollution. The capitalist logic of accumulation is another driver of resource depletion and deforestation, since large portions of Lake Victoria's basin have long been commercially enclosed for purposes of timber production and cash-crop plantations. In keeping with the expansionary logic of accumulation, transnational operators of logging companies and large-scale agrocommercial plantations of sugarcane, coffee, tea, cotton, tobacco, pyrethrum, and sisal have subjected the Lake Victoria basin's forest to uncontrolled clearance. Forests are auctioned to logging companies, which engage in selective logging to pick the "lowest hanging fruit" first. The supposed legal logging operations are supported by illegal felling of prime forests by individuals who seek to improve their lot by selling the timber to middlemen. The direct exploitation of forests as commodities is compounded by other activities, such as commercial charcoal production, the construction of hydropower stations, transmission lines, roads, expansion of urban centers, mining operations, and exploitation of hydrocarbons.

The current rush to speed up the transition to the carbohydrate economy has made forests even more vulnerable to destruction in two ways. First, the forests in the tropics and subtropics must be cleared to make way for bioenergy crops to feed the agrofuel processing plants in the global north. Second, wood chips and pellets have now become highly valued inputs for biofuel production. In 2008, over 10 million metric tonnes of wood pellets were globally traded and that figure was projected to double in four years (ETC Group, 2010). The Canadian MagForest company, operating in the DRC, was exporting 500,000 metric tonnes of wood chips to Europe per year as of writing; IBIC Ghana Ltd. also boasts that it has the capacity and the resources to export 1.2 million metric tonnes of hardwood and softwood every year from Ghana to biofuel factories in Europe (ETC Group, 2010).

Having broadly outlined the drivers of deforestation in the Lake Victoria basin, by which rainfall generation, food production, watershed protection, wood extraction, and the provision of other goods and services are being eroded, now I turn to particularize the discussion by looking at how each riparian country has been managing its forest resources.

All of the countries covered in this work have long been ravaging their forest resources for a variety of uses. In the last century, Uganda lost more than half of its forest cover and, at the current rate of forest depletion of 92,000 ha per year (2.2 percent per year compared with the global average of forest depletion rate of 0.2 percent per year), Uganda may lose 80-90 percent of its remaining 3.6 million ha of forests over a 30-year period. Between 1990 and 2005, the net forest loss in Uganda was 1,297,000 ha. Unless the current rate of deforestation and climate change is comprehensively addressed, 80 percent of Uganda may become desert by the end of this century. This will increase the pressure on Lake Victoria and other natural freshwater reservoirs. Since Uganda's forest resources contribute a total of $546 million (direct and indirect) per year to the economy, forest depletion of this magnitude and the consequent entropic degradation of the ecosystem would be a devastating blow to all Ugandans (FAO, 2005; BBC News Monitoring Africa, Oct. 13, 2008).

The threat to Uganda's forest resources comes from many fronts, including tourism developers, subsistence and commercial cultivators, agrofuels producers, tobacco growers, firewood harvesters, and charcoal producers. In order to give a broader scope to its highly trumpeted export-led growth, for example, the Ugandan government granted BIDCO Oil Refineries Ltd. 6,449 ha of forested land on Bugala Island in Lake Victoria. In 2003, the Ugandan government granted the same company another 30,000 ha of forest land to develop a palm oil plantation and processing facility. In 2007, these land grants were followed by the government's decision to make 7,100 ha (25 percent of the Mabira Forest, which is the largest forest reserve in the country) available to the Mehta Group to develop sugarcane plantations for purposes of biofuels production (*Africa News*, May 8, 2007).

Forest resource harvesting for private consumption or sale is another serious threat to Uganda's forests. Various reports show that forest resource utilization by private actors in Uganda has increased three-fold in recent years; by 2002, the annual wood consumption was already 25 million metric tonnes or 1.1 metric tonne per person, 95 percent of which was used as firewood or charcoal (*Africa News*, May 8, 2007).

Tobacco production (which requires cultivable land space and massive amounts of wood for curing) is also a major culprit in Uganda's forest destruction. The West Nile District, where 70 percent of Uganda's tobacco is grown, is emblematic of the problem. The tobacco growers there have almost wiped out the native miombo trees in the district, and are increasingly threatening the Merchison Falls National Park as they desperately look for trees to fell. In 2006, the West Nile District produced more than 10 million kg of tobacco, a major source of its status as the most-wood-deficit area in Uganda. Even though a number of NGOs have echoed their concerns over the linkage between tobacco production and deforestation, the pirate state bourgeoisie have turned a deaf ear to these pleas. Just as poor farmers are desperate for income to make ends meet, the pirate state bourgeoisie are equally desperate to keep the accumulation process going. After all, tobacco is the third largest foreign exchange earner after coffee and fishery; in 2007, tobacco exports registered $66 million (*Africa News*, May 15, 2008).

Another threat to Uganda's forests comes from landless peasant encroachers. Under circumstances of no option to eke out a livelihood, poor Ugandans have been flocking to the edges of forests and water bodies in growing numbers in the last 20 years. The number of forest encroachers increased from 180,000 in 2006 to more than 300,000 in 2008 (*New Vision*, July 10, 2008). The threat to Uganda's forests from encroachment is proportional to the stark demographic conditions. The country's present population of 36 million is projected to reach 103.2 million by 2050; by then, the per capita forest resources in the country will be zero (*New Vision*, Sept. 15, 2009). This suggests that even the 1.9 million ha of protected reserves may be gone by midcentury as a result of commercial logging and encroachment. The majority of Uganda's protected forest areas are found in Lake Victoria Crescent, a pristine region straddling 13 districts covering over 1.5 million ha. Therefore, the disappearance or fragmentation of the forests in this area may represent a huge loss, particularly for Lake Victoria, as the consequent soil erosion and pollution may hyperaccelerate the sedimentation and eutrophication of the lake.

Although Tanzania enjoys superior endowment in terms of forest resources over its immediate neighbors, the net forest loss in the country now stands at 1.2 percent per year and is growing rapidly for the same reasons as in Uganda. Tanzania has 35.2 million ha (39 percent of its total land area) under forest cover. Between 1990 and 2005, the net forest loss in Tanzania was 14.9 percent of forest cover, which translates to 6,184,000 ha (FAO, 2005a; World Bank, 2008; Milledge et al., 2007). Of the 35.2 million ha of forest resources, 15 million ha are in state-managed reserves, of which 13 million ha are characterized as production forests. Most of the remaining forest resources are classified as general resources, defined by an open access regime (World Bank, 2008).

This classification means that competing uses for industrial development, agriculture, human settlement, livestock grazing, timber, and charcoal production make Tanzania's forests vulnerable to heavy deforestation, exacerbating water and wind erosion, physical and chemical degradation, and desertification. Of the 40.4 million cubic meters of fuelwood consumed nationwide in 1999, rural households consumed 26 million in charcoal, while urban dwellers consumed another 13.4 million cubic meters (World Bank, 2008). Moreover, almost all rural-based small operations rely exclusively on fuelwood to process their products, including smoking fish, curing tobacco, brewing beer, making bricks and pottery, distilling liquor, making bread, building boats or canoes, and drying tea and lime. Since 95 percent of energy supply in Tanzania is in the form of firewood use, it would require massive investments in alternative sources of energy to halt or reverse the current rate of deforestation.

Forest resources in Tanzania are also under assault as they are heavily targeted by the pirate state bourgeoisie themselves for purposes of export revenue generation. According to official accounting, the forestry sector contributes 2-3 percent to GDP and 10-15 percent to foreign exchange earnings (World Bank, 2008). These official statistics understate the value of forest extraction, since illegal forest harvesting and timber exports never enter official bookkeeping. By some estimates, 96 percent of commercial tree felling and 85-90 percent of timber export from Tanzania is considered illegal. In May 2007, Traffic International produced a scathing report on the rate of deforestation taking place, particularly in the pristine coastal forests and miombo woodlands of southern Tanzania, which are among the 10 global biodiversity hotspots (Milledge et al., 2007). The report reveals the extent of the nexus that exists between commercial vandalization of the country's forest resources and official corruption. Indeed, the social and economic arrangements under which commercial forest vandalization is taking place in southern Tanzania resembles the Nile perch-dominated fishery mode of production in Lake Victoria.

At the bottom of the intricate relationship in forest resource extraction are 10 or so local facilitators, generally composed of village chiefs. Above them are 5 or so receiving middlemen, connected to a particular logging or timber exporting company. At the top of the pyramid are 15 permanent hardwood sawmills present in southern Tanzania alone, and some 40 timber exporting companies. These companies have cozy institutional relationships with senior government officials and foreign interests, in terms of patronage, shareholding, or membership in corporate boards. The result of this interlocking relationship is undervaluation of Tanzania's precious hardwood, environmental devastation, and gross misdistribution of forest benefits. According to Traffic International, local timber harvesters receive less than 1 percent of the value of hardwood extracted. District and central government entities also lose millions of dollars each year to illegal subterranean channels. For example, the revenue loss to the Forestry and Beekeeping Division in the Ministry of Natural Resources during 2004 and 2005 was a staggering $116 million. These losses were due to illegal forest extraction and illegal timber export. The anomalies in official trade statistics help to expose the extent of forest destruction and loss of royalties from timber export.

The officially recorded revenue from timber export from Tanzania to all destinations between 2003 and 2005 was 19,300 cubic meters, while the statistics of timber export to China alone was a whopping 108,500 cubic meters. It is a small wonder that timber export from Tanzania to China increased by 1,400 percent since 1997 (Milledge et al., 2007). In the view of Traffic International, at the 2003 rate of forest extraction, all commercially valuable trees in the coastal forests and miombo woodlands in southern Tanzania will be wiped out in 20 years. For reference, in 2003, between 91,000 ha and 98,000 ha of pristine coastal forests and miombo woodlands were lost in 2 southern districts alone due to primarily illegal logging (Milledge et al., 2007).

Tobacco production, which accounts for 1 percent of Tanzania's total land area and 12 percent of its foreign exchange earnings, is another driver of deforestation in the country (Mayunga and Kapooria, 2003). Following liberalization, Tanzania intensified tobacco cultivation; between 1984 and 1992, the area under tobacco cultivation grew by 66 percent to reach 90,000 acres, and curing the tobacco produced on these acres required the destruction of 99,000 acres of forest each year (Hammond, 1997). From the point of view of Tanzania's pirate state bourgeoisie, tobacco production is too lucrative to be given up in the interest of forest conservation. Even though national tobacco production sharply increased by 167 percent over a period of 8 years—from 22,000 metric tonnes in 2000 to 58,702 metric tonnes in 2008—the long-term goal of the regime has been to increase tobacco production to over 100,000 metric tonnes per annum. This makes tobacco the second largest cash crop in Tanzania, contributing $143.3 million to the economy and $41 million in foreign exchange earnings (Kiishweko, 2009).

The Tabora region in central Tanzania, which produced 26.1 million kg of tobacco in 2006-2007, has been the focus of tobacco farming; the region is richly endowed with miombo trees, a natural condition that has made it exceedingly attractive to tobacco growers (*Africa News*, April 4, 2008). A case study by Mangora (2005) in the Urambo District of the Tabora region highlights the relationship between tobacco cultivation and forest destruction. Because tobacco production in the district rests almost exclusively on shifting cultivation, 42,738 ha of woodlands are cleared each year to plant tobacco. This figure represents 3.5 percent net forest loss per year. Moreover, the average farmer who grows tobacco on 1.3 ha would use 23 cubic meters of stacked wood to cure his tobacco. Therefore, this step adds another 3 percent of net forest loss. In total, Urambo District loses 6.5 percent of its miombo woodlands each year as a result of tobacco cultivation and tobacco curing.

A major driver of shifting cultivation is the fact that the first cultivation of tobacco is veritably followed by sharp soil macronutrient depletion, necessitating further miombo woodland destruction or a resort to chemical applications required to restore the soil macronutrients. Fertilizers are not easy to come by for poor tobacco growers, since they must be imported at constantly rising price mark ups. In 1997, for example, tobacco growers paid $20 for a bag of fertilizer, which cost $6.60 3 years earlier (Hammond, 1997). Lack of access to fertilizers to develop an intensive system of cultivation thus compels poor farmers to continually hack their miombo woodlands to make way for tobacco farms. As the carrying capacity of the tobacco- growing district shrinks, farmers have begun to encroach on the forest reserves in the area, which extend over 994,000 ha (Mangora, 2005).

As a percentage of its total land area, Kenya is less endowed with forest resources than its neighbors. The country has some 3.7 million ha under forest cover, amounting to 3 percent of the country's total land area. The net annual loss of forests stands at 0.3 percent per year. Between 1990 and 2005, Kenya lost 5 percent of its forest cover, amounting to 186,000 ha. Most of this loss is accounted for by tobacco cultivation and curing, agrocommercial plantations, timber and nontimber production, and fuelwood consumption. Of the 20 million cubic meters of fuelwood consumed annually, 95 percent is harvested from forests and woodlands. Moreover, 2.9 million Kenyans who reside within 5 km of the forests depend primarily on forest products to eke out a living; in some areas, households earn up to 70 percent of their cash incomes from harvesting forest products (FAO, 2005a; UNEP, 2009).

Runaway population growth, contextualized by deepening poverty and government-sanctioned capitalist agrarian accumulation, are the principal drivers of deforestation in Kenya. As agrarian accumulators grab the fertile and well-watered land, the poor are pushed further and further to marginal areas, forced to depend entirely on nature.

Among Kenya's 8 provinces, Nyanza Province has witnessed widespread deforestation over the last 4 decades, since it has been identified (among other things) as suitable for tobacco production. Ecological degradation of this province is particularly disconcerting because it hugs Lake Victoria, which means that its wetlands and forests are the last line of defense against silts and sediments flowing to the lake. In any case, in the late 1960s, British American Tobacco (BAT) introduced tobacco cultivation into the province on account of the region's abundant rainfall, fertile soil, and available forest resources. These variables are all deemed essential for growing and curing tobacco. Prior to tobacco cultivation, agropastoralism (alongside with fishing) was the dominant mode of production in most of Nyanza. The rapid expansion of tobacco production, however, put the region on a trajectory of intensification and commodification of agriculture.

A case in point is Kuria District in Nyanza Province, where colonialism had barely left its imprint on the precapitalist mode of production. Well until the early 1970s, people still relied on livestock keeping as the primary source of livelihood. But the introduction of tobacco cultivation into the district changed not only the food ecology of the inhabitants, but also the entire landscape of the area. BAT managed to lure some 20,000 agropastoralists in the various districts of Nyanza to shift to tobacco cultivation on the basis of outgrower schemes. Under this arrangement, the company would provide farm inputs on credits to these contracted farmers, who would sell their tobacco leaf to the company. As a result, tobacco cultivation took off rapidly. In the Kuria District alone, annual tobacco production rose from 30 metric tonnes in 1975 to 4,386 metric tonnes in 1998 (*The Nation*, July 21, 1999). The traditional sources of food such as the cattle, millet, maize, and sweet potato were soon put by the wayside. Correspondingly, such commercial crops as rice, coffee, and sugarcane took their place. All of these commercial activities required clearance of vast forested areas, to make way for the farms and to collect wood to cure the tobacco.

Obviously, specialization in the production of tobacco has had 2 implications. First, continuous removal of indigenous forest vegetation has exposed the landscape to water and wind erosion, leading to increased siltation and sedimentation of Lake Victoria. Second, contrary to the promised bonanza from the sale of tobacco, contracted farmers have grown poorer over the years. For example, over 70 percent of people in Nyanza Province currently live in extreme poverty, even though the province accounts for 80 percent of Kenya's tobacco production. Furthermore, the tobacco companies literally hold tobacco growers hostage. This is so since their farm fields are devoted to tobacco cultivation and the growers are tied to the companies by the loans they receive in the form of fertilizer, seeds, and insecticides. Furthermore, the growers must repay these loans rates fixed by the companies.

Some of the inputs growers receive from the tobacco companies pose serious health risks, since they contain such harmful chemicals as Aldrin and Diedrin, which are banned in the advanced capitalist countries. In addition, the smog from tobacco curing generates a slew of respiratory illnesses among tobacco growers. In years of bumper crops, the companies also violate their contractual obligations to the growers by not buying all of what is produced. In 2001, for example, BAT stopped buying the tobacco leaf from Kenyan farmers. The company argued that the supply of raw tobacco exceeded the company's projected capacity to buy tobacco leaf by 1.85 million metric tonnes. As a result, Kenyan tobacco growers lost hundreds of millions of shillings (*The Nation*, August 1, 2001). Ironically, as nature and the people in tobacco growing regions have grown poorer, tobacco companies and the pirate state bourgeoisie have continued to prosper. In 1996, for example, BAT earned $200 million from the sale of Kenyan tobacco while the pirate state bourgeoisie collected another $110 million in taxes from the tobacco sector (Kweyuh, 1997).

Other regions severely affected by deforestation are the country's five water towers. Kenya's forest and water resources are principally concentrated around Mount Kenya, Mount Elgon, Cherangani Hills, Abedare Range, and the Mau Complex. These water towers provide drinking water and hydropower supply to almost 80 percent of the country.

These water towers are not only the sources of virtually all of Kenya's rivers, but are also home to the country's largest forests. They provide water for irrigation, modern industry, urban centers, and hydropower energy supply; 60 percent of Kenya's installed hydropower energy is derived from rivers originating in the water towers. For example, the Abedare Range forests, covering an area of 250,000 ha, are critical to the Tana River flow, which Nairobi depends on for its drinking water (UNEP, 2009). Precisely because Kenya's forest and water resources are primarily located around the 5 water towers, they have become the magnet for dense urban development, intensive agriculture, and heavy timber and nontimber production. Combined, these objective material conditions have made the forests around the 5 water towers highly vulnerable to degradation and depletion.

The present state of the Mau Complex illustrates the deep contradiction stemming from the necessity of nature, the necessity of human survival, and the capitalist logic of destructive creation. Occupying a vast forest area of 403,000 ha in southwest Kenya, the Mau Complex is a composite of 12 forest reserves and constitutes the largest canopy forest in Eastern Africa. The Mau Complex catchment is where significant amounts of rainfall production takes place, feeding 5 Rift Valley lakes and 12 rivers. Of these river flows, Lake Victoria is among the chief beneficiaries. Such important rivers as the Nzoia, Yala, Mara, Sondu, and Nyando flow to Lake Victoria. However, in the face of state predation and encroachment by dispossessed peasants, critical ecological functions and services provided by the Mau Complex are rapidly dissipating. In the last ten years alone, the Mau Forest Complex has lost 125,224 ha of forest cover, as the government continued to degazette portions of the complex for distribution to well-connected cronies (UNEP, 2009).

A government report, released in April 2009, exposed 25 powerful individuals and companies that have received tracts of land in the Mau Forest Complex between 1996 and 2005. This report also uncovered instances of double and triple allocations of land to the same individuals. The beneficiaries include local chiefs, district and provincial commissioners, heads of land and forestry departments, directors of parastatal organizations, sitting parliamentarians, government department heads, senior cabinet ministers and permanent secretaries (BBC Monitoring Africa, April 3, 2009). The cover for such land grab has been a regulation that the Moi regime put in place to ostensibly resettle the dispossessed Ojiek people, who were driven out of their forest dwellings following inter-ethnic clashes. Even though more than 60,000 ha of the Mau Forest have been excised from the complex in the name of resettling the Ojiek people, no land actually went to them (BBC Monitoring Africa, April 3, 2009). The result has been that the 15,000 people who invaded the Mau Forest have had to rely on their own devices to make ends meet. Out of desperation, these people have reportedly felled the equivalent of 104,000 ha of trees from the Mau Forest Complex over a period of 15 years (Vidal, 2009).

Moreover, as part of the overarching 2030 vision, Kenya's pirate state bourgeoisie dream is to grow the economy by 10 percent per annum to meet the economic and social needs of the country's burgeoning population. As a component of this vision, the Kenyan government has targeted the Mau Complex as integral to its grand strategy of national development and poverty alleviation. In the government's estimation, the hydropower generation potential of the Mau Complex is 535 megawatts per annum, equal to 57 percent of Kenya's current total annual hydroelectricity production. At present, the hydropower projects either already installed in, or under construction around the Mau Complex, are expected to supply 190-mw per annum (UNEP, 2009). So, hydrodevelopment on the rivers flowing from the Mau Forests is bound to interfere with the watershed outflow to Lake Victoria.

It is helpful to bear in mind that the microclimate of the area is intimately linked to the maintenance of the Mau Complex. This is due to such ecosystem services as evapotranspiration, cloud formation, precipitation, temperature regulation and moisture retention, soil formation and nutrient cycling, water storage and groundwater recharge, water flow regulation and flood mitigation, carbon sequestration, and erosion prevention. All of these services are provided by the canopy of the Mau Forest. Given this importance, the grand vision intent on accelerating industrialization of hydropower energy production and furthering human settlement around the Mau Forests can have serious implications; in particular, these implications concern the long-term integrity of the forest ecology and also the communities in the region, the state of Lake Victoria, and the Kenyan economy, in general (UNEP, 2009).

After all, the Mau Complex is also projected to constitute one of the pillars of tourism development. In 2007, the tourism sector directly employed 400,000 Kenyans and another 600,000 people indirectly benefited from this sector. According to Vision 2030, the country has plans to raise the number of foreign visitors from 1.6 million in 2006 to 3 million by 2012, thereby generating KSh80 billion. In order to realize this vision, Kenya has plans to construct a network of airstrips near or around forested regions and hotel accommodations are projected to increase from 40,000 beds at present to 65,000 by 2012 (UNEP, 2009).

Deforestation in Kenya is not limited to the Mau Complex. The forests on the slopes of Mount Kenya are in similar predicaments. From base to summit, the mountain has 8 different vegetation bands corresponding to altitude. The lower slopes are covered with thick forest, housing rare and endemic tree species. The natural wealth of these slopes is of such extraordinary magnificence, that the 715 km^2 Mount Kenya National Park is recognized as a UNESCO world heritage site. Mount Kenya is also a catchment area for many of Kenya's water systems, including the Tana and Ewaso Nyiro rivers. Even though the Tana River begins its 800-km-long journey in the Abedare mountain range, it receives half of its water from Mount Kenya. However, a combination of legal and illegal timber and nontimber production, unauthorized cultivation, charcoal production, clear cutting of the slopes for marijuana cultivation, and increased human settlements has placed the long-term hydrobiological integrity of Mount Kenya in serious jeopardy. Additionally, climate change has an impact on the glaciers. In the last century, Mount Kenya lost 7 of its 18 glaciers, and the remaining 11 glaciers are fragmenting and thinning at an alarming rate (UNEP, 2009).

At present, the logic of accumulation and the necessity of survival appear to be heading for permanent clashes over Kenya's forests, water resources, and energy supply. The national accumulators are poised to realize Vision 2030 through the construction of a network of reservoirs for purposes of energy production, agricultural irrigation, and tourist villages. Assuming that the elite are able to supply the needed capital for the construction of a network of modern projects, the Vision is inevitably bound to require conversion of extensive portions of the forest reserves for airstrips, roads, hotels, lodges and much more. There exists the question of technical competence, as to whether the predatory state will be in a position to efficiently build and manage these modern projects in accordance with the ecological sustainability requirements of the areas involved. The present state of the Turkwel hydropower dam is notable here.

The Turkwel River rises in Mount Elgon on the border of Uganda and flows to Lake Turkana. A hydroelectricity project on the river started in the late 1980s with French financial support for 4 objectives including generation of hydroelectric energy, fishery development, promotion of agriculture through extensive irrigation, and transformation of the North Rift into a tourist haven. Twenty years later, even after the cost exceeded the original projected outlay by a factor of 5, only the hydropower-generation component of the ambitious project has made a modest contribution to national energy production. However, this has been without necessarily benefiting the communities in the region. In fact, the dam has done more harm than good to local communities which now have to deal with the added burden of coping with the dangerous proliferation of crocodile population and malaria infestation associated with the standing waters of the dam. As one grieving resident father put it,

> This dam has consumed 4 of my children after drowning. It has become a home to crocodiles and a breeding place for mosquitoes. It has converted a river that was once a source of water for our livestock into a jinx and thrown us into a dilemma on what to do with it (*The Nation*, July 13, 2007).

Moreover, ethnic clashes have now become a common occurrence in the area over water and pasture because the dam has disrupted the natural flow of the river and the associated biosocial patterns of herding livestock.

The second major threat to Kenya's water towers and forest reserves comes from the natural necessity of survival. Socially marginalized, economically dispossessed, and politically unrepresented, millions of Kenyans are bound to increasingly rely on their own devices for survival. In search of food, pasture, water, and firewood, the poor will have no choice other than to invade the reserves, which they have already started doing. The crisis of accumulation and the crisis of survival are thus colliding to produce a general ecological crisis. This situation is bound to sharpen the class struggle, compelling the predatory state to use its levers of power to tilt the balance in favor of accumulation, at the expense of the broad masses.

As of this writing, the Kenyan government is planning to erect thousands of electric fences around 26 forest reserves in the Mau Complex, Mount Kenya, Mount Elgon, and the Cherangani hills to keep dispossessed peasants and herders out. At the same time, the number of armed guards protecting the national parks is expected to double over a five-year period. The model is the 250-mile fence already erected around the Abedare

Mountain range (Vidal, 2009). However, the magnitude of the ecological crisis and the extent of indigence among the masses are so stark, that the considered militarization of the national forest reserves and the natural water systems is unlikely to avert broad social confrontation between the predatory state and the dispossessed masses. In the face of widespread dislocation and mass pauperization, the various communities adjacent to the forest reserves have only 2 options: fight the predatory state for access to the forest and water resources, or migrate to Kenya's burgeoning towns and cities to further swell the reserve army of slum dwellers.

In order to fully grasp the enormity of this challenge, the preceding discussion should be put in a broader context of Kenya's geographical profile. As indicated in chapter 2, 83 percent of Kenya is characterized as semiarid or arid. These conditions make the majority of Kenyans vulnerable to the impacts of drought, water scarcity, famine, and structural poverty. With the degradation and depletion of forest resources, and the associated microclimate change, these conditions are likely to worsen. Already, the mean annual temperature in Kenya has risen by 1.0 percent since the 1960s, and is expected to rise by another 1.0 to 2.8 degrees Celsius by the 2060s. If the current trend holds, the mean annual temperature in Kenya is bound to rise by another 4.0 percent by 2100 (UNEP, 2009). These rises in temperature will mean low rainfall, frequent drought, water shortages, and mass dislocation and migration, probably to larger water bodies such as Lake Victoria or even to the cities.

WETLANDS ECOLOGY AND SUPPLY CONDITIONS

Historically, wetlands have been viewed as impediments to capitalist agricultural, industrial, and urban expansion; removal of these impediments through conversion of wetlands to other uses has been the logical response. Because they constitute the interface between natural water systems and the terrestrial ecology, wetlands are pivotal to the long-term maintenance of the geochemistry, hydroecology and physiology of lakes. That is why removal of riparian vegetation and degradation of wetlands today pose grave threats to Lake Victoria. Indeed, John Balirwa (1998) hypothesizes that the ecological change in Lake Victoria and the consequent loss of fish diversity and abundance in the lake and its fringing wetlands may be attributable to substantial human modification and degradation of wetlands. Human population pressure-driven changes in land cover, removal of riparian vegetation for agriculture, urban sprawl, and runoff from rural and urban activities appear to have eroded the buffering functions of the fringing wetlands of Lake Victoria. As a result, the refugia, spawning, breeding, and nursery grounds of fish have been reduced, the lake has undergone rapid eutrophication followed by increased turbidity, massive algal bloom, hypolimnotic deoxygenation, suppression of underwater plants, and fish kill (Balirwa, 1998). The hypothesis appears plausible since, consonant with the laws of gravity, negative externalities resulting from actions performed on terrestrial ecosystems are naturally bound to find their pathways into aquatic ecosystems via multiple water channels of rivers, streams, and creeks.

In order to fully appreciate the pertinence of wetlands to Lake Victoria, it is useful to move the analysis of wetland ecology from general to particular levels. At the most general level, wetlands must be conceived as constituting complex and diverse components of an indivisible ecological unit -- one whose ecosystem functions and services on the macrospatial and temporal scales are inseparable from one another and other components of the global ecology in terms of the synergistic contributions they make to overall ecological maintenance, sustainability, and climatological stability. Wetlands cover 8.2 million km^2 or 6.4 percent of the earth's surface; yet they hold 20 percent of terrestrial carbon (Gronewold, 2008).

At the regional level, riverine wetlands, marshes, and floodplains in the wider catchment areas of the lakes can be regarded as the first line of defense against insertions of unwelcome loads into the lakes. By removing or retaining, absorbing or transforming sediments, silts, nutrients, and toxicants long before they enter water bodies and wetlands, riverine wetlands and riverine vegetation help to prevent or mitigate the effects of nutrient-enriching actions performed in the catchment areas of the lakes. These may include nitrogen, phosphorus, pesticides, fungicides, herbicide, and other residues.

The microlevel aspect of the analysis pertains to the wetlands, marshes, and floodplains that immediately fringe the lakes. These ecological assets are the only natural boundaries that exist between lacustrine and terrestrial ecosystems, constituting the last line of defense against aggressions of terrestrial origins. In this sense, wetlands are to lakes what kidneys are to humans. Wetlands play a vital role in the filtration process by blocking or mitigating the effects of contaminants and pollutants before they enter the lakes; in effect, they are the natural sink for excess nutrients. Although their retention efficiency rates vary from region to region, wetlands are a vital player in the regulation of mineralization and sedimentation processes as well. Studies have shown that wetlands retain between 23 and 91 percent of incoming sediments (Dodd, 2002).

The economic imperative, however, to drain wetlands for agriculture, industrial, and urban development and human settlement is short-circuiting the vital functions from wetlands. In the absence of wetlands, sedimentation, siltation, deterioration in water quality, and accumulation of pollutants combine to create permissive conditions for eutrophication of lakes. Taken together, the various components of the wetland ecology perform diverse and complex functions, and, in the process, they provide invaluable goods and services. Even though quantitative market and nonmarket valuation of wetland services is inordinately difficult, in 1997, scientists attempted to put a monetary value on global wetland ecosystem services at $14.9 trillion, accounting for 45 percent of the estimated $33 trillion for the total global ecosystem services provided annually (Costanza et al., 1997).

To learn more about the full range of benefits gained from wetlands, the reader is referred to the official website of the Ramsar Convention on wetlands at the website www.ramsar.org/info/values_intro_E.htm. The brief discussion here on the benefits of wetlands draws from this website, unless reinforced by additional citations from other sources. The discussion should help us to recognize the importance of wetlands to lakeshore communities and national societies at large in the Lake Victoria basin context.

To begin with, wetlands play pivotal roles in stabilizing shorelines, controlling flood and erosion, providing storm protection, retaining sediment and nutrients, and detoxifying harmful substances. By anchoring soils, sediments, and riparian vegetation in place, and by diverting water flow from river channels or by slowing the velocity of floodwaters, wetlands defend lakes against violent surges, wind actions, and from being overwhelmed by excessive external loads. By slowing the passage of water, wetlands intercept sediments and nutrients and force their deposition within the wetland ecology. Since the sediments and nutrients usually contain agrochemicals and industrial effluents such as nitrogen, phosphorus, pesticides, and detergents, their retention within the wetland ecology plays a double role of protecting the lakes against nutrient enrichment (a major source of eutrophication), on the one hand, and enriching the productivity of the wetlands themselves, on the other.

Moreover, the ecological processes and biochemical actions occurring in wetlands serve to transform deposition of alien substances into useful elements that are either neutralized or removed by wetland vegetation uptake. With respect to the regulation of nitrogen and phosphorus, wetlands play a particularly important role, since the quantity of these limiting elements determine the levels of primary production and productivity of living aquatic resources. Serving as sinks, wetlands also accumulate nitrogen in plant biomass in the form of organic matter. As we shall see later, the thick papyrus fringing Lake Victoria is an excellent illustration of this point. The loss of such ecosystem functions means an acceleration of entropic degradation of lakes and their influent natural water channels. The socioeconomic consequences can also be catastrophic. The 1998 great flood in China, for example, severely affected 230 million people, displaced more than 20 million. It killed more than 3,500 persons, destroyed 7 million homes, and swept away more than 15 million farms, raising the economic cost of the flood to a staggering $32 billion.

Even though total prevention of the great Chinese flood might not have been possible, the severity and effects of the flood could have been substantially mitigated if the natural conditions of the catchment and wetland areas were undisturbed. A number of causes are attributed to the ferocity of the flood, including concentration of human settlements in low-lying areas to ease the demographic pressure, drainage of wetlands to grow food, and deforestation in the upper reaches of the natural water systems causing excessive siltation of water channels, lakes, and wetlands. These adverse conditions cumulatively reduced the water absorption capacity of the entire watershed ecology. It is worth noting that 0.4 ha of wetland can hold as much as 6,000 cubic meters of flood water.

Carbon sequestration is another pivotal function performed by wetlands. The various types of wetlands in the world hold between 202 and 535 billion tons of carbon, and peatlands store between another 400 and 500 billion tons of carbon. This fact alone makes wetlands and peatlands key players in the regulation of climate at all levels global, regional, and local (Gronewold, 2008; Keddy et al., 2009). The presence of well-watered wetlands has always been a temptation to conventional economists in the context of growing demand for food, industrial growth, and urban sprawl. Using their standard model, they are quick to make a trade-off between preserving the wetlands to insure ecological sustainability and using them to ostensibly promote overall economic growth. The trade-off has consistently been in favor of the latter.

Conventional economists, however, conveniently gloss over the fact that conversion of wetlands to agricultural expansion, industrial growth, and urban sprawl entails two long-range and unwelcomed consequences. First, wetland destruction inevitably leads to the release of inordinate amounts of greenhouse gases into the atmosphere and, as a result, climate change accelerates. In addition, the total terrestrial ecology's capacity to capture and store greenhouse gases will be severely curtailed, leading to the same effect of climate change. Second, degradation and destruction of wetlands deprive lakes of their last line of defense against insertions of unwelcomed loads, potentially leading to their eutrophication.

Wetlands are also vast reservoirs of biological diversity. Even though freshwater wetlands cover a mere 1 percent of the earth's surface, they are home to 40 percent of all plant species and 12 percent of all animal species on earth. There are at least 100,000 animal species making their residence in inland wetlands alone (Keddy et al., 2009). Africa's wetlands, covering 341,000 km^2, are considered rich sources of fisheries. Indeed, Africa's rivers and their associated wetlands and floodplains are said to account for half of the continent's inland-capture fisheries (Welcomme, 2003; Welcomme et al., 2010).

As we shall observe later, the expansive wetlands in Africa's Rift Valley are fabulously endowed with vast numbers of aquatic plant, fish, herbivorous animal species, and avifauna of both resident and migratory water birds. Millions of humans depend on these wetland resources for fish, grain, vegetables, fruits, livestock grazing, fuelwood, fiber, timber, traditional medicine, building materials, and much more. For example, the inner Niger delta wetlands support more than 1 million goats and sheep. Likewise, the 150,000 km^2 wetlands in the southern Sudan (the largest wetland system in Africa) support an even larger population of livestock and other wildlife (Gottingen, 2006). The economic value of wetlands in the Zambezi River basin in terms of fisheries, food crops, and maintenance of grassland for livestock production is estimated at more than $200 million per annum (UNEP, 2002). Moreover, wetlands provide habitats for countless pollinators and refugia for a variety of fish species from predators. The transition areas between the open water of a lake and adjacent wetlands are crucial sources of food, shelter, and protection for species endangered by predation. A study by Chapman and others (1996), for example, reveals that the wetlands, marshes, and riparian vegetation around Lake Victoria have provided refugia for a number of fish species following the introduction of the Nile perch into Lakes Victoria, Nabugabo, and Kyoga. According to their findings, the fish species that could tolerate low oxygen levels have migrated to adjacent wetlands, marshes, and papyrus vegetation in order to avoid the perch. The species, long gone from open waters, are now found in wetland lagoons, in streams separated from the lake by dense papyrus swamps, as well as in the transition zone between the open waters of the lake and vegetated areas. The structural and hypoxic conditions prevailing in the wetlands have effectively blocked the Nile perch from expanding their spatial hold, thereby guaranteeing the biological security of the endangered species.

Wetlands also perform indispensable functions in the storage, retention, and purification of freshwater, as well as recharging or discharging groundwater resources for domestic, agricultural, and industrial uses. Water resources stored in wetlands are particularly indispensable during dry seasons or drought since people resort to them for drinking water, watering their livestock, and irrigating their farms. By recharging or discharging aquifers, wetlands raise the water table and thus make water resources accessible to people and animals. After all, more than 97 percent of global unfrozen freshwater is stored in aquifers, which is drawn out for consumption or moves laterally to other areas as discharged to water other wetlands or to be used by people. Indeed, 17 percent of the world's agriculture, which produces 40 percent of global food, is irrigated by water from aquifers. The water that is stored in aquifers is clean because wetland vegetation and soil filter out agrochemicals from surface runoff, and remove, retain, or recover organic matter, toxic chemicals, and excess nutrients leached out from catchment areas before entering natural water systems. It is believed that Florida's cypress swamps remove 98 percent of the nitrogen and 97 percent of the phosphorus inflow before water carries these elements underground (Hails, 2000). The removal of agrochemicals, such as nitrogen and phosphorus, by wetlands before entering the natural water systems is particularly important for the ecosystem health of lakes, without which, the deposition of these elements in a lake can accelerate de-oxygenation and eutrophication, fostering rapid algal growth.

Moreover, the quality of wetlands as repository of information is another crucial nonmarket value. People depend on the scenic beauty of wetlands for recreation, spiritual companionship, bird watching, and boating.

The above discussion of the benefits that could be had from wetlands invites 2 central questions. Given the objective material conditions prevailing in the countries covered in this study, what are the chances that the wetlands in the countries concerned will be conserved and used wisely? And, if the wetlands are not conserved or wisely used, then what are the potential economic, social, hydrological, and ecological consequences of destroying them?

If the past is a useful guide to making predictions, the answers to these questions would not inspire optimism. The problem lies in the fact that the universal model being adopted in the region is the one imported from the advanced countries, which have almost wiped out their natural wetlands to make way for agricultural expansion, industrial growth, and urban sprawl.

As stated before, there are 2 logics of necessity competing for the same natural water systems and riparian resources. On the one hand, there is the logic of sheer survival on the part of the impoverished masses; for the poor, tomorrow is a theoretical time, which the masses may experience if they survive until then. Given this harsh reality, preservation and/or conservation of wetlands resources is an abstract notion to the poor. Many writers erroneously attribute the unsustainable utilization of wetland resources to ignorance about the potential future benefits of those natural resources on the part of the masses. In reality, the masses fully know the value of their natural ecology. They have grown in it; they live in it; they still depend on it; and they have used its natural resources efficiently and sustainably since the beginning of time -- until capitalist penetration and commercial enclosure began to despoil nature. A mid-1990s study, for example, revealed that 91 percent of fishers on the Kenyan side of Lake Victoria acknowledged that fish catch per unit effort had drastically declined over the years, and 62 percent of them attributed the decline to reductions in fish stocks and fish diversity, and catch of immature species. When asked why they continued to stay in the sector in light of fish depletion, 70 percent of them cited lack of employment opportunities in other sectors of the economy, and another 18 percent cited their proximity to the lake or family traditions as reasons (Ikiara and Odink, 2000).

On the other hand, there is the logic of capital accumulation at work, which requires the further alienation, commodification, and incorporation of nature for the continued expansion of capital. Since 1900, the world has lost half of its wetlands, primarily associated with capitalist expansion. Europe has lost more than 90 percent of its original wetlands; the United States has converted 50 million ha of its 90 million ha of wetlands to other uses. Ten of the 50 states have lost 70 percent of their wetlands, while California, Iowa, Ohio, and Florida have lost 90 percent or more of their wetlands. Louisiana, which contains 40 percent of the wetlands in the 48 contiguous states, today loses between 100 and 150 km^2 of its wetlands each year. Canada does not fare better than the United States in protecting its wetland resources since the various components of its wetland ecology have suffered losses ranging from 65 to 80 percent (Cox, 1993; Moser et al., 1996; Postel and Carpenter, 1997; Gronewold, 2008).

Asians have similarly used up much of their wetlands, mainly for rice cultivation. Vietnam has completely wiped out the wetlands of the Red River delta, which once had covered 1.75 million ha. The 1 million ha of floodplain wetlands in Bangladesh's Sylhet basin are gone. The destruction of the 6 million ha of floodplain wetlands in central Myanmar is total. The 1.9 million ha of rice cultivation in central Thailand is the result of wetland conversion. Most of the 40 million ha of rice fields in central India are also the result of wetland conversion (Moser, 1996).

In comparison to other regions, Africa's wetlands remain relatively intact precisely because of the limited scope of capitalist penetration on the continent. Sub-Saharan Africa has over 200 million ha of wetlands with only 3 percent of these currently under rice cultivation. Historically, people in the African continent generally used wetlands marginally to supplement their primary source of livelihood, be it livestock production, fishery, farming or harvesting natural foods and vegetation. Lately, however, both private developers (backed by government bureaucrats) and impoverished citizens are making aggressive forays into the domain of wetlands. The reasons are apparent most terrestrial ecosystems have undergone significant entropic degradation and are in the process of collapsing throughout the continent. After all, Nigeria alone imports 1.2 million metric tonnes of rice annually. Africa, as a whole, imports 40 percent of its annual rice requirements. This amount is equal to one-third of the globally traded rice, a factor that makes the continent vulnerable to global supply disruptions and price shocks

(Appropriate Technology, Sept. 2008). This objective material condition alone provides a powerful incentive to turn attention to wetlands with an eye to convert them to rice fields. Indeed, experimental projects are proliferating throughout Africa to popularize what is termed "NERICA (New Rice for Africa)," which is a cross between high yield Asian and low-yield African varieties.

One rice experimental project currently under way in West Africa is being held as a potential model for the rest of the continent. Imported from Asia, the new rice cultivation method is known as "sawah." The method involves transforming the wetland landscape by leveling the ground using heavy machinery so that fertilizers, pesticides and water can be evenly applied to the rice fields. This amounts to industrialization of rice production with heavy labor, capital input and planting high-yield rice. According to the projects, rice farmers using the sawah method can grow from 3 to 5 metric tonnes per ha compared to 1 metric tonne per ha under the traditional method. Under this optimistic scenario, the sawah system can save the rice importing nations of West Africa $2 billion a year as ten million farmers would be deployed to the sawah rice production. By converting its wetlands to sawah rice fields, Nigeria alone is expected to be in a position to produce 10 million metric tonnes of rice annually when the experiment completes its development cycle by 2020 (Appropriate Technology, Sept. 2008). Indeed, the integration of wetlands and flood plains into the overarching strategy of agrarian growth (under the slogan of poverty alleviation and food security) has received official government and donor sanction.

A general review of some cases that have become focal points of attention should supply ample illustrations, as we begin assessing the state of wetlands in the countries surrounding Lake Victoria.

Wetlands in Kenya cover between 2 and 3 percent of the country's land surface. Of these total wetlands, 37 percent fringes Lake Victoria (Awange and Ong'ang'a, 2006; Hughes and Hughes, 1992). Although this percentage may appear small when compared to the wetland coverage of its immediate neighbors, Kenya's wetlands play a critical role in terms of hydrology, agroecology, biodiversity, and socioeconomics. After all, 83 percent of Kenya's land surface is characterized as arid and semiarid; only 17 percent is classified as moderate to high fertility and considered suitable for farming (UNEP, 2009). Moreover, Kenya is prone to drought and water stress. Indeed, only the Lake Victoria and Tana River basins enjoy water surplus. The remaining 3 basins (Rift Valley, Athi, and Ewaso Ngiro) basins are water-deficit regions, which rely on transbasin water transfers (FAO, 1998).

In this context, Kenya's wetlands have become a focus of contention between indigenous users and external intruders. On the one hand, wetland-dependent communities view the wetlands as biosocial assets, which are essential for survival in terms of providing livelihood, income, employment, traditional medicine, water, fuelwood, building material, and grazing pasture for their livestock. On the other hand, land grabbers, with full government sanction, are increasingly encroaching on the wetlands. The government believes that wetlands are the answer to the country's food insecurity and poverty. The government puts the estimates of the national irrigation potential at 540,000 ha; of these, 300,000 ha are wetlands. Of the total irrigation potential, 118,000 ha are in the Lake Victoria basin and another 52,500 ha are in the Kerio Valley basin, which can supposedly be irrigated by transferring water from Lake Victoria. In 1998, Kenya had 52,000 ha under irrigation, which rose to 110,000 ha by 2003. Thus, by draining 300,000 ha of wetlands and by converting another 340,000 ha of forests to agriculture, the country could possibly have 592,000 ha under irrigation (FAO, 1998; Aquastat, Kenya, May 2006).

It is against this drive for agrarian accumulation that the prospects of the wetlands bordering Lake Victoria loom large. The Lake Victoria basin in Kenya covers over 8.4 million ha. Since 12 rivers drain the lake's basin, the wetlands bordering the lake are essential gatekeepers. Lake Victoria's wetlands on the Kenyan side of the lake occupy 78,000 ha, of which 21,000 ha are taken up by papyrus vegetation, phragmites, and other plant species (Awange and Ong'ang'a, 2006; Gottingen, 2006). The Yala and Nyando wetlands are 2 of the largest of their kinds on the Kenyan side of the lake.

The Yala Swamp Complex, fed by waters from the Yala and Nzoia Rivers and from the backflow of Lake Victoria itself, occupies the northeastern Kenyan portion of the lake straddling the Siaya District and a small portion of the Bondo District. The Yala wetland complex is 25 km from east to west and 15 km from north to south. The size of the wetland is variously defined as covering between 17,500 ha and 21,765 ha. Others put the size even higher at between 38,000 ha and 52,000 ha (Hughes and Hughes, 1992; Kenya Wetland Forum, 2006, April 27, 2008).

This wetland complex hosts 3 highly productive small freshwater lakes: Kanyaboli, Sare, and Nmboyo. Together with these small lakes and the thick riparian vegetation, the Yala Swamp represents one of the most delicate hydroecological tapestries in East Africa. Not only does it host the largest papyrus vegetation, by far on the Kenyan side of Lake Victoria, but it also represents the highest papyrus endemism in East Africa. It is a small wonder that the Yala Swamp is described as a "living museum" because many of the species that had long disappeared from Lake Victoria have found refuge in the Yala Swamp and its lakes. The endangered Sitatunga antelope, for example, is sheltered in the thick papyrus forest of the swamp. So, too, are waterbuck, bushbuck, reedbuck, and many water birds. It also hosts countless number of resident and migratory water birds. In fact, Birdlife

International (n.d.) has categorized the Yala Swamp as one of Kenya's great ornithological sites of national and international importance (Kenya Wetland Forum, 2006; Global Response, 2007).

In addition to acting as natural filters by removing silt, biocides, and other pollutants from waters of the surrounding catchment area before being inserted into Lake Victoria, the wetland papyrus, phragmites, and other plant species provide a natural wealth to the indigenous Luo, Luhya, and other ethnic groups. These resources provide materials for building homes, thatching roofs, and making mats, fish traps, and baskets. They also serve as fodder for cows, goats and sheep, fisheries, and farm plots. Children and widows traditionally used to make a living by harvesting papyrus to make baskets and mats for the local market.

To sustain themselves, the communities around the Yala Swamp combined such horticultural production as cassava, millet, sorghum, beans, peas, tomatoes, potatoes, maize, and onions with hunting wildlife, catching birds, and fishing activities. As Schuyt (2005) compellingly notes, the dependency ratio of the communities on the Yala Swamp is incredible: 100 percent of the local people use it for irrigation, washing, cooking, and drinking; 74 percent use the swamp to graze their livestock; 81 percent harvest wood for subsistence and another 28 percent gather wood for sale; 66 percent use the swamp for subsistence fishing and another 34 percent fish commercially; 46 percent catch birds and hunt wildlife for subsistence; and 66 percent of the communities use the waters of the Yala Swamp for transport. For centuries, the indigenous communities have used the wetland sustainably for fishery, farming, medicine, energy, transport, and building materials. Seventy percent of what has been produced on the swamp has been for household consumption, while the rest has been traded to earn some income (Gottingen, 2006).

Since the late 1960s, however, the Yala Swamp Complex has been in the process of rapid transformation. Several drivers have been at work behind this transformation. First, local population growth in need of more wetlands has been one of the driving forces. The size of the wetland taken up by agriculture progressively increased from 1,564 ha in 1973 to 5,939 ha in 2001 commensurate with the rise in population (Gottingen, 2006).

Also, general degradation of the catchment area served to generate several push factors. Rising population density in the upper reaches of the Yala Swamp altered or modified the regenerative capacity of the catchment area. As a result, the Yala and Nzoia Rivers increased the amounts of sediment and silt, which they brought from the degraded landscape for insertion into the swamp and finally, into Lake Victoria itself. More importantly, however, in response to decreasing farm sizes and decline in soil fertility in the catchment area, more and more people have drifted to the

Yala Swamp in search of livelihood. This influx meant conversion of more wetland to farming, human settlements, and livestock production.

Finally, government policy to drain significant portions of the Yala Swamp for modern agrarian accumulation has perhaps been the most harmful factor behind the transformation.

Beginning in 1965, the Kenyan government formally launched the Lake Basin Development Authority (LBDA) with a master plan to put a significant portion of the Yala Swamp under irrigation. In the first phase of the plan, the LBDA drained 2,300 ha for agriculture. An additional 10,000 ha have been slated for drainage under the second and third phases of the plan (Kenya Wetland Forum, 2006; *The East African Standard*, April 27, 2008). However, the LBDA was unable to transform the 2,300 ha into modern agricultural enterprise, let alone implement the second and third phases of the master plan. The LBDA project, from the start, has been bedeviled by corruption, mismanagement, and technical incompetence, which rendered the 2,300 ha already under irrigation an agro-ecological fiasco. This failure compelled the Kenyan government to place the Yala Swamp on the international auction block. The highest bidder was Dominion Group Ltd., a U.S. multinational corporation based in Oklahoma.

Dominion's president, Calvin Burgess, is said to have taken just one trip to Kenya to discover that the thick riparian vegetation, streams, and lakes of the Yala Swamp were long-ignored hidden gold. A technocentric optimist by temperament, Burgess convinced himself that there was a lot of money to be made by transforming the swamp into a giant agrocommercial enterprise. But he had to put the social infrastructure in place first, before embarking on the neocolonial penetration of the swamp. He knew that bribing state and local authorities alone was insufficient to achieve his vision, since the 600,000 people who inhabit the area would not voluntarily acquiesce in their own dispossession, so he had to combine seemingly humanitarian work with cold business calculations. To do so, he gleaned a few important lessons from the colonial model. As the late Kenneth Kaunda of Zambia used to say, European colonists did not exclusively depend on the gun to colonize the continent; they always carried the Bible in their right hand also.

That is what Burgess did. He got involved first in philanthropic work with some church groups in Kisumu town bordering Lake Victoria. Then, he befriended a certain local preacher in Siaya District by the name of Were, who played an important role in cajoling local communities. Among the many promises dangled to the communities was a reduction of poverty and unemployment by 50 percent. Those who were not cajoled with promises of schools, clinics, jobs, and much more were threatened with the Bible. Were and Burgess warned the communities that opposition to the project was tantamount to opposing God's will; even Jesus' own crucifixion was said to have been used to threaten community members with excommunication and damnation (Boggs, 2007; Lackmeyer, 2007). After all, Burgess quickly learned from his recent experience in Nyando District next to Siaya District. When he attempted to take over the Ahero rice irrigation scheme in Nyando with government support, local farmers who worked the scheme chased him out (*The East African Standard*, April 27, 2008).

At any rate, after 2 years of haggling, Burgess put together a high-caliber team that included Professor Keith Taylor of the University of Arkansas (an expert on rice cultivation), Professor Tim Streeter of the University of Missouri (specialist in earth moving), and the Grahm Vetch Group of the United Kingdom was hired to bring the farm operations to fruition. Burgess also obtained the services of Bob Green, his long-time confidant and contractor, as the head of Dominion Farms in Kenya (Lackmeyer, 2007).

Dominion started its operation in 2003 to introduce a high-tech rice cultivation and fishing operation, on the promise that its modern procedures would create jobs, alleviate hunger and poverty, generate hard currency for the country, and make Siaya the "breadbasket" of Kenya. By the end of 2006, the corporation hired 350 permanent and 1,000 contingent workers (Boggs, 2007; Lackmeyer, 2007). According to the original stipulation contained in the 25-year lease, Dominion would grow rice for domestic consumption, thereby saving the country significant amount of hard currency that had been used to import rice. Since Kenya annually produces only between 35,000 and 50,000 metric tonnes of the 250,000 metric tonnes of rice the nation consumes annually, the production of 100,000 metric tonnes of rice annually on the Yala Swamp would certainly make a huge dent into the country's rice import bill (*The East African Standard,* April 27, 2008).

In short order after penetrating the Yala Swamp, however, Dominion proceeded to exceed its original intent to grow not only rice, but also maize, cotton, sunflowers, groundnuts, and atanasia. Thus, by 2006, Dominion already drained an additional 6,900 ha to expand its holdings, not only for agriculture but for a variety of additional projects. In its long range vision, Dominion seeks 65 percent of the Yala Swamp for what it calls "integrated development" to supposedly produce 100,000 metric tonnes of rice a year and a constant supply of fish for both national and international consumption, making the Yala project a model agricultural plan for other regions to emulate (Kenya Wetland Forum, 2006; Global Response, 2007).

Calvin Burgess' extravagant promises to make the area the "breadbasket" of Kenya notwithstanding, the corporate agrocolonization began to negatively impact the lives of the communities. Not only did the corporation begin commercially enclosing communal land, but also private property without proper legal authority and adequate compensation. In fact, the company exposed its neocolonial stinginess by offering each household affected by its plunder a mere one-time compensation of $64. Those who refused to take the offer and move away found their homes and property drowned. For example, when the company submerged whatever she owned in water after she refused to leave, a certain grandmother (Salome) lamented, "I grew cabbages, I made mats, I planted maize and millet. Now all my fields are flooded" (Boggs, 2007). Today, people like Salome either depend on handouts from nongovernmental organizations or have migrated to Nairobi to join the reserve army of slum dwellers.

To make matters worse, the construction of elaborate networks of roads, a hydroelectricity dam, a weir, airstrip, diversion canals, dikes, flood controls, a fish farm, a fish-processing plant, and a fishmeal factory began to undermine the biosocial environment of the indigenous communities. For example, the Yala River used to flow through the eastern portion of the swamp into Lake Kanyaboli before the river waters spilled over the main body of the swamp and into Lake Victoria. However, because of canalization, the eastern part of the swamp is reclaimed and the Yala River flows directly into the main swamp, depriving Lake Kanyaboli of important freshwater. The long-term effect of this deprivation on the physiology, geomorphology, trophic profile, hydrochemistry, and functioning of the lakes and swamps is predictable. In combination, these operations have altered the flow from the Yala and Nzoi Rivers. Contamination of soil from oil spills, loss of grazing pasture, and aerial spraying are already having adverse impact on the hydroecology, biology, socioeconomics of the Yala Swamp, and the health of the communities.

As private property holdings, homes, and farms are flooded, dwellers have to relocate against their will; fishers are cut off from communal fishing grounds, and agrochemicals are contaminating the vital water upon which communities and their livestock populations depend for drinking. With the rapid privatization and commodification of the swamp, the vast network of projects is blocking people from having access to the lake, schools, health clinics, and markets. With the standing water rising, water-related diseases, such as bilharzia, malaria, and typhoid, are increasing. Even though villagers reported losing livestock to aerial spraying in 2006, Dominion requested permission to spray its fields with DDT, a chemical that is banned internationally (Boggs, 2007). Complaints fall on deaf ears and protests are met with tear gas, harassments, and arrests.

> A certain John Ayila, a delegate who represented the Yala Swamp communities at the World Social Forum in January 2007, put the case against the corporation in these terms:"We worked hard growing maize and bananas, raising cows and bees. I inherited my farms from my father and grandfathers. I have lived here all my life. One day I

> saw surveyors taking measurements on my land, and then came the bulldozers. They slashed all my crops and they fenced me out of the lake where we used to fish. They built a dam and now the reservoir has submerged all my land, my beasts, my houses. When I protested to my MP [Member of Parliament], he said I should take the money and move away. I will not take their money. I want my home back. (Global Response, 2007)

In early April 2008, the class struggle gathered storm as even those who agreed to give their holdings in exchange for a better future rose up. Villagers from the Siaya and Bondo Districts stormed the premises of the corporation and defiantly released 150 head of cattle on the company's fields. One of the placards they carried read: "Burgess must go. He promised us honey but we are starving because he has taken all of our land. He has made us squatters in our ancestral homes" (*The East African Standard*, April 27, 2008).

The communities accused Dominion of not only failing to keep its promises, but also of destroying their entire livelihoods. They blamed Dominion for polluting their water sources; firing members of local communities and replacing them with outsiders; for using biocides that damaged their tomatoes, kale, and onions and harmed their animals; and, for blocking them from accessing the lakes and local markets. Furthermore, the villagers accused state and local authorities of colluding with Dominion at their expense.

Burgess placed the blame of the gathering social movement against his operations squarely on local politicians. In his view, the protesters have been primed, agitated, and led by politicians who wanted more kickbacks from Dominion. He alleged that his company paid out KSh6 million to local authorities in 2006 without any trace of where the money went. Burgess challenged his hosts:

"I cannot operate in a situation where a clique is perennially intimidating my staff. I will never be intimidated through corruption and blackmail, which are serious offences" (*The East African Standard,* April 27, 2008). Then he threatened to pull out from Kenya and relocate his operations in neighboring countries. He began to play hardball by suggesting that Kenya was going to be the loser big time if he closed his fish farm operation that was expected to export more than 20 million kg of fish to the European market, and if he suspended the KSh1.5 billion International Youth Center (supposedly a philanthropic project) that gives vocational training to 20,000 young adults.

Burgess was probably correct in some of his accusations. Through its own independent investigation, *The East African Standard* discovered that some politicians regularly refueled their cars at Dominion's fuel stations free of charge and traded their support for the corporation for employment of their family members and friends by the company. It is possible that these politicians received numerous of kickbacks (*The East African Standard*, April 27, 2008). However, the main point is how Calvin Burgess got the Yala Swamp as a trophy for the kickbacks with which he lavishly cajoled his host in the first place. In any case, Burgess was not alone in his pretenses of anger and defiance; his national, provincial, and local clients were angered, too, by the threat from the grassroots social movement. In open support of Burgess, more than 20 church leaders in Kisumu town ingratiated themselves by marching in the town. Not only did they curse those who tried to intimidate him, but they also lavishly praised his good work (*The East African Standard*, April 27, 2008).

The Yala Swamp is not the only site under threat. Virtually all sites on the Kenyan side of Lake Victoria are facing a bleak future. For example, Dunga, Koguta, and Kusa papyrus swamps in the vicinity of Kisumu have lost most of their papyrus since 1969 due to cultivation, grazing, burning, human settlement, and papyrus harvesting for making mats, baskets, and fuel. According to a recent study, if the present trends in human settlement and land reclamation persist, Dunga and Koguta will lose all of their remaining papyrus by 2020. In the case of Kusa, the area under papyrus cover will be 19 percent of its original size (Owino and Ryan, 2006). In addition to the fact that the structural characteristics, density, and physiological functions of the papyrus on these swamps are crucial to buffering Lake Victoria against insertions of urban and terrestrial runoff, they are pivotal in providing important refugia, breeding grounds, and food to a host of fish, reptiles, and resident and migratory birds.

Probably no other East African country rivals Uganda in terms of the abundance and spatial distribution of its wetlands. The country has more than 20 million ha of wetlands, accounting for 15 percent of the country's total land surface. Over 2.6 million ha of Uganda's total wetland area of 20 million ha fringes Lake Victoria. According to a recent study by the World Resources Institute, the International Livestock Research Institute, and the Ugandan Ministry of Water and Irrigation, more than 5 million rural Ugandans get their daily freshwater from wetlands, which translates to $25 million a year. In addition, 2.7 million Ugandans are directly or indirectly dependent on wetlands for full-time or subsistence employment. Moreover, Ugandans use the wetlands for purposes of water collection, livestock production, and tree harvesting simultaneously. The study estimates that the annual value of wetland goods and services to the Ugandan economy ranges between 9 and 18 trillion Ugandan shillings. This translates into between $300 and $600 per ha of wetland per year. If the nonmarketed value of such ecosystem services as carbon sequestration, water purification, flood protection, soil formation, and sediment retention provided by Uganda's wetlands is included in the computation, the total value from 1 ha of wetlands reaches $10,000 per year (World Resources Institute, 2009).

The aforementioned study draws a parallel between the proper utilization of wetlands and the prospect of poverty reduction, which the government can make part of its overall development strategy. According to the computation of the authors of the study, Ugandans can arguably sustainably harvest 400 bundles of papyrus per year from each ha of papyrus wetland, which translates to $170 per ha per year or $51.8 million in yearly aggregate (World Resources Institute, 2009). In this view, the papyrus wetlands along the Victoria/Kyoga/Albert axis can be harnessed sustainably for purposes of poverty alleviation. According to this study, 514 subcounties of Uganda's 958 subcounties have papyrus wetlands that can be put to good use. Of the 514 subcounties with papyrus wetlands, 210 subcounties have rich papyrus resources to harvest, and they can sell raw papyrus and generate revenue sufficient to close the poverty gap in their administrative units. While the remaining 304 subcounties with papyrus wetlands may not be able to generate revenue by simply harvesting and selling raw papyrus, they can, nonetheless, generate revenue by adding value to raw papyrus, which means producing mats, fish traps, baskets, and hats instead of selling raw papyrus (World Resources Institute, 2009).

While such computations look promising on paper, the suggestion that Uganda's papyrus wetlands can be harvested sustainably under the prevailing social conditions actually portends danger for the papyrus wetlands and, by extension, for the country's water bodies. What the authors of the document inadvertently recommend is the wholesale commercialization of Uganda's 3.6 million ha of papyrus swamps fringing Lakes Victoria, Kyoga, George, and other minor water bodies. It is useful to bear in mind that Cyperus papyrus, widespread and very developed in East Africa, has historically been integral to the ecosystem health of the region's freshwater bodies. This perennial herbaceous plant occurs as free-floating mats rooted in loose, detritus, emergent, and stranded complex vegetation fringing water bodies (Toowoomba, 2000). Because papyrus swamps have the reputation for high tolerance for pollution, they are effective purifiers of urban and agricultural runoff, regulating, removing, and recycling nutrient flows. They provide fixed nitrogen to animals and plants fringing bodies of freshwater. Furthermore, the papyrus swamps around Lake Victoria provide habitat, refugia, spawning, nesting, and breeding grounds as well as food to a host of fish and other animal species. Hypoxia-tolerant fish species in particular make effective use of the papyrus swamps of Lake Victoria. A study by Chapman and others (1995), for example, discovered 12 hypoxia-tolerant fish species making good use of these swamps. Reliant on the refugium provided by the papyrus swamps, these species have adapted themselves to habitats characterized by low dissolved oxygen in order to avoid predation by the Nile perch.

However, in the face of high population growth, shrinking arable land area, and terrestrial degradation, more and more Ugandans are increasingly encroaching on the nation's wetlands. In particular, the wetlands fringing Lake Victoria on the Ugandan side are under serious threat not only from the drive for land reclamation but also from raw sewage, solid wastes, and pollution originating in the city of Kampala itself. The city used to contain 12 important wetlands, which have all but disappeared as a result of the physical expansion of the city. Kampala has spatially grown from just 170 acres in 1902 to a metropolis of 25 hills, encompassing 386 km^2. These hills are characterized by steep slopes, separated by vast tapestries of wetlands, swamps, and flood plains with enormous water purification and nutrient-retention capacity. Correspondingly to the spatial expansion of the city, Kampala's population has grown from 330,000 in 1969 to 1.8 million in 2008. Currently hosting 40 percent of Uganda's urban population and still growing at over 5.6 percent per annum, Kampala's population is projected to reach 2.4 million by 2015 (Nyakaana et al., 2007; Giddings, 2009). Consequently, hunger for urban land has become extremely acute in all zones of the city. Today, Kampala has 274,000 housing units with average occupancy rates of 6 persons per unit. The gross housing deficit stands at over 300,000 units and is projected to reach 1 million units by 2025 at the present rate, presaging a total "slumization" of the city. After all, 1.5 million of Kampala's 1.8 million residents are already classified as slum dwellers (Giddings, 2009; Nyakaana et al., 2007).

In the face of rising demand from the burgeoning urban population for housing, and from commercial establishments for offices, land developers and individual encroachers have been set loose to reclaim and drain the city's wetlands, formally and informally, legally and illegally. It is a small wonder that the size of the city's wetlands has shrunk from 13 percent of the city's land area some 3 decades ago to just 3.3 percent today, which continues to degrade and shrink further (Nyakaana et al., 2007). The danger of Kampala's unplanned growth, spatial expansion and slumization inheres in the city's immediate proximity to Lake Victoria. Since the lake is the natural drainage for Kampala and its entire environs, the insertion of Kampala's raw sewage, solid waste and water pollution into the lake is unavoidable.

The state of Nakivubo wetlands illustrates the danger that wetland degradation poses to Lake Victoria. This historic wetland, with 40 km^2 in catchment area and covering an area of 5.29 km^2, is a tropical wetland on the northern shores of Lake Victoria, separating Kampala from Merchison Bay. It is fed by the Nakivubo River and 4 tributary streams. The wetland vegetation is dominated by papyrus, Typha, phragmites, and Viscanthidium grass (IUCN, May 2003a). The spatial configuration of Nakivubo, running from the central district of Kampala all the way to Lake Victoria, coupled with the presence of a railway line, has created conditions for invasion for human settlements as well as industrial and commercial establishments. This geographic condition allowed human settlements and business establishments to encroach upon the wetlands from all directions. As of 2002, there were over 100,000 slum dwellers on the fringes of Nakivubo wetlands drawing such important natural resources as woodfuel, papyrus, fish, water, and wild food, or growing sweet potato, cocoyams, cassava, cocoa, and sugarcane on the wetlands (Schuyt, 2005; IUCN, 2003a). On the northern side of Nakivubo wetlands, there are over 200 industries of all sizes, including breweries, battery and paint factories, abattoirs, oil and soap factories, soft drink facilities, fish- processing factories, tanneries, plastic and foam factories, metalworks, pharmaceutical factories, sawmills, paper-making facilities, and distilleries. The wetlands continually receive untreated industrial effluents and wastewater from the industries and residents in the area before discharging the load into Lake Victoria. More than a third of the industrial establishments are known to have no waste-treatment facilities at all; they directly dump their effluents and solid wastes into the wetlands. These waste products are known for containing such elements as nitrates, phosphates, detergents, acids, lubricants, and oil, as well as a variety of heavy metals (Schuyt, 2005).

The ecological importance of the Nakivubo wetlands inheres in the fact that the wetlands constitute part of the ecological complex that had historically guarded Lake Victoria against insertions of terrestrial runoff via the mouth of Merchison Bay. The wetlands used to act as a buffer or sponge to absorb both falling rain and effluents from Kampala itself and, then, filter the water before sending it to Lake Victoria. Today, however, the functional capacity of the wetlands is severely eroded since the wetlands have given huge ground to encroachers, real estate developers, and business establishments. As the size of the wetlands has dangerously shrunk, so has its capacity to arrest and process waste-water and pollution from households, industries, and municipal sources. When the half million of Kampala's slum dwellers release their latrines and dump their solid waste, feces, manure, and other organic matter find a pathway into the wetlands, overwhelming the environment's capacity to process, assimilate, store, and transform the waste and pollution. Complicating this matter is the fact that engineers have built a channel in the wetlands to receive inflows from the city and its environs and to carry the water to Lake Victoria in order to presumably reduce potentially harmful flooding. However, the channel has now become an open sewer that carries industrial and human wastes.

The substantial reduction in the size of the wetlands and the excessive external load have eroded Nakivubo's ability to perform its biological, physical, and chemical functions to process the pollutants, contaminants, and metallic elements through reticulation before they are inserted into Lake Victoria.

The destruction of the papyrus, which has historically played critical role in water purification and nutrient retention, is of particular concern. Because papyrus mat is thin and open, permitting vertical penetration of wastewater, the permissive interaction between the biochemical elements in the papyrus mat and the wastewater in the water column make the processes of decomposition, mineralization, and sedimentation highly efficient in purifying water, removing or reducing the potency of harmful substances, while retaining nutrients. Kansiime and van Bruggen (2001) have found that Nakivubo's papyrus wetlands have been responsible for effectively neutralizing fecal coliforms on the wetland. This ecosystem service has been invaluable since 40 percent of Kampala's raw sewage enters Nakivubo's wetlands on its way to Lake Victoria (IUCN, 2003a). The problem is compounded by the fact that partially treated sewage from a nearby sewage treatment plant is continually added to the wetlands. The water of the inner Merchison Bay in the area where the Nakivubo channel dumps its load today turns black and the stench is said to be unbearable.

Moreover, thanks to external loads of organic matter, the bay area is covered with blue-green algae. In the past, the algae lasted for only a few days. They now persist for weeks at time before they disappear or die off and sink to the lake floor. Because the algae have depleted the dissolved oxygen in the area, fishers have not been able to harvest fish. Additionally, since the people near the bay rely on raw water from the lake for drinking, cooking and bathing, they are regularly exposed to dangerous health hazards (New Vision, March 7, 2007). In June 2010, Uganda's Wetland Commission declared the inner section of Merchison Bay a biological desert. By all estimates, cleaning up the area, which receives 14 metric tonnes of raw sewage daily, would take 40 years and tens of millions of dollars. The cost of water treatment at Kampala's main waterworks has increased fourfold in the last 3 decades because the capacity of Nakivubo wetlands to filter and purify the water is severely eroded. It is instructive to note that the water purification services from undisturbed Nakivubo wetlands is in the range of $1,500 to $2,000 per ha per year, which is the opportunity cost of not having the wetlands to do the job of removing nutrients and purifying the water free of charge (World Resources Institute, 2009; Candia, 2010).

Another area in Uganda that has received increasing attention in recent years is rice production, which naturally presents opportunities and perils Insofar as wetland ecology is concerned. Historically, Uganda was not known as rice country; today, however, this image is rapidly changing. With assistance from China, Uganda first started the Kibimba and Doho rice schemes in the 1970s in the Lake Victoria region. By 1998, the 2 schemes had a combined hectarage of 1,600 under rice irrigation. In the 1990s, the FAO also started 3 small rice schemes between Lake Victoria and Lake Kyoga, primarily for demonstration purposes (Aquastat, Uganda, June 2006).

After a long gestation period, rice production in Uganda has suddenly begun to soar; in the past 10 years, it has increased by 400 percent. As a result, rice imports have been falling by half every year. In 2001, Uganda's rice output barely reached 35,000 metric tonnes, which steadily rose to 135,000 metric tonnes in 2006 and 180,000 metric tonnes in 2008. Further yearly increases are likely if Uganda continues to aggressively convert its wetlands to rice production (*The Monitor*, Oct. 28, 2008; Appropriate Technology, June 2008). Today, there are 5 private companies that sell NERICA seed to rice producers, and 9 private firms are in rice milling business (Appropriate Technology, June 2008). Encouraged by the prospect of making Uganda the regional rice basket, the government at present views rice production as the touchstone of its "poverty alleviation intervention" strategy. By definition, rice is a wetland plant, inviting irrigation works. In the past, Uganda's irrigation potential was estimated at between 200,000 and 400,000 ha (Aquastat, Uganda, June 2006). Given the current popularity of commercial rice production, it would appear that Uganda is more likely to aim toward realizing further expansion of rice production through a more aggressive conversion of wetlands.

Serious concerns over the future of Uganda's wetlands around Lake Victoria also arise from the growing popularity of floriculture. Ugandan export of flowers grew from $313 metric tonnes worth $1 million in 1993 to 7,239 metric tonnes worth $35 million in 2005. Excited by the prospect of more revenue from the sector, flower exporters and the government have been poised to imitate Kenya by luring foreigners to invest $200 million in the sector with a view to increasing flower export to 16,000 metric tonnes, thereby generating $72 million annually (*East African Business* Week, January 7, 2008). The wetlands around Lake Victoria are considered suitable for large-scale floriculture operations. Lutembe Bay, on the northern shores of Lake Victoria between Entebe and Kampala, where flower growers are concentrated, is one of the wetlands facing rapid degradation. Agrochemicals from 5 flower farms in the area are having visible impact on the ecological integrity of the bay. Moreover, more and more people are unsustainably harvesting the bay's papyrus and other wetland vegetation. Lutembe Bay is known to host over 207 species of birds, both resident and migratory. On certain months, up to 1 million birds use the bay and its wetlands for food, water, breeding and resting (Boere et al., 2006).

CONCLUSION

By situating the current state of Lake Victoria in macrolevel discussions, this chapter has shown how the lake has become vulnerable to extreme ecological stressors resulting from human-induced climate change, deforestation, and wetland depletion or fragmentation. The chapter has also shown how any discussion on the preservation or conservation of forest and wetland resources must be situated in a broader socioeconomic context and development dilemmas. The stark demographic condition and the degree of absolute poverty currently obtaining in Victoria's littoral countries, coupled with the rising affluence of the pirate state bourgeoisie and the commercial class, have made abundantly clear that appropriation of plant biomass as well as livestock and fishery products in the scale witnessed so far is unsustainable.

The relevance to the state of Lake Victoria of the discussions widely aired in the chapter boils down to 5 essentials.

First, using fire to clear forests, woodlands, shrubs, and bushes for farming or charcoal making has mobilized numerous substances and elements that have been inserted into the lake by wind and water. For example, the insertion rates of ash (known for its rich phosphorus content) into Lake Victoria are among the highest in the world, contributing to the eutrophication of the lake in major ways (Hecky et al., 2006).

Second, forest fragmentation and wetlands depletion veritably leads to the modification of macro- and micro-climatic conditions, in which precipitation patterns are adversely affected. The role of forests and riparian vegetation is vital in terms of providing watershed protection, oxygen production, carbon sequestration, and regulation of temperatures. These are all requisite conditions for the precipitation regime. Substantial reductions or modifications of these critical ecosystem functions, arising from forest-depletion-related climate change at both micro- and macro-scales, are having an enormous bearing on the state of Lake Victoria and its basin. This observation has particular resonance with Victoria's riparian countries, since they are all vulnerable to varying degrees of climate change in light of their proximity to the equator.

The third major implication of forest degradation and wetlands depletion for Lake Victoria involves the siltation and sedimentation of the lake, precipitated by widespread soil erosion throughout the basin. Forests and wetlands are pivotal players not only in soil formation but also in landscape stabilization and sediment retention. In light of the extensive deforestation and wetlands destruction taking place in its watershed, Lake Victoria is today choking from external loading, and becoming increasingly less productive. Several studies suggest that nutrient loads transported by water from denuded areas to the lake have increased in recent decades by between sixfold and nine-fold, and sediment yields have increased by between tenfold and fortyfold, depending on the topographic conditions of landscapes and the intensity of flows (Hecky et al., 2006). Apart from the obvious depletion of natural nutrients critical to farms, the deposition of external nutrients and sediments in the lake has adversely affected primary production, overenriching some aquatic primary producers at the expense of other aquatic living things. Owing to anthropogenic factors, 45 percent of the land in the Lake Victoria basin is today prone to water and wind erosion. Between 1963 and 2008, roughly 3.2 billion metric tonnes of soil are believed to have been inserted into Lake Victoria from throughout its catchment areas, whereby sedimentation and siltation of the lake have become problematic (UNEP, 2009). In consequence, the lake has increasingly become turbid and murky, reducing access of food by aquatic living things.

The fourth impact on Lake Victoria is socioeconomic in character since the lake is inexorably linked to the importance of forests and riparian vegetation to rainfall generation, watershed protection, food production, wood extraction, and the provision of other goods and services that are foundational to the food ecology of rural communities. Terrestrial systems are the mainstay of every economy in terms of agriculture, livestock production, supply of freshwater, collection of traditional medicines, timber and nontimber production, and bush meat. Terrestrial vegetation is the driving force behind these benefits. Deforestation or degradation and habitat fragmentation thus erode the very foundation of an economy, compelling forest-dependent communities to migrate to lakeshores to make a living as fishers as witnessed in Lake Victoria. Moreover, in the context of high meteorological variability in Victoria's littoral countries, a major concern of forest depletion is the potential for desertification, precipitating mass migration of people from the terrestrial systems to the lake and its environs in search of water, food, fodder, and riparian vegetation.

Finally, there is the vexing question of finding the right solution(s) for the crisis of deforestation and wetlands destruction. There is a case of classic catch-22 facing Lake Victoria's riparian communities. On the one hand, aggressive transformation of forest and wetland resources to simultaneously serve the necessity of survival and the logic of accumulation risks deepening the degradation of critical components of the environment, whereby the societies in the region may lose all the benefits that could be had from forest and wetland ecologies.

On the other hand, however, given the hegemonic ascendance of the logic of destructive creation, it would be naïve to think that the natural resources would not be exploited in full for purposes of facilitating capitalist accumulation by the pirate state bourgeoisie and their hangers-on while, at the same time, attempting to improve the debilitating poverty of the broad masses in the interest of retaining a modicum of legitimacy. For example, take the 2009 famine in East Africa when more than 4 million Ugandans and Kenyans were on life support with food aid from foreign entities. Even under normal conditions, domestic food production in the littoral states of Lake Victoria is so poor that they are often unable to improve the conditions of those who are already in stark poverty, much less to keep up with population growth. Average landholdings in all 3 countries are shrinking fast; the soils are exhausted, and cycles of drought hit the countries with remarkable regularity. This explains in part why the east African states see Lake Victoria, its fringing wetlands, and the basin's forests as natural default option.

Herein lie the contradictory hopes and dangers, potential benefits and potential costs, depending on whether the water, forest, and wetland resources are wisely used for the benefit of the poor or the resources are misused simply to enlarge the already large pockets of the pirate state bourgeoisie. In the context of recurring drought, entropic degradation of terrestrial and riparian ecosystems, widespread stark poverty, and agrarian fragmentation associated with rising population, the integration of forest and wetland resources into processes of subsistence cultivation and commercial agrarian production is likely to continue. This means that the consequences of forest and wetland depletion and degradation in combination with climate change are likely to further accelerate the human-induced state transition of Lake Victoria. Thus, human actions taking place outside the lacustrine ecology are bound to worsen the integrity of supply conditions in Lake Victoria. Deforestation, removal of riparian vegetation, wetland destruction, soil erosion, agrochemical deposition from agrocommercial plantations, tailings from mineral extraction, and urban sources of contaminants and pollutants all are more likely to exacerbate the entropic degradation of the lake, perhaps beyond restoration.

The crux of the problem inheres in the fact that national and foreign research agencies forward manifold proposals on how to restore Lake Victoria to its previous state without examining whether the policy options they propose are achievable within the existing social order. The reality is that the managerialist quest for solutions within the prevailing capitalist mode of production is elusive. The great conundrum facing those deeply concerned about the future of Lake Victoria and lakeshore communities must be how to create subjective conditions for meaningful change. So long as policymakers and researchers in the region continue to see nature as infinite supplier of low entropy throughput and natural sinks, reversing the state of Lake Victoria does not appear to be in prospect. Moreover, contrary to the lip service frequently paid to the importance of natural resource conservation, there is blind faith in the notion of unlimited economic growth without fully grasping nature's limited capacity to supply the throughputs and sinks necessary for capitalist reproduction. For example, since the adoption of the Ramsar Convention in 1971, more than 1,400 wetlands, covering over 186 million ha, have been designated as sites of international importance worthy of protection; the number of contracting parties to the convention has also risen to 141 by 2009 (Wetlands International, Feb. 2009).

However, even these wetland sites, presumably under protection in accordance with the requirements of the convention, are not immune to encroachment and degradation. A 1993 study, for example, revealed that 84 percent of the wetlands in contracting countries of the Ramsar Convention had been threatened by varying degrees of ecological changes resulting from human activities (Daly and Farley, 2003; Moser et al., 1996). Thus, the greatest challenge to researchers and policymakers is how to find alternative strategies of survival for the broad masses of people by vigorously contesting the master narrative of global capitalism, and by providing secular vision to those who are at the forefront battling the pirate bourgeois state.

4 | Lake Victoria: The Supply Conditions (Part II)

Fortunately, the great lakes are vast and because of their remoteness are as yet unspoiled by industry and all that this entails in the way of pollution and interference. Such hazards may, however, yet be in store. If and when they become a reality, it should be remembered that these lakes are the repositories not only of potentially valuable fisheries but of unique faunas that each contains assemblages of cichlid fishes, which represent the end points of one of the most remarkable of all the feats of evolution and that each contains many fishes which are to be found nowhere else in the world. …As for the introduction of predators such as the Nile perch, this is not simply a stupid but a criminal act and such it should be treated. …We would also condemn the naïve belief that pilot studies can enable one to predict the outcome of introductions. This belief is as dangerous as it is foolish. …Those who believe that, and who call themselves biologists, are unworthy of the name. So little do they appreciate the complexity of the situations with which they interfere. (G. Fryer and T. D. Iles, 1972: 592-593).

THE IMPACT OF INTRODUCED SPECIES ON SUPPLY CONDITIONS

Forty years later, if Fryer and Iles were to rewrite the observation quoted above, the epigraph would have read like a requiem over what has befallen Lake Victoria due to human biomanipulation of its species composition and the continuous insertions of pollution into the lake, which have triggered its state transition through widespread eutrophication. In a way, this chapter is about how what they had feared has actually happened. In this section, I present how the introduction of exotic species into Lake Victoria has radically transformed predator-prey relations and consequently caused the disintegration of trophic structures.

Quite a few writers have publicly indicted the Nile perch with having driven or, at least, with having become accessory to the extirpation of some 300 haplochromine cichlids from Lake Victoria (Barel et al., 1985; Welcomme, 1988; Lowe-McConnell, 1997; Witte et al., 1992; Goldschmidt et al., 1993; Chapman et al., 1998; Kitchell et al., 1997; Crul, 1995; Ogutu-Ohwayo, 1990). Before we look at the material evidence supporting this indictment, some prefatory words are in order on the opportunities and risks associated with species introductions to new hydroecology. Since 1900, some 1.5 million introductions involving 291 species have occurred in 148 countries as part of the supply augmentation strategy (Pitcher, 1995).

While this worldwide phenomenon of fish introductions appears to have made some freshwater bodies productive, the practice may have compromised the integrity of many other aquatic ecosystems. For example, introductions of *Limnotrissa miodon* from Lake Tanganica into Lakes Kariba (Zambia), Bahora Bassa (Mozambique), and Kivu (DRC and Rwanda) have made both artisanal and commercial fisheries in those water bodies viable. In contrast, the 1958 introduction of Micropterus salmoides into Lake Atitlan in Guatemala completely wiped out the native species, on which indigenous Indian communities had depended over the centuries for critical protein and other essential macronutrients (Zaret and Paine, 1973). As a matter of principle, introductions of fish species into new environments tend to involve significant risks for lack of capacity to predict the effects of species introductions on endemic faunal composition, genetic diversity, fecundity, species fitness, adaptability to local conditions, trophic structure, and the biophysical and ecological conditions of the freshwater bodies in question. Introducing or relocating alien animal species from wetlands, rivers, and lakes into other aquatic systems brings the dangers of competition, predation, introgression, hybridization, stuntedness, habitat modification, and the transmission of parasites and other pathogens. There are five aspects to this problem.

First, depending on the biological characteristics of the introduced species in terms of fitness, resilience, and fecundity, competition between the exotic and native species for food, habitat, and breeding ground may disadvantage indigenous populations, which are at risk for decline and even potential extinction. Indeed, in situations where there is an overlap of food and habitat requirements, the intensification of competition between introduced and native species may result in habitat alteration or disruption. For example, competition for breeding and nursery space in Lake Victoria between the introduced *T. zillii* and the endemic *O. variabilis* is regarded as the cause for the extirpation of the latter from the lake (Welcomme, 1988; Pitcher, 2001; Hall and Mills, 2000). Likewise, the introduction of the American comb jellyfish into the Black Sea has long been held accountable for the extirpation of 26 commercially valuable fish species (Millennium Ecosystem Assessment, 2005).

Second, bringing together previously isolated allopatric species can create the opportunity for gene flow between the guest and host species, leading to hybridization or introgression. Over time, this may lead to genetic pollution or extirpation by hybridization and introgression (Rhymer and Simberloff, 1996; Welcomme, 1988; Hall and Mills, 2000). For example, the introduction of rainbow trout in the United States, Canada, Italy, and Sweden has resulted in extensive hybridization and introgression between the guest and host trout species. Ultimately, this has eroded the genetic diversity and distinctiveness of trout species. Similarly, the introduction of smallmouth bass in Texas has led to its hybridization with the native Guadalupe bass, threatening the very existence of the latter (Rhymer and Simberloff, 1996).

Third, if the introduced species is at the top of the food chain, the extent of predation on native species is difficult to predetermine, in which case the native populations may crush under combined predation and fishing pressure. Exemplary of this situation was the fate of Victoria's *Haplochromis*. It is said that the introduction of Micropterus salmoides into Lake Naivasha has been the principal cause for the disappearance of several cichlid species in that lake (Welcomme, 1988; Hall and Mills, 2000). Similarly, in the 1940s, commercial fishers used to harvest 6 million pounds of lake trout annually from Lake Michigan. However, the introduction of sea lamprey into the lake completely wiped out the lake trout by the mid-1950s. The *sea lampry* is a voracious predator, which quickly spread to all Great Lakes and aggressively descended upon native species (Fryer, 1960; Cox, 1993).

Fourth, the addition of exotic species to a new hydroecology may generate its own unexpected dynamics. The resulting interaction between biotic and abiotic entities may change all or some important environmental parameters. The resulting change or modification in habitat conditions, food web, and the erosion of ecosystem services can have the effect of changing the faunal composition, diversity, abundance, and distribution. Further effects may distress the resilience of individual populations. Indeed, the exotic species may destroy the habitat of native populations (Pitcher, 2001; Hall and Mills, 2000). As we saw in chapter 2, introduction of the common carp into Lake Naivasha has substantially modified the habitat of such important fish as the *T. zillii*, which depend on vision and water clarity to gather their food and reproduce. In addition to uprooting aquatic macrophytes, the common carp stirs up much bottom sediments. The carp muddies the water column, choking aquatic invertebrates, and releasing phosphorus from bottom sediments, factors believed to be responsible for much of the eutrophication of Lake Naivasha. The net result has been an unwelcome change in the faunal composition, abundance, and distribution of resident species.

Finally, there is the problem of stunting entailing deterioration in the quality of fish stocks, which may result from the effects of guest species on host species. As part of the evolutionary strategy to adapt to sudden change in the natural conditions resulting from introductions, native species may suddenly expand their populations so rapidly that individuals of the species mature at much reduced size. In this event, the species may not be commercially fishable. *Oreochromis niloticus*, *Tilapia rendalli*, and *Oreochromis mossambicus* are said to have displayed the characteristics of stuntedness in Lake Victoria and other Rift Valley lakes (Welcomme, 1988).

It is in the context of these considerations that proponents of the biomanipulation-driven extinction hypothesis attribute the demise of the more than 300 haplochromine cichlids of Lake Victoria. More specifically, they attribute the demise of the cichlids to the introduction of the Nile perch from Lakes Turkana and Albert into Lakes Nabugabo, Kyoga, and Victoria. Obviously, because sequencing the impacts of climate change, eutrophication, and Nile predation is a hazardous exercise, making a definitive attribution of blame for the largest case of vertebrate extinction in modern times to a single event could be overstretching. However, the fact remains that the perch (as the top predator), has certainly played a major role in the extirpation of the haplochromine cichlids, as well as many other native fish species. The controversy surrounding the relationship between the introduction of the perch and alien *Tilapiine* cichlids into Lake Victoria and the disappearance of a large number of native haplochromine cichlids and tilapias point to the uncertainties and surprises involved in the interplay of the laws of evolution and thermodynamics. However, the unfortunate episode of vertebrate extinction to this magnitude appears to validate the concerns of those who had the wisdom and the gift of farsightedness to caution against playing Russian roulette with nature. Of course, so long as the logic of destructive creation remains the dominant organizing principle of capitalist accumulation, precaution will not be regarded as an asset, but rather as a liability standing in the way of destructive creation. When retrospection transforms precaution into an asset, it becomes too late to reverse what is already foregone.

As we saw in chapter 2, concerned over the rapid depletion of native fish stocks, British colonial authorities had already begun entertaining the idea of relocating large piscivores from Lakes Albert and Turkana to Lake Victoria as early as the second half of the 1920s. The supposition then was that large predatory piscivores, like the Nile perch, could gorge on the small bony haplochromine cichlids. The belief was that this would produce dining-table fish to support commercial fishery. In order to find out the causes of the depletion of the commercially valuable *esculentus* species, while simultaneously exploring the prospect of reinvigorating commercial fishery through the possible introduction of alien species, British naturalist Michael Graham took upon himself to make a lakewide survey of Lake Victoria's fish ecology in 1927-28 (Graham, 1929; Worthington and Worthington, 1933). Graham was mindful of the importance of ascertaining the lake's fish population dynamics, biological characteristics, trophic profile, and ecological requirements as requisite conditions for any course of action that might follow (Graham, 1929; Worthington and Worthington, 1933). Upon completion of his survey, Graham saw great opportunities and potentially perilous risks in the potential introduction of alien species into Lake Victoria. On the one hand, there were the haplochromine cichlids, huge in abundance, varieties, sizes, shapes, and feeding habits. Their abundance was so incredible that he reportedly caught 2,600 *Haplochromis* in just one hour of trawling. The *Haplochromis* were not considered palatable for human consumption because of their bony characteristics and small sizes. However, Graham thought that the haplochromine populations could potentially support a large fleet of trawl vessels. This is how he expressed his excitement:

> The number of individuals is almost incredible. I have taken comparatively large catches of these with a small trawl almost anywhere except in the very deepest part of the lake. So great are their numbers that I had contemplated suggesting trawling for them in order they may be used for manure in Kenya colony. I believe that the stocks can withstand an industry consisting of 200 trawlers, that is provided that it became economic to use trawlers at all. (p. 20)

However, Graham shared the concern with his assistant E. B. Worthington that trawl fishery could be damaging to other species. Trawlers would naturally trample upon other valuable fish, such as *esculentus* and *variabilis*. Therefore, both Graham and Worthington took up the matter of introducing large piscivores, such as the Nile perch and the tiger fish, from Lake Albert into Lake Victoria. As a possibility, these fish would be good and easy to catch.

The introduced predators would potentially convert the haplochromine cichlids into commercially palatable fillets, thereby averting the danger of trashing the fish ecology by the introduction of trawl fishing. However, both naturalists agreed that introductions of predatory species into Lake Victoria could be disastrous, without making a thorough and complete scientific inquiry into the risks involved before introducing the Nile perch into the lake. They proposed to use Lake Nabugabo as a natural laboratory for the introduction of the Nile perch. In their view, Lake Nabugabo was similar to Lake Victoria in terms of faunal composition, trophic characteristics, water temperature gradients, and habitat conditions. After all, Nabugabo was once part of Lake Victoria before the accumulation of sands broke off the small lake from the main body of Victoria. In their scientific minds, this made sense because observations of how the Nile perch behaved and performed in Nabugabo would reveal the potential opportunities and risks involved in species introduction. Even if the experiment in Nabugabo failed, the magnitude of faunal and ecological disruptions was bearable. The 20 or so fishers then making a living in Nabugabo could easily be relocated to Lake Victoria if the ecological disruptions made their livelihoods there untenable. The Worthingtons (1933: 173) put the case in these terms:

> Supposing that the Nile perch was introduced and that he survived the difference of the Lake Victoria environment, the cooler water and the different food, one of two things must happen: either the Nile perch would strike a natural balance with the other fish of the lake or he would find the Lake Victoria fish unused to its voracious habit too easy prey. Consequently, he would breed too fast until the entire food supply was exhausted and then would die out himself with disastrous results to the existing fishery. In fact, although the introduction might prove of great value, it must not be made until it has been known exactly what the results will be.

The ambivalence contained in the above statement underpinned the proposal to use Nabugabo as natural laboratory, in which the progress of the Nile perch could be closely monitored and evaluated. In the end, though, the cautionary recommendation of the British naturalists was fraught with contradictions. This largely reflected their naïve understanding of how the social world was constructed and used. On the one hand, the naturalists were cognizant of the fact that species introductions in the past had produced disastrous results, causing extinctions of species and the spread of pests, parasites, and diseases worldwide. In the words of the Worthingtons (1933: 177): "Indeed, a strong case can be made out against man's tampering with nature for his own end, for he may so easily go too far. Man is a greedy creature and all too often in the past his greed had proved too strong for his discretion."

Despite this acknowledgement of man's greed, however, the British naturalists seemed to propose that preservation of nature and colonial accumulation could go hand in hand, if guided by caution and the right mind. Placing their faith in the religion of managerialism, the Worthingtons (1933: 177-78) were so convinced of the potential domestication of man's greed that they wrote:

> Against these arguments, however, we can say that civilization cannot stand still...a lake like Naivasha or Victoria, lying in populous neighbourhoods, should not be left unfarmed if there is a reasonable chance of making a product of food or even a place of harmless recreation.

In the midst of hesitation and the dearth of research products, the idea of introducing a large piscivore into Lake Victoria remained dormant for over two decades, until the precariousness of commercial fisheries roused renewed interest in the 1950s. By the early 1950s, commercial fisheries in Lake Victoria were on the verge of total collapse. The *Oreochromis esculentus*, *Oreochromis variabilis*, *Labeo victorianus*, *Clarias*, *Bagrus*, *Propterus*, *Synodontis*, *Mormyrus*, and other species, which were the mainstays of the commercial fishery, had befallen enormous stresses because of modernization-driven overfishing.

At any rate, the objective conditions of dwindling fisheries, threatening both the necessity of survival and the logic of commercial accumulation, reenergized proponents of species introduction. The captains of colonialism finally determined to introduce the large piscivorous Nile perch to Lake Victoria to convert the *Haplochromis* into a palatable table fish. Furthermore, four presumably vegetarian *Tilapiine* species would be introduced, which would feed on macrophytes that were being inefficiently utilized by other native fish species. Proponents now vigorously honed their basic contention by stretching the benefits of introduction and by downplaying the risks involved, using four arguments

First, the foundational proposition was that, since a mature perch required larger mesh sizes to catch, fishers would presumably use large nets, which would translate into easing the pressure on inshore *Esculentus*, *Variabilis*, *Victorianus*, and other native species that had become the target of small mesh sizes. It was further argued that, since the Nile perch was an offshore deep-water dweller, fishers would venture out into the deeper portions of the lake to chase the perch, allowing inshore fish species to recover.

Second, proponents of introduction even suggested that, since the Nile perch was probably native to the Lake Victoria region before the desiccation of the lake some 14,600 years ago, re-nativization of the Nile perch would not pose a danger to the species already in the lake. This observation was reinforced by the supposed empirical evidence derived from the evolutionary coexistence of the Nile perch and *Tilapiine* species in Lake Albert, where the Nile perch had supposedly struck a natural balance with their prey species. This supposed natural equilibrium was said to have been empirically supported by the fact that two-thirds of the fish landed in Lake Albert comprised *Tilapiine* and other species, while the Nile perch represented the remaining one-third. Furthermore, the stomach contents of the Nile perch were said to have shown that the predator was flexible in its dietary profile. This suggested that the predator would be content to gorge on the superabundant *Haplochromis*, without necessarily affecting the larger *Tilapiine* species, thereby assuring the continued flourishing of tilapia-based commercial fishery.

Third, even if the abundance of native fish populations declined in response to species introduction, the abundance of large high-quality fish like the Nile perch would compensate for any decline in catches of other species (Anderson, 1961; Fryer, 1960; Achieng, 1990; Beverton, 1959).

Fourth, using data gathered since the introduction of the perch into Lake Victoria, Reynolds and Greboval (1995) painstakingly made the empirical case to buttress earlier contentions. In their view, the gains of $560 million made between 1975 and 1989 due to species introduction validated the importance of the perch to the fishery mode of production. Moreover, because of perch fishery, the number of people engaged in the primary and secondary fishing activity and their dependents rose to 1,266,000 in 1989 from 475,000 in 1975. This further indicated the growing importance of the Nile perch as a source of both livelihoods for indigenous communities and hard currency generation for the riparian countries.

There were some British ecologists, who understood the myopia of the policy soon enough to raise strenuous objection to the introduction of the Nile perch, since little to nothing was known about the lake's ecomorphology, hydrobiology, primary production, and trophic profile. After all, the arguments in favor of introduction were more deductively rationalized than empirically supported. Indeed, Fryer and Iles (1972: 592) were so disturbed by the rush to introduce exotic species into Lake Victoria and other Rift Valley lakes that they compellingly wrote:

> The introduction of foreign species to these lakes is to be deplored and should be strictly be forbidden. While the introduction of herbivorous fishes such as *tilapia zillii* into Lake Victoria may have

> been made for the very best of motives mainly in the hope of increasing production of food for human consumption, we would submit that even in cases such as these is never possible to predict with certainty what the outcome might be. The history of animal introductions, including that of fishes, provides more than sufficient examples of disasters, some on huge and costly scale, which have resulted from this practice.

In the opponents' view of the introduction of exotic piscivores into Lake Victoria, the cost of the trophic requirements of the Nile perch would outweigh its value in the longterm. In other words, the amount of *Haplochromis* or tilapia the perch would eat to produce a given amount of dining-table fish food could be less than what a herbivorous tilapia might produce. By shortening the food chain, the perch might promote grossly inefficient energy use. From a biological point of view, it would thus be fanciful to suggest that the perch would exclusively limit its feeding habit to the *Haplochromis*. Suppose that the *Haplochromis* were crushed due to unanticipated events, then what would the perch do? Accept voluntary extinction or diversify its food sources to *Esculentus*, *Variabilis*, and other native species? Furthermore, since the perch does not consume phytoplankton and detritus, Nile perch dominance in the lake, by reduction or elimination of phytoplanktivorous and detritivorous populations, could negatively alter the lake's trophic structure. After all, the Nile perch is known for its exceptionally high fecundity. A female perch can lay up to 17 million eggs in a single spawning.

There were early studies that warned that removal of too many herbivorous species (such as native tilapias) from Lake Victoria could potentially spell disaster for biological production in the lake, since herbivores are ecologically most efficient in energy use. If introduction of exotic species were to be considered at all, then detrivores and herbivores would be the most viable candidate for introduction into Lake Victoria. Some had even suggested that promoting and developing aquaculture based on species native to Lake Victoria, such as tilapias, was a safer bet than risking an unknown outcome by introducing exotic predators (Fryer, 1960; Achieng, 1990; Barel et al., 1985; Anderson, 1961; Witte et al., 1992; Goldschmidt et al., 1993; Fryer and Iles, 1972). In the final analysis, the Nile perch represented a conundrum for both proponents and opponents of introduction. For proponents, the commercial promise of the Nile perch was too attractive to ignore; weighing up to 300 or 400 pounds, the perch could supply massive amounts of fillets that could be exported to generate hard currency. From the viewpoint of opponents, the size of the perch was precisely the reason why it should not have been introduced into Lake

Victoria, because the ecological and biological consequences of its introduction into the lake remained elusively indeterminate. In the words of Geoffrey Fryer (1960: 267):

> Ecology is a complex subject and the elucidation of the interaction of one animal with the other members of the community in which it lives constitutes some of its most complex problems. This calls for the application of more than empirical knowledge. Moreover, as the fishery of Lake Victoria is worth several million pounds annually and is of incalculable worth as the source of protein to the indigenous population, it behooves those responsible for its continued prosperity to give attention to the information which science can provide and to take heed of the advice which it has to give. It is, however, just as much a dictate of commonsense as scientific principle that predators can never be as abundant as nonpredators.

In any case, caution was put by the wayside, as the logic of destructive creation won the day and the captains of colonialism rushed to introduce the Nile perch and four *Tilapiine* species into Lake Victoria. Now the jury is in, and the outcome has not turned out as proponents of introduction projected. Prior to 1980, 80 percent of the fish biomass in Lake Victoria comprised haplochromine cichlids and 90 percent of the fish species were endemic to the lake. Fifteen years later, however, half of the 450 native species were said to have become extinct and the Nile perch was among the major factors answerable for their extinction; the perch now accounts for 80 percent of both fish biomass and fish catch (Chapman et al., 1995; Cohen et al., 1996).

The epistemological orientation of the promoters of exotic species introduction was fatally erroneous from the start. The metaphysical certitude with which they approached the question of species introduction was replete with intellectual arrogance or ignorance, as well as myopia, since they excised the dimensions of uncertainty and surprise inherent in nature from the fundamentals of evolutionary biology.

In the first place, as historically shown, the use of Lake Albert as a model was an excellent case of the fallacy of displaced concreteness. Even though Lakes Albert and Victoria may share some important limnological conditions, biophysical characteristics, and ecological parameters, they are never identical in terms of size, species diversity, trophic structure, temperature, or physical and chemical properties.

Second, the projection of the *Haplochromis* as a mere source of food for the introduced Nile perch, without fully understanding the ecological functions performed by the species, equally betrayed the epistemological shortcomings of those who promised a biological rose garden. The proposition that the Nile perch would exclusively gorge on *Haplochromis* was illusory. It never occurred to them that the *Haplochromis* might crush under the combined weight of heavy human fishing, predation by the perch, and/or from other stochastic events.

As some writers have suggested, if the Nile perch populations had reached 120,000 tons by 1980, the introduced species might have consumed 650,000 tons of fish that year, with haplochromine species comprising 85 percent of the diet (Moreau, 1995).

Third, the early optimism that the productivity and high quality of Nile perch fishery would compensate for any decline in the catches of the other species has also come to naught. For example, annual yields of the Nile perch in Lake Kyoga sharply dropped from 70,000 tons in the 1970s, to 15,000 tons in the late 1980s. This compelled a radical downward revision of MSY of the Nile perch to 3,000 to 15,000 tons per annum, while *O. niloticus* comprised 80 percent of the total catch from Kyoga. Similarly, the estimated MSY for the Nile perch in Lake Victoria was scaled down to between 71,000 to 327,000 tons per annum (Pitcher, 1995; Ogutu-Ohwayo, 1990).

Of course, defenders of the Nile perch-based commercial fisheries have been quick to point to the temporary bonanza in perch catches as evidence of their prediction. They have frequently pointed to the large quantities annually harvested from the lake and the increased number of fishers, processors and traders involved in the perch-dominated fishery sector as empirical support for their contention. However, such presentation obscures the long-term impacts of the outcomes on intragenerational and intergenerational equity, on the one hand, and on the biology and ecology of the lake, on the other. The Nile perch-dependent commercial fisheries have undeniably increased the quantity of fish supplied in the short run. The fundamental question worth pondering is whether the benefits accruing from species introductions have improved the lots of the toiling masses, or have simply expanded the supply of fish to the global haves to the detriment of local equity and ecological integrity. In aggregate terms, the supply of fish from Lake Victoria has no doubt appreciably increased since the introduction of the Nile perch. For example, as a result of the exponential growth of the Nile perch, the total annual catches from the whole lake of Victoria rose from 100,000 metric tonnes in the 1970s to between 300,000 and 500,000 metric tonnes in the 1990s. The Nile perch now constituted 80 to 90 percent of the total catch. Similarly, 19,000 fishing boats, along with 100,000 operators and crews were directly deployed to Victoria's fisheries by the mid-1990s. Another 250,000 persons were involved in secondary and tertiary activities. If dependents are included in the statistics, the lake was supporting roughly 2 million people by 1995 (Lowe-McConnell, 1997; Jeheb et al., 2008; Kitchell et al., 1997; Hall and Mills, 2000). However, these statistics do not escape the unsavory conclusion that the supply of fish has been for the affluent global consumers and wealthy nationals who are hooked to the global grid of consumerism. The vast majority of lakeshore communities are condemned to consuming carcasses, heads, tails, and other organs of the Nile perch. Even these items, previously unwanted by the fish factories, are now becoming rare to the poor since fishmeal factories are converting them into fish feed.

Insofar as the biological consequences of species introductions are concerned, opponents of species introduction appear to be proven right in their prediction. The faunal composition of Lake Victoria is probably altered forever. When British colonialism made its first inroads into the lake region, local fishery was based on some 300 cichlid fish species and 14 cyprinid fish species, which were integral to local food security (Balirwa, 1998; Ogutu-Ohwayo, 1990; Kitchell et al., 1997; Kudhongania and Chltamwebwa, 1995). However, introduction of the perch and four *Tilapiine* cichlids into Lake Victoria has simplified the lake fish ecology, since now only three species dominate commercial fishery, namely: the perch itself, *O. niloticus*, and the cyprinid *Rastrineobola argentea*. The population explosion of the Nile perch has transformed Lake Victoria's commercial fishery in terms of volume and commercial value. With the virtual dominance of the Nile perch throughout Lake Victoria, now constituting more than 80 percent of the lake's fish biomass, the faunal composition and trophic structure of Lake Victoria have begun to evolve in an unwanted or unexpected direction. The biological characteristics and ecological requirements of the Nile perch are at the root of this new evolutionary trajectory. As a top piscivorous predator, the Nile perch has the widest possible food spectrum and can switch its feeding habit from large and abundant to small and less abundant prey species, including its own juvenile.

Given this hydrobiological reality of the high fecundity of the Nile perch, it was no coincidence that native species rapidly declined or disappeared in the wake of perch propagation and predation. Between 1984 and 1987 alone, over 200 *Haplochromis* are believed to have disappeared from the sublittoral and offshore areas of the lake (Kaufman 1992; Witte et al., 1992 and 2007; Seehausen et al., 1997; Kitchell et al., 1997; Lowe-McConnell, 1997; Chapman et al., 1998; Ogutu-Ohwayo, 1990; Kudhongania and Chltamwebwa, 1995; Moreau, 1995; Cohen and Ogutu-Ohwayo, 1996). In addition to the number of Haplochromine cichlids driven to extinction, the native O. *esculentus* and O. *variabilis*—which historically had been of significant commercial value—have also disappeared from the main body of

Lake Victoria. The *Rastrineobola argentea* is the only native fish that remain commercially viable.

Furthermore, the material evidence strongly suggests that the Nile perch has been the principal driver in the disappearance of a large number of haplochromine cichlids in areas where there has been habitat overlap between the two species. The haplochromine species make their domicile in the offshore pelagic zone, deep water, and sublittoral areas, and have been the most vulnerable to perch predations, resulting in their extirpation. Over 200 haplochromine species used to reside in the sublittoral, deep, and/or open offshore waters of the lake—also favored by the predatory Nile perch. It appears only natural that introduction of a top predator into Lake Victoria could have a determinate impact on the abundance and distribution of these cichlids species, which happened to share the same habitats with the voracious predator. The larger piscivorous, insectivorous, and molluscivorous *Haplochromis* appear to have been the perch's favorite dish. Thus, they were the first to show signs of stress and eventually disappeared from the sublittoral and offshore regions of the lake.

In the Mwanza Gulf on the Tanzanian side of the lake, of the 123 haplochromine species that were caught before, 80 species disappeared overnight following the upsurge of the Nile perch. In fact, in some sublittoral fishing areas, the entire flock of haplochromine species was wiped out. The surviving haplochromines are the ones that reside near littoral vegetation and rocky inshore areas, areas less visited by the Nile perch.

In recent years, following the decline in Nile perch biomass, some resurgence of *Haplochromis* in sublittoral and offshore areas has been witnessed. However, diversity is still lacking. For example, zooplanktivorous *Haplochromis* have recovered in significant abundance, but their diversity has sharply declined from twelve to three species in the Mwanza Gulf. Similarly, the number of detrivorous *Haplochromis* has fallen from 13 to 4 species. The fact that some haplochromine species have shown a promising resurgence following the decline in perch biomass suggests that there has been a strong relationship between predation by the perch and extirpation of the *Haplochromis*. The above observation by no means exonerates the role that human exploitation and anthropogenic eutrophication played in the extirpation of the *Haplochromis*. For example, with the operationalization of commercial trawl fishing beginning in 1976, long before the upsurge of the Nile perch was noticeable, Tanzanian trawlers had been hauling 10-15 tons of *Haplochromis* per day in the Mwanza Gulf (Witte et al., 2007). Moreover, the widespread use of seine and mosquito nets by subsistence fishers was already having an impact on the *Haplochromis* in many fishing areas of the lake.

In any case, with the haplochromine populations having rapidly declined, the Nile perch has diversified its food sources. It has expanded its feeding habit to include *Cardina nilotica* (freshwater shrimp), *R. argentea*, lungfish, catfish, elephant snout fish, and even the tilapias. In the context of a rapidly eutrophied lake, however, the simplification of the lake's biotic community has soon manifested its profound implications for the Nile perch itself, as the perch is forced to make auto-predation part of its survival strategy. Today, larger perch have descended upon smaller perch as part of their menu of necessity. This may explain in part the alarming reduction in perch biomass in Lake Victoria and the abrupt fall in Nile perch catch as well. Indeed, 98 percent of catches of the species are juvenile; catches of Nile perch have overshot recommended MSY by 27 percent as of 2007, suggesting that exploitation of the species must be reduced by 50 percent if restoring a natural balance is a conservation priority (Knaap et al., 2007). This is a powerful illustration of how the utilitarian strive for more from nature, driven by the logic of destructive creation, could produce many unintended harmful consequences.

Along with the Nile perch, four *Tilapiine* species (*O. niloticus*, *O. leucostictus, T. zillii*, and *T. rendalli*) were introduced into Lakes Victoria and Kyoga. The introduction of *O. niloticus* and *O. leucostictus* into Lake Victoria was intended to relieve the fishing pressure on O. *esculentus* and O. *variabilis*. Furthermore, the *T. zillii* was expected to feed efficiently on macrophytes not being utilized by other commercial species (Ogutu-Ohwayo, 1990; Kudhongania and Chltamwebwa, 1995; Hall and Mills, 2000). Contrary to intentions, however, only the *O. niloticus* has survived in the lake, but only at the expense of the native *Tilapiine* species. It appears that the native *Tilapiine*s were outcompeted by the *O. niloticus* for the same ecological niches. The much smaller native *O. variabilis* was outcompeted by *T. zillii* for food and nursery grounds, while the *O. niloticus* appears to have outrun the O. *esculentus* in the competition for the same biological necessities. Moreover, the changes in ecology following the introduction of the exotic species may have engendered permissive conditions for hybridization between the guest and the host *Tilapiine*s (Kudhongania and Chltamwebwa, 1995; Ogutu-Ohwayo, 1990; Witte et al., 1992; Hall and Mills, 2000).

The relatively successful coexistence of the *O. niloticus* with the Nile perch in Lake Victoria is probably due to three important factors.

The strong showing of the *O. niloticus* in Lake Victoria suggests that both the perch and the *O. niloticus* had developed a coevolutionary relationship in Lakes Albert and Turkana long before their joint relocation to Lake Victoria. As a result, the latter might have already refined its survival maneuvers vis-à-vis the perch. Moreover, its predation-avoidance strategy appears to have been strengthened by its omnivorous feeding habit. Contrary to the earlier understanding of the *O. niloticus* as a vegetarian species, its food sources now comprise crustaceans, detritus, mollusks, insects, fish, and a variety of phytoplankton species.

Next, in comparison with the native *Tilapiines*, the *O. niloticus* has high fecundity, grows to a large size, and has long life expectancy.

Finally, the primary residency of the *O. niloticus* is in the littoral and sublittoral areas of Lake Victoria, having little or incidental habitat overlap with the perch (Ogutu-Ohwayo, 1990; Witte et al., 1992; Balirwa, 1998; Kudhongania and Chltamwebwa, 1995; Knaap et al., 2007; Kitchell et al., 1997). Indeed, the *O. niloticus* appears to have thrived in Lake Victoria, since catches of this species jumped from 13,000 metric tonnes in 1975 to more than 105,000 metric tonnes in 2000 (LVFO, Strategic Vision 1999-2015).

The *O. niloticus* may have played a minor role in the status of some haplochromine species through competition for food, especially in the rocky shores of Lake Victoria. It is here that *O. niloticus* were seen to have outnumbered haplochromine algae grazers. Since *O. niloticus* and algae-grazing *Haplochromis* enjoy a vegetarian diet to an equal extent, competition may thus have placed the *Haplochromis* at a disadvantage in the Darwinian struggle for survival (Seehausen et al., 1997).

However, the *O. niloticus* is showing signs of stress, since the species has reduced its size at first maturity and its maximum mean size has fallen. This is perhaps an early indication of stuntedness. Of the native species, the *R. argentea* is the only species that has acquired commercial prominence today. Well until the third part of the 20th century, the *R. argentea* had little commercial value beyond casual exploitation for personal consumption by local communities (Ogutu-Ohwayo, 1990). The *R. argentea*'s commercial value in Lake Victoria appears to have gained from the disappearance of the *Haplochromis*, where its own trophic requirements overlapped with those of the *Haplochromis*. In this sense, the *R. argentea* has benefited from the spontaneous proxy services provided by the Nile perch in removing the competitor zooplanktivorous *Haplochromis* from the sublittoral waters. Now, adult *R. argentea* are at liberty to monopolize the macrobenthic invertebrates previously held by the *Haplochromis*. Indeed, with this new trophic niche fully at its disposal, the *R. argentea* has effectively made bottom dwelling a permanent feature of its habitat by making important morphological adaptations to cope efficiently with decreased oxygen concentrations in near bottom habitats. Adult *R. argentea* now alternate their migration between bottom dwelling during the daytime, and near surface waters at night, when they chase zooplankton and *Caridina nilotica*.

This habitat and trophic niche diversification, coupled with its evasive maneuvers to avoid predation, may explain the relative success and rapid population growth of the *R. argentea* (Wanink and Witte, 2000; Witte et al.,

1992; Kitchell et al., 1997). Indeed, this cyprinid pelagic species is now second to the Nile perch in terms of fish biomass in Lake Victoria. This has made the *R. argentea* fishery important to low-income people, since the fish is generally considered of low commercial value on the international market, and is relatively inexpensive. Therefore, the *R. argentea* is consumed domestically for the most part, even though fishmeal processers are now looking at the species with covetous eyes. The standing crop of *R. argentea* in the early years of this century was rapidly growing, with fishable potential yield of 580,000 tons per annum (Knaap et al., 2007; Tumwebaze et al., 2007). In 2001, the estimated harvest of *R. argentea* in Tanzanian waters was 167,789 tons, while the estimated harvest was 70,333 tons and 66,620 in Ugandan and Kenyan waters, respectively (Tumwebaze et al., 2007).

The *R. argentea* has today improved its trophic flexibility, since it has diversified its food sources from being predominantly zooplankton to include *Caridina nilotica*, insects, and fish. Thanks to predation by the Nile perch, disappearance of the zooplanktivorous *Haplochromis* (which used to constitute 25 percent of the total pelagic haplochromine biomass and whose food requirements overlapped with those of the *R. argentea*) has also strengthened the food security of the *R. argentea*. This explains, in part, why the biomass of the *R. argentea* has substantially increased in the open waters of the lake. Furthermore, the *R. argentea* is known for having exceptionally high fecundity (Goldschmidt et al., 1993; Reynolds and Greboval, 1995; Knaap et al., 2007). However, probably in reaction to overfishing and predation by the perch, the *R. argentea* has decreased both in maximum size and in size at first maturity (Knaap et al., 2007). Whether this behavior is an early manifestation of stunting or a selective survival strategy, only the future will tell.

Finally, the ecological and socioeconomic dimension of perch introduction has proven to be highly damaging, if not catastrophic. Indeed, the Nile perch may have been one of the important surrogates for the eutrophication of Lake Victoria and, hence, for its state transition. As noted earlier, the phytoplanktivorous and algae-grazing *Haplochromis* were known for their efficiency in removing massive amounts of algae from the lake. Through their vertical descent and horizontal migration, the *Haplochromis* also used to move and recycle huge amounts of nutrients (Kaufman, 1992; Hall and Mills, 2000). Their absence from the lake may explain why blooms of cynanobacteria have become a permanent feature of Lake Victoria, and why dead algae now sinks to the bottom of the lake where it rots, leading to the deoxygenation of the lake floor (Kaufman, 1992; Hall and Mill, 2000; Bootsma and Hecky, 1993; Hecky et al., 2010). Therefore, because of the disappearance of important haplochromine species, Lake Victoria may have lost irreplaceable ecosystem services that may not be fully captured by any quantification and modeling processes.

To point out the vital importance of biodiversity, biologists use such metaphors as rivet popping and portfolio effect to highlight the functional unity of species in an ecosystem. The idea of rivet popping describes how the succession of losses of species can have an exponential effect on the performance of an ecosystem. If several keystone species are lost from an ecosystem in rapid succession, without the chance to recover, the efficiency of the ecosystem is likely to erode in many unquantifiable ways and the ecosystem may eventually collapse. The reason is that the elimination of some species may lower the resilience of the ecosystem as a whole, since resilience may be a function of species diversity and collective capacity to respond and adapt to external perturbations. In this sense, biological diversity resembles a portfolio of stocks in a market, in which wise investors diversify their stockholdings to insure themselves against unexpected shocks. In this case, they will be in a better position to absorb the loss without going bankrupt. In the same way, biodiversity in the ecology allows the ecosystem to effectively respond to perturbations and preserve critical services because of the collective effort of species to respond differentially to external shocks. In the final analysis, ecosystem stability, productivity, and resilience may depend on biodiversity. The more complex and diverse an ecosystem, the more likely that the ecosystem will contain several species that are superior in their ability to complementarily strengthen and influence ecosystem functions. Perhaps it is tautological to state that since ecological systems are both nondeterministic and irreducible to their particular components, the components of an ecosystem depend on everything else within the system for their continued existence and performance (Millennium Ecosystem Assessment, 2005; Jorgensen and Szirezhez, 2004; Tilman, 1997; Costanza et al., 1997; E. Wilson, 2002; Daly and Farley, 2003).

In terms of socioeconomics, introduction of the perch has proven to be disruptive to the subsistence economy of lakeshore communities. The large size of the perch requires huge amount of firewood to smoke the fish. When large numbers of fish harvesters descended upon the riparian vegetation to collect firewood, the result has inevitably been extensive deforestation. The consequential soil erosion of the landscape and siltation of the lake may have significantly contributed to the erosion of the food security of aquatic organisms, while the eutrophication and deep-water deoxygenation of Victoria forced the oxygen-deprived *Haplochromis* to migrate to shallower areas, where they became vulnerable to Nile perch predation (Bootsma and Hecky, 1993; Hecky et al., 1994). Since differential sensitivity to cultural eutrophication between the Nile perch and the vision-dependent *Haplochromis* appears to have placed the latter at disadvantage in the struggle for survival and adaptation, lakeshore communities have contributed to the fate of many haplochromine cichlids by destroying riparian vegetation. Moreover, subsistence fishers, who traditionally relied on *Haplochromis* and *Oreochromis Tilapiine*s to make ends meet, have been driven out from their biosocial environment since the large-scale perch-based fishery is now monopolized by owners of capital (Barel et al., 1985).

In summary, the debate over the radical alterations in the species composition, trophic profile, and hydro-ecological parameters of Lake Victoria are functions of predation, eutrophication, or climate change, and is very likely to continue far into the future. The issue is what is to be done about it. The fact is that the once-rich multispecies fishery regime, now replaced by three species, is probably irreversibly extinct. Moreover, the biological characteristics, spatial distribution, and abundance of these three species have also changed. In the 1970s, *O. niloticus* constituted a mere 1 percent of catches, whereas in 2005, it accounted for 25 percent. By the same token, the dominance of the perch has dropped to 58 percent of the catches in 2005 from 95 percent at its peak in the early 1990s. Furthermore, more than 98 percent of perch catches are immature, and 70 percent of *O. niloticus* catches are immature (Njiru et al., 2008). What the story of the Nile perch and all its relationships with the native species tells us is that human biomanipulation of nature can be a perilous undertaking.

THE IMPACT OF SEDIMENTATION, POLLUTION, AND EUTROPHICATION ON SUPPLY CONDITIONS

Lake Victoria and all Rift Valley lakes are not only natural water reservoirs and habitats for countless aquatic species, but are also nutrient sinks for their entire catchment areas. Solid organic and inorganic matter, soils, silts, sediments, phosphorus, nitrogen, sulfur, mercury, and other trace elements from the terrestrial and urban ecologies are regularly inserted into the lakes via direct precipitation, influent rivers, and/or dry deposition. Interactively, these external loads have the effect of altering the limnology, biodiversity, and productivity of freshwater lakes. The above description of the conditions that can interfere with the health of lakes is particularly pertinent to Lake Victoria. In the watershed of Lake Victoria, changes in land cover are pervasive, degradation of terrestrial resources is extensive, and rates of urban and periurban sprawl are alarmingly rapid. Following the 1994 Rwandan genocide, for example, the Kagera River transported 40,000 human corpses to Lake Victoria (Rabi, 1996; Chege, 1995).

Owing to changes in land cover in catchment areas associated with too much clearing of forests and drainage of wetlands for human settlements, subsistence cultivation, agrocommercial plantations, overgrazing of rangelands, extraction of firewood, and infrastructure construction, sedimentation is increasingly becoming highly pronounced in Lake Victoria. For example, the insertion of sediment loads into Lake Victoria via the Nyando River alone increased by seven and half times during the 1990s (Odada et al., 2004; Allogo and Odada, 2007). In the context of widespread denudation of hills, mountains, and plains, as well as the destruction of wetlands, the rivers and streams are busy transporting heavy loads of soils and silts to Lake Victoria. These loads are oversaturated with agrochemicals—exacerbating the siltation of the lake, decreasing water quality, and increasing turbidity. Because sediment deposition predominantly takes place in littoral and inshore zones of the lake—which fish species use for habitat, feeding, spawning and nursery—sedimentation reduces available living spaces, as well as the supply of essential nutrients for the species involved (Cohen et al., 1996; Odada et al., 2004; Ollago and Odada, 2007).

Today, even inflowing rivers are experiencing intense siltation, worsening the reduction of living space and food availability for fish and other aquatic species. For example, 13 anadromous fish species from Lake Victoria are known to swim up the rivers annually for spawning. Now, because of pervasive siltation and sedimentation of the rivers, these species have found migration up the rivers strenuously difficult. For example, the Nyando River, which flows into the Nyanza Gulf, has become so turbid that the once-abundant spawning fish species have become either extinct or rare. Over 46 percent of the Nyando River basin is said to be completely denuded, making soil erosion a permanent feature (Odada et al., 2004; Awange and Ong'ang'a, 2006; Ollago and Odada, 2007). Moreover, since the vast majority of people who live in towns around Lake Victoria depend on adjacent forested areas and wetlands for biomass primary energy, brick making, construction material, sand mining, and farms, the resulting inflows of soils, sediments and mud have exacerbated the siltation of Lake Victoria. In sum, the biological impacts of sedimentation on the supply conditions of Lake Victoria include disruption of light penetration necessary for photosynthesis, erosion of the nutritional importance of detritus, physical damage to aquatic organisms, interference with the feeding and respiration patterns of organisms, and adverse changes in food web interaction resulting from releases of nutrients from the stirring up of soil particles (Cohen et al., 1996).

Another growing problem in Lake Victoria is pollution. The lake is known to be hypersensitive to pollution, in light of its high pollution retention time, and prone to high evaporation rates relative to its tropical location, low oxygen saturation potential, and higher oxygen consumption rates occurring during periods of high temperatures (Cohen et al., 1996). Transformation of the urban ecology, livestock production, soil erosion, combustion, mining operations, and the promotion of chemical-dependent agriculture are the major sources of pollution in the region. Transformation of the urban and periurban ecology, in particular, is the principal contributor to the microorganic, metallic, and chemical pollution of Lake Victoria. For many decades, point and nonpoint sources of pollution and contamination—containing PBCs, pesticides, heavy metals, and untreated industrial wastes and agrochemicals—have been rapidly proliferating throughout the watershed areas of Lake Victoria. Livestock production, coffee and tea processing plants, tanneries, breweries, sugar factories, fish-processing facilities, soft drink and food-processing plants, abattoirs, oil-processing mills, soap-making facilities, distilleries, mining operations, pulp and paper mills, textile factories, battery and paint production, metallurgical works, agroprocessing facilities, fertilizer and pesticide factories, energy plants, and other allied establishments today discharge their effluents untreated directly into Lake Victoria or into satellite water channels and wetlands in its wider environment (Crul, 1995).

In terms of microorganic pollution, the 40 cities and towns bordering Lake Victoria—where urban growth rates are in the order of 5-10 percent per annum—are the principal source of microorganic pollution. Lakeshore cities like Mwanza, Bukoba, and Musoma in Tanzania; Kisumu and Homabay in Kenya; and Kampala, Entebbe and Jinja in Uganda have antiquated or no sewage treatment plants and, as a result, raw sewage is simply dumped into the lake on a daily basis, creating permissive conditions that support the growth and spread of water-borne diseases (Odada et al., 2004; UNEP, 2002; Chege, 1995). For example, Kisumu's municipality collects a mere 1 percent of the city's 500,000 tons of solid waste, responsible for air and water pollution, lecheate generation, spread of infectious diseases, and blockage of sewer and drainage works. In Musoma, only 10 percent of the liquid waste generated annually is collected by the municipality. In the early 1990s, Tanzanian towns were believed to have been dumping 2 million liters of untreated liquid waste into Lake Victoria daily (Chege, 1995; Kinuthia-Njenga, 2008).

Moreover, the 40 million people who live in the rural regions of Lake Victoria have no sewer treatment facilities. Human and animal wastes in the watershed ecology are washed into the natural water channels, and finally into Lake Victoria during rainy seasons, amplifying the effects of microorganic pollution and the concomitant outbreaks of cholera, typhoid, paratyphoid, and other water-borne diseases, as well as the spread of malaria-carrying mosquitoes, bilharzias, and river blindness (Crul, 1995). Between June 1997 and March 1998, for example, cholera outbreaks along Kenya's Lake Victoria resulted in 547 deaths and 14,245 hospital admissions (Odada et al., 2004).

The magnitude of insertions of these pollutants into the lake is closely associated with the rates of urbanization, industrialization, and human biomass growth in the region, which are among the highest in the world (Nweke and Sanders, 2009). Cyanobacteria dominance of the phytoplankton community, resulting from overfertilization of the lake with nitrogen and phosphorus, has worsened microorganic pollution in and around Lake Victoria. For example, cyanobacteria dominance in the lake is believed to have led to the proliferation of microcystis in Lake Victoria, which produces a potentially dangerous toxin known as microcystin. Today, this toxin threatens the health of fish and people, especially the lakeshore communities that depend on this water for drinking and on the fish protein (Njiru et al., 2008).

The fact that Lake Victoria is an international body of freshwater magnifies the effect of insertions of microorganic pollution into the lake. There are 40 towns and cities around the Lake Victoria basin including such big cities as Kampala in Uganda with two million residents, Kisumu in Kenya with 360,000 people, and Mwanza in Tanzania with half a million residents, that increasingly discharge their effluents and solid wastes into Lake Victoria and its subsidiary channels (Nweke and Sanders, 2009). Townspeople wash their clothes directly in the lake using detergents and bathe themselves in the lake using soaps. The number of open-air garages and car- wash stations are multiplying along the shores of the lake; the dirty water from the cars together with gasoline and diesel leaks from cars in disrepair then enter the lake (Awange and Ong'ang'a, 2006).

There is also growing evidence that insertions into the lakes of heavy metals, originating in manufacturing establishments in urban centers, are growing. Metallic pollution thus represents yet another category of threat to the long-term health of Lake Victoria. Quarrying and mining operations (both large-scale industrial and small-scale artisanal) in the ecoregions of the lake are growing in number and intensity, making enormous contributions to lacustrine contamination. While global corporations control industrial mining operations, local individuals or groups engage in artisanal or small-scale operations that are dependent on nonmechanized operations. Of the countries covered in this volume, Tanzania, in particular, is a major producer of a variety of minerals, with huge ecological and socioeconomic significance for Lake Victoria. Mining operations pose two crucial challenges.

First, of the 92 known elements, 70 are metallic and are subject to quarrying and mining. Producing, processing, consuming, and disposing of these metallic elements strictly obey the thermodynamic laws of conservation and dissipation. From the perspective of the law of conservation, these metallic elements cannot be created or destroyed. From the point of view of the entropy law, however, the metallic elements are transformed from low-entropy throughput to high-entropy wastes. After production and consumption, the final destinations of the metals are terrestrial and lacustrine ecologies, where they do more harm to the environment and humans after their transformation than when they are in their natural state.

Second, in a context in which profit maximization and utilitarian considerations determine the scope of natural resource mining operations, control or regulation of pollution resulting from mining operations becomes exceedingly difficult. This is more so in rent-dependent countries such as those covered in this volume. In countries where the pirate state capitalists are held unaccountable, one needs to make a leap of faith to believe that mining corporations would be held accountable to live up to their obligations to further the cause of social responsibility. Even in the United States, where redundant formal oversight and regulatory institutions exist, there are still over 570,000 abandoned mines awaiting major clean ups at the cost of between $32 and $72 billion to the American taxpayer (Ali, 2009; Daly and Farley, 2003). In addition to the class limitations of the pirate state capitalists in the countries covered in this work, the states in the region lack institutional capabilities to monitor and regulate mining operations to come down in favor of environmental protection and social promotion. Furthermore, these states lack the institutional capabilities to make a difference in the relationship between mining corporations and the imperatives of ecological sustainability. For purposes of illustrating the point I have already made above, I shall now consider the impact of gold mining operations on the state of Lake Victoria.

A growing sector of importance in the world of metals is gold production. The greatest lure of this precious metal is its soaring global price. With respect to gold, the pirate state bourgeoisie and global mining corporations not only share the same vision, but they have also begun to speak the same language. The conventional vision is that greater accumulation through gold mining results in greater national economic growth through the injection of foreign investment, technology transfer, infrastructure development, employment generation, and greater earnings from the export of this particularly precious metal.

As a result, Tanzania has targeted gold production as a growth sector. In 2009, in a World Gold Council-commissioned study, Maureen Upton (2009) presents Tanzania as exemplar in demonstrating how large-scale infusion of foreign capital into the gold production sector has made a difference in GDP growth and revenue generation. She draws a spurious connection between the supposed rapid GDP growth in the Tanzanian economy and the six large capital-intensive gold mining operations that have come on stream since 1998. Since Tanzania rented out gold production to global corporations, the country's stock of foreign direct investment grew from $2.78 billion in 2000 to $5.94 billion in 2007. Gold accounted for two-thirds of the increase, reflected in the $770 million that Tanzania earned from gold exports in 2008. In 2002, Tanzania exported 900,000 ounces of gold, which jumped to 1.75 million ounces in 2007. Gold exports eclipsed all traditional exports by eight-to-one in 2004 and the earnings from gold export were reportedly 700 percent higher than the receipts from coffee exports. Thanks to global mining corporations led by Barrick of Canada and AngloGold Ashanti of Britain, Tanzania is today the third largest gold producer in the world (Upton, 2009). To placate popular opposition to natural resource mining, global mining corporations have found it easy to offer bundles of social promises such as the offer to build schools, health clinics, and roads.

Contrary to conventional claims, however, mining corporations are neither generators of employment nor cleaner than artisanal gold miners. In fact, the official rush to rent out gold mining has spurred the growth of artisanal mining operations worldwide. As we shall see below, while the environmental and socioeconomic impacts of industrial operations are well known, that of artisanal mining operations is less covered and, hence, less known. Poverty is driven by social marginalization, economic exclusion, and political disenfranchisement. People who lack capital and technology, proper training, and proper health and safety measures carry on artisanal mining operations. According to the International Labor Organization, there were 13 million people (including 1 million child miners) in 30 developing countries directly engaged in small-scale mining operations in 1999. Another 80-100 million were deriving their livelihoods indirectly from artisanal mining operations. However, artisanal miners are not limited to mining gold. They actively engage in the extraction of gemstones, diamonds, and tin.

In the DRC, for example, artisanal miners account for 70 percent of national diamond output (*Business Wire*, Sept. 7, 2009). However, the most serious environmental and health risks originate in the artisanal gold mining sector because of the heavy reliance on mercury amalgamation. Artisanal gold miners produce between 500 and 800 tons of gold per annum and, at the same time, release between 800 and 1,000 tons of mercury, which poses serious occupational, health, and environmental risks (Spiegel, 2009). In 1999, the DRC, Zambia, Malawi, Mozambique, Tanzania, and Zimbabwe combined had 1.2 million artisanal miners, women making up a significant percentage. For example, of the 60,000 artisanal miners in Mozambique in 1999, 18,000 were women; in Tanzania, of the 550,000 artisanal miners, 137,000 were women; and of the 350,000 miners in Zimbabwe, 153,000 were women. Since these figures relate only to the 1999 estimates, the present absolute number of women in the sector is very likely to be higher. For example, the number of artisanal miners in Tanzania is estimated to exceed 1 million today (Hentschel et al., 2002; Spiegel, 2009).

From both environmental and socioeconomic considerations, artisanal gold mining operations pose a very serious challenge, since they exist outside of formal regulatory machinery of the state and defy statistical accounting. More important, since the pirate state bourgeoisie generally define artisanal gold miners as "illegal," they are excluded from international assistance, national policy considerations, and public health programs (Spiegel, 2009). As a result, artisanal miners are largely reliant on mercury to amalgamate and concentrate gold because the use of mercury is a simple process, which does not require large capital investment, advanced technology, or great managerial skills. This explains why the artisanal gold mining sector uses more mercury than any sector in the mining industry (Spiegel, 2009). In order to recover the gold from the mercury, miners simply burn the amalgam using charcoal, without containment of emissions. Once the gold is released, the mercury is simply left to contaminate the air, soils, farmlands, and water systems, or the residue is left in the open shaft and washed away later by rain into rivers and lakes. Contaminated tailing material is stored in places near farms and wells. The final destination of the mercury is the natural water systems, nearby houses, schools, food stalls, and farms.

Two important characteristics make mercury pollution particularly dangerous. Mercury pollution is not visible, and thus its presence, long-term health impacts, and environmental impacts are not immediately noticed. For this very reason, mercury has earned the dubious distinction as the "chemical time bomb." Furthermore, since 70-80 percent of mercury is released to the atmosphere during the initial stages of concentration and amalgamation, the resulting pollution from gold mining operations travels far and wide in keeping with the laws of global biogeochemical cycling. Because of this reality, Lake Victoria is vulnerable not only to mercury pollution originating within the region itself, but also to atmospheric mercury deposition originating in such countries as South Africa and Zimbabwe. Since 1.2 to 1.5 grams of mercury are lost to the environment for every gram of gold produced, the dispersion, bioavailability, and bioaccumulation of mercury in soils, sediments, streams, rivers, and lakes are proportional to the size and scope of gold production in the environment.

Also, mercury is known to undergo transformation into methylmercury, through a process known as biomethylation. Microorganisms that live in anoxic conditions in lake-bottom sediments, such as bacteria, transform inorganic mercury into organic mercury compounds of which methylmercury is one form. Once transformed, methylmercury bioaccumulates in aquatic organisms and enters the food chain. Methylmercury is a neurotoxin, which causes many deformities in humans, especially pregnant women, children and diabetics. While the health of miners and communities who live close to gold mining areas is compromised through inhalation of mercury vapor and direct exposure to mercury, people who live far away are also adversely affected by consuming fish caught in mercury-contaminated bodies of water (Straaten, 2000; Spiegel, 2009; Nweke and Sanders, 2009). Several studies on the Geita District, in Tanzania, illustrate the double jeopardy involved in artisanal gold mining.

One of the eight districts in the Mwanza region and with a population of 712,000, including 150,000 artisanal gold miners, the Geita District is Tanzania's "King Solomon's mines." Constituting the heartland of the Lake Victoria green stone belt where precious metals are extracted, two factors make the district crucially important from ecological and socioeconomic considerations. The Geita District is located within the watershed areas of both Lakes Victoria and Tanganyika. The Rwamagasa area within the Geita District is populated by some 27,000 households and is just 37 km from Lake Victoria, but the area drains into Lake Tanganyika. And the Geita District accounts for 95 percent of Tanzania's national gold production. This distinction has made the area a magnet to both large-scale and artisanal gold mining operations. Recent sample tests on drainage sediments, soils, water, and fish from local ponds, as well as Lake Victoria, have shown significant elevation in mercury concentrations. In fact, in 1999, the European Union prohibited importation of Tanzanian fillets into European markets because of public health concern over mercury contamination and cholera outbreaks; in early 2008, Turkey declined over 16,000 kg of Tanzania's fish after discovering that the fish contained high levels of mercury (Spiegel, 2009).

Global corporations, at the invitation of the Tanzanian pirate state bourgeoisie, and defenseless artisanal miners are intensifying gold digging and processing throughout the Lake Victoria region with two profound socioeconomic, ecological, and public health ramifications. Because foreign corporations are privileged over artisanal gold miners to cast a wide net to excavate the country's gold in the Mwanza region, the traditional artisanal gold miners are being displaced. Since 1998, an estimated 40,000 artisanal miners have been displaced by giant global mining corporations, dedicated to the growth ideology (*The Citizen*, Dec. 5, 2009). The resulting intensification of the rush for gold has been more scarring of the lake region's landscape while, at the same time, deepening poverty in the area. When the Geita Gold Mine was opened by AngloGold Ashanti a decade ago, local communities were awash with corporate promises of job creation and social development. Within two years of the operation, however, the local communities accused AngloGold Ashanti of converting their land into another Ogoniland, a reference to the country of the Ogoni people in the Niger Delta, where Shell Oil Corporation transformed the pristine ecosystem into a biological desert. Furthermore, the Ogoni's water was polluted, the fish on which they depended for generations were contaminated, and the air they breathed became choking, all without any benefits to the people. As early as 2002, Geita District communities similarly began to endure the consequences of deforestation, acid rain, arsenic contamination, and cyanide pollution; AngloGold Ashanti boasted about spending $5 million in two years on social services such as schools, roads, and health centers, yet it earned $330 million from the area gold production over the same period (Mgama, 2003).

Also there is a common myth surrounding large-scale mining operations that regard them as cleaner than artisanal mining operations, because of the switch they have made from the use of mercury to the application of cyanide in their operations. Indeed, cyanide can be just as deadly if mine wastes from industrial operations are not properly disposed. Given the history of corporations, the gold mining companies in the Lake Victoria region are more likely than not to dump their wastes recklessly in order to avoid costs. To be sure, the cumulative impact of industrial and artisanal mining operations on terrestrial and lacustrine ecologies in the Lake Victoria basin is not hard to imagine, given the history of the mining sector. Even in the advanced capitalist countries, where politicians brag about having transparency, accountability, and redundant regulatory institutions, mining corporations constantly evade responsibility for proper disposal of waste and closure. Today, deforestation, soil erosion, river siltation, acidification, cyanide and mercury pollution, and alluvial degradation associated with mining operations are threatening Lake Victoria. After all, mining operations around Lake Victoria release 1.2 tons of mercury into the environment for every ton of artisanal gold that is mined (World Bank, 2008). Since the so-called Lake Victoria green stone belt is emerging as the new "el dorado," the increase in the bioavailability and bioaccumulation of methylmercury in the aquatic food chain appears very much in prospect. Even though mercury contamination in Lake Victoria has not yet reached threshold levels, there is mounting evidence that the elevation of methylmercury in the lake is growing. What is of grave concern is not only the mercury already in the lake, but the prospect of continual addition of mercury pollution to the lake. Whether originating in the urban ecology or in the mines, the insertions of multiple metallic agents into Lake Victoria pose serious threats to aquatic life and human health. These agents include mercury, uranium, radium benedium, chromium, copper, zinc, nickel, tin, cadmium, arsenic, and other toxic metals.

As Dodd (2002) notes, the first step in the process of metallic contamination of aquatic food sources involves bioconcentration of toxins in aquatic plants, in which potentially toxic compounds enter the tissues of organisms, followed by bioaccumulation of these compounds in the food. When the increases in the bioconcentration and bioaccumulation progressively move up the entire food web, biomagnification of the toxins becomes unavoidable. These contaminants are accumulated in the tissues of fish species with consequent biomagnification, which, in the end, affects the health of people who ingest contaminated fish. Owing to the long water cycling time, Lake Victoria is particularly sensitive to the bioaccumulation and bioconcentration of dangerous organic and inorganic pollutants.

Destitution in institutional capacity to monitor deposition of pollution in the lake simply worsens the situation. After all, even advanced countries know very little about the more than 72,000 chemicals that are commercially circulating. In fact, only 10 percent of these chemicals are reportedly screened for toxicity. Even in the United States, the most advanced country, only 0.5 of the chemicals in commercial use are subject to regulation (Dodd, 2002).

Petroleum prospecting is yet another source of pollution that may have far-reaching influences on the general ecology of the Great Rift Valley and the biolimnology of Lake Victoria. The recent finds of multibillion-barrel oil reserves in the Lake Albert basin (between Uganda and the DRC) have excited both Uganda's pirate state bourgeoisie and global oil corporations. Estimates of these finds range between 2 and 6 billion barrels on the Ugandan side of the Lake Albert basin alone, and more barrels of oil are awaiting discovery on the Congolese side of the basin (Westerkamp and Houdret, 2010). As of this writing, three blocks around Lake Albert had been under development with combined daily intermediate capacity of 150,000 barrels, which may potentially rise to 350,000 barrels per day by 2018. It is expected that the pirate state bourgeoisie will have $2 billion annually at their disposal from oil export by 2015 (*The Economist,* April 3, 2010*)*. The leading oil multinational was the U.K.-based Tullow Oil PLC, which was planning to invest $5 billion in the sector by 2015. At the time of writing, Tullow was negotiating with China's CNOOC to accelerate exploitation of Uganda's petroleum jointly (*Oil and Gas Journal*, Feb. 8, 2010).

Petroleum exploitation in the fragile ecosystem in the Lake Albert basin has both biosocial and security dimensions. To begin, the oil finds are in areas where the Bunyoro people consider all natural resources within their dominion integral to their biosocial possessions, essentially off limits to the central government without their consent. They had already questioned the legality of the secretive deals that Museveni struck with oil multinationals. If the proprietary status of the resource is not resolved to the satisfaction of the Bunyoro people, or if Museveni does not offer the Bunyoro elite a big cut of the oil bonanza, we may see a replay of the Niger Delta situation with far-reaching social, political, and environmental implications for the Lake Victoria basin.

Next, some areas in and around Lake Albert are disputed by the DRC. The politicization and securitization of oil may thus lead to a reckless pattern of military escalation between Uganda and the DRC. After all, in August 2007, there was an incident in which Congolese soldiers on the disputed Rukwanzi Island in Lake Albert killed oil prospectors carrying out survey work on the Ugandan portion of the lake. Even though 90 percent of the islanders are Congolese, Uganda contends that the colonial map places the island within Uganda (Westerkamp and Houdret, 2010).

In anticipation of potential internal rebellion and a regional imbroglio, Museveni had already created an elite army, supported by a lacustrine police unit equipped with six modern speedboats to protect the oil fields. Museveni's own son is now placed in command of this elite force. Moreover, Uganda and Russia were on the final stage of negotiation, as of this writing, in which Uganda would receive six Sukhoi jet fighters at the cost of $300 million; the Sukhoi jets are considered among the top fighter planes in the world (Watkins, 2010).

Finally, the drilling, production, refining, and transportation of petroleum in the Lake Albert basin would inevitably entail a significant flare of associated gas spills, deforestation, soil erosion, and water pollution. This carries wider implications for the biolimnology of Lakes Victoria, Kyoga, Albert, Edward, and George, as well as other water bodies. Only 40 percent petroleum is actually oil by weight and the rest is natural gas, composed primarily of methane. In the advanced countries, oil companies are required by law to capture the associated gas, which they either use for other purposes or reinject into the wells to pressurize them. In developing countries, however, it has been the practice of oil companies simply to burn off the associated gas because doing so is cheaper than installing modern equipment to capture the gas. The Niger Delta is a textbook case of this practice. In addition to dumping toxic water waste and mud into the natural water channels and lakes, the oil companies are likely to burn off the gas, polluting the air, soils, and water. Uganda has plans to stretch a pipeline from the point of production to the Indian Ocean, while China has offered to build a refinery to produce gasoline and petroleum products for Ugandan domestic consumption and regional distribution. These activities are likely to entail spills, affecting the lakes directly or indirectly. Indeed, the environmental situation can worsen since Tanzania and the DRC have also jointly agreed to prospect petroleum around Lake Tanganyika.

Lake overfertilization, known as eutrophication, has been another growing threat to Lake Victoria. Indeed, many authors who use the eutrophication-driven change hypothesis to explain the present limnological condition of Lake Victoria attribute the sudden banishing of some 300 haplochromine cichlids in Lake Victoria to anthropogenic eutrophication (Hecky et al., 1994; Kaufman, 1992; Seehausen et al., 1997; Deweerdt, 2004). Even though lakes undergo varying degrees of natural eutrophication over time, owing to changes in the biogenic and geomorphological conditions, what concerns us here is why cultural eutrophication of Lake Victoria has accelerated since the 1980s. Also of concern are the cumulative impacts that this phenomenon has had on the biochemistry of the lake and the composition of the aquatic species therein. Since cultural eutrophication is generally a result of human activities, it appears that the reactive mobilization of nitrogen, oxygen, phosphorus, carbon, hydrogen, and sulfur (the key elements that make up 95 percent of all living things) in the urban and terrestrial ecologies of Lake Victoria, may have provided the triggers for the process of eutrophication. Two particular human-induced changes in the urban and terrestrial ecologies of the lake are in noteworthy of consideration.

First, Lake Victoria is surrounded by vast fertile plains, where extensive agrarian and livestock production occurs. These processes release nutrient-rich residues that ultimately end up in the lake, enhancing the conditions for overfertilization and, hence, eutrophication. Ever since the introduction of colonialism into the region, agrochemical-dependent commercial plantations of sugar, tea, coffee, cotton, tobacco, and other allied cash crops have proliferated throughout the catchment areas of Lake Victoria.

Residues of such nutrient-rich chemicals as fertilizers, pesticides, and herbicides have been increasingly inserted into the lake for well over a century. This suggests that the lake may well have been on the trajectory of eutrophication for a long while. It is estimated that the Lake Victoria basin annually receives approximately 22,000 metric tonnes of phosphorus from its catchment. In addition to the synthetic fertilizers widely used throughout the basin, domestic waste and animal manure contribute 4,000 metric tonnes of nitrogen per year to the lake's total nitrogen budget (Awange and Ong'ang'a, 2006; Ochumba and Kibaara, 1989).

It thus stands to reason that the promotion of increasingly chemical-dependent agriculture, as a major source of fertilization and contamination of Lake Victoria, offers a partial explanation for its growing eutrophication. In addition to the reactive mobilization of phosphorus and nitrogen in the lake's basin, organochlorines, DDT, dieldrin, aldrin, and other persistent chemicals are believed to be used throughout East Africa to boost agrarian production, as well as to control malaria and sleeping sickness. This suggests that residues of these toxic chlorinated organic pesticides and chemicals may well have been inserted into the lake from eroded soils transported by rivers, precipitation, and wind. After all, high levels of organochlorine concentrations in the pelagic and benthic food webs of the East African lakes have been recently observed (Karlsson et al., 2000; Nweke and Sanders, 2009; Kweyuh, 1997).

The second partial explanation for Victoria's eutrophication is associated with the population pressure exerted on the lake and its catchment areas for living space, raw materials, and food. As indicated in previous chapters, the number of people deriving their livelihoods from the basin's natural resources has increased from 4 million in 1932 to more than 40 million today. The resulting extensive clearing of forests and riparian vegetation, as well as the drainage of wetlands for cultivation, human settlements, wood extraction, and commercial charcoal production appears to have reinforced the mobilization of eutrophying elements. Moreover, since terrestrial vegetation and wetlands are the natural barriers to the insertions of eutrophying nutrients, it is reasonable to hold the view that the absence or substantial degradation of terrestrial and riparian vegetation in the lake's basin may have compounded the biolimnological vulnerability of Lake Victoria to terrogenous nutrient loading and, hence, eutrophication.

In light of the preceding observations, I argue that insertions of phosphorus and nitrogen-rich deposits (the two elements generally seen as the major culprits) into Lake Victoria may have indeterminate influences on the geochemistry and hydrobiology of the aquatic ecosystem, relevant to the eutrophication dynamics. It is helpful to bear in mind that Lake Victoria's ecology, driven by solar radiation, generally depends on seasonal mixing and nutrient supply as a function of internal loading in order to keep natural primary production going. However, this process appears to have been adversely affected by the insertions of too much phosphorus and nitrogen into the lake, which are the limiting factors. Even a modest increase in phosphorus input into the lake, followed by high rates of denitrification in deep waters, can trigger an assortment of undesirable events, including deoxygenation of the lake floor, algae blooms, accompanied by the death of fish and other aquatic vertebrates (Ochumba and Kibaara, 1989; Containers, 2010; Smith and Smith, 2001).

For such freshwater bodies as Lake Victoria, phosphorus can be either a blessing or a curse, depending on the quantity of the element present in the lake. As a limiting nutrient in most freshwater bodies, the right amount of phosphorus is essential for primary production, but if too much of it is inserted into freshwater bodies, it causes eutrophication (Cox, 1993). The problem is that both the mobilization of reactive phosphorus fertilizer and phosphorus accumulation in agricultural soils worldwide have increased threefold since the 1960s (Millennium Ecosystem Assessment, 2005). It is reasonable to infer from this empirical observation that Lake Victoria could not have been immune to insertions of phosphorus beyond what the biological processes in the lake require to carry out all essential functions. With phosphorus-associated eutrophication, the composition of the phytoplankton community shifts markedly from the dominance of useful diatoms to the dominance of harmful blue-green algae, many of which are not harvested by herbivores efficiently. Often, scums of these algae accumulate on the surface and are blown by wind to shore, where they pile up and rot. Large quantities of these dead algae also sink to the bottom waters, where bacterial action makes the supply of oxygen during stratification even scarcer, deepening the eutrophication process.

As indicated earlier, the increases in the accumulation of phosphorus in Lake Victoria may come from direct atmospheric deposition and runoff from soils and rocks, agricultural and livestock production, urban operations, waste-water treatment plants, drained wetlands, and commercial cleaning preparations. In addition, when animals and plants die and sink to the bottom of the lake, bacterial action releases the phosphorus from the dead plant and animal tissues to keep the phosphorus cycle going. Thus, the abundance and bioavailability of phosphorus is an important indicator of algae bloom in Lake Victoria. In the context of growing urbanization around Lake Victoria, future deposition of more and more phosphorus in the lake from urban sewage is very likely to increase. After all, primary sewage treatment removes only 10 percent of the phosphorus, and secondary sewage treatment may remove only another 30 percent of the phosphorus (Smith and Smith, 2001). This means that, even if the urban centers around Lake Victoria are equipped with modern wastewater treatment plants, substantial amounts of phosphorus are going to be inserted into Lake Victoria, worsening the eutrophication of the lake. The persistence of phosphorus is such that it can remain in soils for decades before it enters water bodies (Millenium Ecosystem Assessment, 2005).

Generally, the principal means by which phosphorus in Victoria's waters is removed is by biogenic uptake and sedimentation by diatoms. However, as stated earlier, the shift from diatom dominance to cyanobacteria dominance in the lake's phytoplankton species composition (resulting from the simplification of aquatic communities due to eutrophication) has increased offshore phosphorus buildup. Additionally, inundation of the lake's shore area following the wet years of the 1960s, the weathering of phosphate-rich minerals from anthropogenic soil erosion, poor agricultural practices, livestock production, deforestation, and biomass burning have facilitated insertions of more phosphorus into the lake (Hecky et al., 1994; Hecky et al., 2010; Holttoman and Lehman, 1998; Lehman et al., 1998; Ochumba and Kibaara, 1989).

Nitrogen, like phosphorus, is also crucial to aquatic health and productivity. Even though nitrogen represents 78 percent of the atmosphere, it is not readily bioavailable without fixation. Here, bacterial and algal action play determinant roles. Lightning also plays an important role in this process, by causing nitrogen and water to combine to form nitrogen compounds. These compounds are carried by rainfall to the ground and to water bodies, for assimilation by plants. However, 90 percent of nitrogen fixation is done by bacteria, particularly cyanobacteria. The amount, bioavailability, and mobility of fixed nitrogen compounds are the major determinants in the health of any aquatic ecosystem. Indeed, as we shall see later, it is the magnitude of biogenic nitrogen-fixation that has become of concern in Lake Victoria, precisely because of the nitrogen-fixing cyanobacteria dominance of the phytoplankton community.

As suggested earlier, modernization of agriculture in the ecoregion of Victoria has unintended harmful impacts on the health and productivity of the lake. After all, 60 percent of global nitrogen fixation today is done artificially to boost food and agrofuel production. The biological problem is that bacterial action converts part of the nitrogen-fixation needed to make fertilizers into nitrates, ammonia, and other forms of chemical compounds. These compounds are inserted into natural waterways and the lakes as part of the runoff from agricultural lands. The magnitude of the insertions of such agrochemicals and other nutrient-rich compounds into the lakes is proportional to the size of the lakes' watershed, as well as to the levels of activity in the watershed. In addition, the depth and residence time of water have huge impact on the lakes' ability to flush out excess nitrogen compounds (Millennium Ecosystem Assessment, 2005; Sachs, 2008; Dodd, 2002).

It appears that excess nitrogen compounds have changed the biochemical composition and physiology of Victoria's hydroecology, having stimulative effects on the production of blue-green algae or cyanobacteria. In Lake Victoria, cyanobacteria are the dominant biogenic agents of nitrogen fixation, since they are equipped with heterocysts that are large and transparent cells along their filaments, allowing them to capture and fix nitrogen. This unique ability to convert nitrogen into various compounds in Lake Victoria makes cyanobacteria dominant players in the biochemical cycle and eutrophication of the lake. To be sure, the anthropogenic actions associated with inappropriate agrarian production and deforestation in the Lake Victoria basin are a major source of reactive nitrogen-rich external loads. Changes in the urban ecology, associated with industrialization and the transformation of urban life itself, have also catalyzed the deposition of numerous nitrogen compounds in the lake. This has reinforced the eutrophication process marked by cyanobacteria dominance. As indicated earlier, forming scums on water surface, cyanobacteria produce toxins that can be deadly to aquatic animal species and humans as well (Dodd, 2002; Ochumba and Kibaara, 1989).

From the point of view of fisheries managers, cyanobacteria are particularly dangerous because they can erode the productivity of lakes by their ability to monopolize sunlight. Because their scums form extensive mats, floating on the surface of the water, they shade out the nonbuoyant primary producers below. This is consistent with the general empirical observation of how excess nitrogen and phosphorus inputs have enhanced phytoplankton production on Lake Victoria's water surface and accounted for the total elimination or suppression of submerged macrophytes, on which aquatic invertebrates and many fish species depend. This sort of change may have far-reaching impacts on the biolimnology and productivity of Lake Victoria, since cyanobacteria dominance of phytoplankton community entails intensification of competition for nutrients and light. High phytoplankton density on the water surface, dominated by cyanobacteria, results in light attenuation, insufficient conveyance of inorganic carbon and oxygen, and impedes photosynthetic actions by underwater macrophytes. Moreover, accumulation of organic matter in sediments following massive algal blooms generates toxic substances during decomposition, which severely handicaps the life cycle of underwater macrophytes and aquatic macroinvertebrates. Since the submerged autotrophs cannot make their own sugar without light interception, light penetration, and light saturation, the overall lacustrine primary production becomes substantially reduced due to cyanobacteria dominance. Consequently, the heterotrophs, which depend on the autotrophs for the production and supply of primary nutrition, inevitably experience food insecurity. This partially explains the growing impoverishment of Lake Victoria.

What makes the self-shading cyanobacteria competitively dangerous is their unique possession of phycobilins, which are pigments that can readily absorb all wavelengths: blue, green, red, orange, and yellow. Other aquatic plants are limited to a very narrow spectrum of wavelengths, mainly to the greenish chlorophyll. In such eutrophic lakes as Victoria, phycobilins give cyanobacteria extraordinary competitive advantage over other aquatic primary producers, by intercepting sunlight and carbon dioxide, on which other aquatic primary producers also depend to carry out their photosynthetic activities.

In view of the fact that cyanobacteria have a gelatinous coating, which is toxic, herbivorous fish (which thrive on a phytoplankton community dominated by green algae) generally avoid consuming blue-green algae. As a result, the lacustrine trophic structure remains impoverished, as has happened in Lake Victoria. Furthermore, the accumulation of suspended particulates, nitrogen, phosphorus, dead organic matter, and other biochemical and physical agents, in combination and interaction, have the effect of reinforcing the eutrophication dynamics in Lake Victoria. In keeping with the second law of thermodynamics, dead algae sink to the bottom of the lake where they undergo the process of organic sedimentation through microbial actions, relying on increasingly scarce dissolved oxygen.

This has increased the biological oxygen demand throughout the lake. The result is an ever-expanding biological desert, in which Lake Victoria finds itself today. The oxygenated zones have shrunk, whereby the suitable fish habitats have been reduced. With the expansion of anoxic regions, the lake's species are limited to an ever-shrinking space; even worse, massive fish kills have been observed during upwelling of anoxic water (Toowoomba, 2000; Dodd, 2002; Johnson et al., 2000).

In light of the arguments advanced so far, scholars who forwarded the eutrophication-driven change hypothesis appear to be on solid ground, insofar as the explanation of the present limnology, biodiversity, and productivity of Lake Victoria are concerned. The available evidence suggests that supply conditions in Lake Victoria are adversely affected, not only by the modernization of the fishing industry and by the increasing number of fish harvesters, but also by the metabolic transformation of both terrestrial and urban ecologies that produce nutrient-rich by-products. It is clear that both the extensification of subsistence cultivation and the intensification of modern agrarian production throughout the catchment areas of Lake Victoria, coupled with the metabolic transformation of the urban ecology, have led to rapid increases in external loading. This contains excess phosphorus and nitrogen, with considerable consequences for aquatic species composition and the lake's trophic profile. The radical shift in the phytoplankton community composition, from diatom and chlorophyte into one favoring cyanobacterial abundance and dominance, is a strong evidence of excessive concentrations of phosphorus and nitrogen in Victoria. These concentrations may have provided the biogenic and physical triggers for state transition of Victoria's ecosystem, with profound ramifications for not only the biolimnology of the lake itself, but also far more for the welfare of riparian communities.

Today, the biochemistry of Lake Victoria is dramatically changed, due to external overload. Phytoplankton production has doubled, algal production has increased fourfold, the dominance of diatoms is replaced by blue-green algae, sulfate in the lake has increased fivefold, and total phosphorus concentration has more than doubled. Silicon concentration has declined by a factor of ten and chlorophyll concentrations have increased fivefold, accounting for 88 percent of the reductions in water transparency. Moreover, higher chlorophyll concentrations on surface waters indicate the unmistakable dominance of the self-shading filamentous cyanobacteria, resulting in oxygen supersaturation of surface waters, followed by widespread deoxygenation of lake bottom. Furthermore, the dominance of invertebrates has shifted from clanoid to cyclopoid copepods, and the benthic invertebrate community has shifted to freshwater shrimp *Caridina niloticus*, on which the Nile perch widely gorge today in the context of much reduced haplochromine populations (Bootsma and Hecky, 1993; LVFO, Strategic Vision 1999-2015; Hecky et al., 2010; Ogutu-Ohwayo and Balirwa, 2006; Njiru et al., 2008; Johnson et al., 2000; Lehman and Franstator, 1994; Crul, 1995; Seehausen et al., 1997; Lowe-McConnell, 1997; Kaufman, 1992; UNEP, 2009).

Before concluding this section, a few sentences ought to be written about whether the eutrophication-driven change hypothesis fully accounts for the sudden extirpation of a large number of haplochromine cichlid species of Lake Victoria. Whether the sudden collapse of the haplochromine cichlid populations and the widespread eutrophication of Lake Victoria in the 1980s was a matter of temporal coincidence or one of causal relationship still remains debatable. However, the fact is inconvertible that the biological characteristics and ecological requirements of haplochromine cichlids are incompatible with eutrophic conditions in Lake Victoria. The 1969-70 trawl survey of Lake Victoria showed that the 500 species of haplochromine cichlids had constituted 80 percent of fish biomass in the lake, a situation that remained unchanged until approximately the early 1980s. By then, the lake had undergone rapid eutrophication, accompanied by massive algal bloom of the blue-green variety. By then, the secchi transparency index of Lake Victoria's deep waters had sharply decreased from 5.5 to 8.0 meters in the 1920s to 1.3 to 3 meters, shortcircuiting the ability of the haplochromine cichlids to freely move throughout the lake in search of food or sexual mates.

We know that Lake Victoria has long been known for the abundance and diversity of haplochromine cichlids, which evolved in short geological time, through remarkably explosive sympatric speciation and extraordinarily rapid morphological radiation, into more than 500 species. Male coloration, water transparency, and trophic specialization appear to have been the determinants driving the evolution and diversification of the haplochromine cichlids. The prevailing view is that female mate preferences are determinant in the sympatric speciation and adaptive radiation of the haplochromine species. This female choice-driven sexual selection is, however, male color-driven. Male cichlids are either bright red, bright blue, or yellow. Such color polymorphism allows females to choose their mates discriminately, thwarting the potential for hybridization. In this case, visual endowment becomes a determinant factor in choosing mates.

Then, if color and sight are crucial determinants for haplochromine cichlid reproduction, water transparency becomes another determinant in haplochromine sexual reproduction. Loss of water clarity resulting from sedimentation and eutrophication short circuits female cichlids' ability to choose proper mates, with the potential to violate the biological integrity of the species. This potentially leads to hybridization. Since eutrophication alters the environmental light gradients, the *Haplochromis* tend to lose this important biological attribute of color polymorphism essential to sexual selection. Because of the loss of short wavelengths, owing to the narrowing of light spectrum, the female *Haplochromis* find it difficult to be choosy about their sexual mates in the absence of color polymorphism (Seehausen et al., 1997 and 2008 Chapman et al., 1998; Deweerdt, 2004).

It seems that even a minor change in the physical properties of water arising from lake sedimentation and eutrophication can disrupt the evolutionary trajectory of the species. Significant reductions in dissolved oxygen in the habitats, where the *Haplochromis* frequent to gather food and sexual partners, works to their disadvantage since freshwater bodies that experience high rates of organic decomposition and sedimentation, high rates of light attenuation, and high thermal stability are known to have low dissolved oxygen (Chapman et al., 2004). Water with hypoxic condition strongly affect demersal and benthic haplochromine cichlids in particular, forcing them to move to sublittoral and inshore waters in search of dissolved oxygen. Here, they become exposed to predation by such top predators as the Nile perch. Indeed, Lake Victoria's turbid waters have given all the advantages to the Nile perch, which does very well in dim light and turbid waters, whereas haplochromine cichlids are known to be poor performers under turbid conditions. To escape hypoxic condition and asphyxiation, demersal *Haplochromis* are forced to abandon their bottom-dwelling places and swim to the surface and/or inshore waters, where they become ready meals for the Nile perch (Kaufman, 1992; Chapman et al., 1995; Seehausen et al., 1997; Bootsma and Hecky, 1993; Hecky et al., 1994; Deweerdt, 2004).

The preceding discussion leads to two conclusions regarding the relationship between the cultural eutrophication of Lake Victoria and the fate of the *Haplochromis*.

First, the eutrophic condition appears to have an immediate impact on the feeding habits, dwelling places, and sexual reproduction of the haplochromine cichlids. They are known to require clear waters to make vertical descents, to swim laterally to avoid predation, and to search for food. This natural freedom appears to have been affected by changes in light attenuation and oxygen depletion, due to eutrophication of the lake. As the eutrophic region of the lake spatially expanded and lake waters became increasingly turbid, the dwelling space and food source of the haplochromines appeared to have substantially shrunk. For example, Kaufman (1992) found that only one species of *Haplochromis* per genus was present in the turbid waters of Lake Victoria, whereas the clear waters of the lake had six species per genus.

In addition, in areas marked by water clarity, the profusion of male fish colors allowed females to be choosy regarding their mates. In contrast, in turbid waters, the males from both species took on dull brown color, probably leading to hybridization as the barriers to gene flow between the species dissipated in response to high turbidity. At the same time, the trophic specialization of the *Haplochromis* dissipated as the species shifted their feeding habits toward homogeneity. For example, prior to the eutrophication of Lake Victoria, it was observed that the red cichlid species used to feed in the water column above rocks, while the food ecology of blue cichlids was in the bottom of rocks; but now, the once different species probably having become hybrid, niche partitioning does not appear to obtain in the lake (Kaufman, 1992; Seehausen et al., 1997; Deweerdt, 2004).

The second conclusion that emerges from the discussion about the impact of eutrophication on the fate of the haplochromine cichlids involves the biological simplification of Lake Victoria. It appears plausible to argue that the rapid entropic degradation of the lake has reduced the complexity and diversity of biotic communities and their food web structures, thereby eroding the co-evolutionary dynamics of the species concerned. As stated earlier, from an evolutionary biological viewpoint, both ecological complexity and biological diversity are crucial for species collaboratively to confront the effects of fluctuations in water levels, meteorological variability, epidemic diseases, and other perturbations (Goldschmidt, 1994).

It follows that simplification of Lake Victoria's faunal composition and biotic structure, resulting from anthropogenic eutrophication, appears to have short-circuited the adaptive mechanisms available to aquatic species. For example, the elimination of the haplochromine cichlids appears to have exacerbated the eutrophication dynamics in Lake Victoria. In their absence, the sinking of large quantities of blue-green algae to the lake floor has transformed large areas of the lake into a biological desert. Before the Nile perch decimated them under the cover of eutrophication, the blue-green algae grazers and detritivorous haplochrominne cichlids, which removed algae from the water surface and sifted decaying organic matter from the lake floor, used to act as vacuum cleaners and kept the algae from proliferating and sinking to the bottom of the lake. This thwarted the accumulation of dead organic matter on the lake floor. Now that the phytoplanktivorous and detritivorous haplochromines are gone or reduced, the dead blue-green algae have nowhere to go except to the lake floor, where they are decomposed by bacteria using scarce oxygen as an energy source. The bacterial activity has increased the biological oxygen demand, thereby reducing available oxygen supply to other species. Today, 70 percent of Lake Victoria is said to be anoxic year round, leading to a 50 percent loss of fishing areas (Goldschmidt, 1994; Njiru et al., 2008).

In summary, sedimentation, pollution, and eutrophication of Lake Victoria appear to have accelerated the entropic degradation of crucial lacustrine functions, resulting from the insertions of urban and terrestrial wastes into the lake. What this suggests is that without addressing the causes of urban and terrestrial conditions of pollution and degradation, the sedimentation and eutrophication of Lake Victoria are bound to continue. More important, without addressing the issues of survival concerning the masses, and the greed of the pirate state bourgeoisie and other wealth accumulators, doing more studies to explain why Lake Victoria has undergone such rapid entropic degradation in a short period becomes an exercise in futility. Although we talk about pollution, sedimentation, and eutrophication as responsible for the entropic disintegration of the lake, the workings of the necessity of survival and the logic of accumulation are actually the reasons for the state transition of the lake. And, so long as the prevailing social order remains ascendant, the further entropic deterioration of Lake Victoria is very much in prospect. Those who ignore this simple fact are peddlers in commercial scholarship.

THE PROMISE OF INDUSTRIALIZATION AND SUPPLY CONDITIONS

The area that has lately captured the imagination of native elites, expatriate advisers, and economic theorists as an alternative to dependency on natural resources is industrialization. The wisdom of the argument that diversification of livelihoods is essential if the ecosystem health of Lake Victoria is to be maintained is not in dispute. Furthermore, industrialization of the economies of Lake Victoria's littoral countries as an important alternative to the vandalization of natural resources is also something that cannot be disputed. Indeed, the 2005 World Bank's categorization of countries into natural resources-dependent and intangible capital-dependent is very instructive. The riparian East African states are too heavily dependent on natural capital, which means that they are depleting their natural resource base beyond the regenerative capacity of the environment and, in the process, destroying critical life-supporting ecosystem functions. This reliance on natural capital stocks will, in the long run, make them invariably and inevitably vulnerable to desertification, droughts, famine, and hunger. The burning issue is, however, industrialization under what conditions, for whom, and under whose control?

Without fully comprehending the type of industrialization required to serve as an alternative to the prevailing social system, industrialization strategy may, in fact, accelerate the metabolic transformation of the general ecology, leading to further entropic degradation of Lake Victoria and its basin. Given the state of development in the nations covered in this study, the word "industrialization" can only mean "dirty." After all, we have witnessed in the last half century that migration of companies from the global north to the global south has been accompanied by dirty industrialization. This migration has been widely supported by politicians and global institutions alike. It is useful to recall the infamous memorandum that someone leaked to the *The Economist* magazine from Lawrence Summers, President Clinton's Treasury Secretary and President Obama's former director of the National Economic Council. When he was chief economist at the World Bank, Summers wrote a memo advising the Bank to consider encouraging the migration of dirty industries to developing nations. The memo in part reads:

> Just between you and me, shouldn't the World Bank be encouraging more migration of dirty industries to LDCs [less developed countries]? I think of three reasons: ... the measurements of the cost of health impairing pollution depend on the foregone earnings from increased morbidity and mortality. From this point of view, a given amount of health impairing pollution should be done in the country with the lowest cost, which will be the country with the lowest wages. I think the economic logic behind dumping a load of toxic waste in the lowest wage country is impeccable and we should face up to that.... The costs of pollution are likely to be non-linear as the initial increments of pollution probably have very low costs. I've always thought that under-polluted countries in Africa are vastly underpolluted, their air quality is probably vastly inefficiently low compared to Los Angeles and Mexico City. (World Bank, Dec. 12, 1991)

Indeed, as we saw in chapter 2, members of the East African Community appear ready to accept Summers' advice since they have developed a master plan to use the Lake Victoria basin as the driving locomotive for industrial transformation of the entire region. Moreover, there is an even bigger and more ambitious plan in conception to create a north-south corridor all the way from eastern to southern Africa. The immediate focus is on several transit corridors: the central corridor running from the port of Dar Es Salaam, Tanzania, to Burundi and Rwanda; the north corridor running from Mombasa to Uganda, Burundi, Rwanda, and the DRC; and, the Lamu corridor going from southern Sudan to Ethiopia.

To this end, the East African Community, the Common Market for Eastern and Southern Africa, and the Southern African Development Community trading blocs held a conference in Lusaka, Zambia, on April 6-7, 2009. The primary aim of the conference was to deepen the regional integration process toward a single free trade area. Because the 24 countries belonging to the three regional trading blocs had a combined GDP of $624 billion by the end of 2008 and a combined population of 500 million potential consumers (not to mention their combined vast natural resources), the conference did enjoy a remarkable international visibility. In attendance were presidents of Kenya, Uganda, Zambia, and South Africa, and representatives from the World Trade Organization, the World Bank, the IMF, the African Development Bank, the European Union, the United Kingdom, the European Investment Bank, and the Standard Charter Bank Group. There and then, $1.35 billion was pledged, including $500 million from the World Bank, $600 million from the African Development Bank, and $150 million from the U.K. This was just the initial installment toward a cluster of $10 billion in projects covering railway links, road networks, port expansion, and energy infrastructure development to which member countries of the trading blocs will be connected. The 8,000 km long highway from Dar Es Salaam to Durban alone would cost $7.4 billion. Considered key to regional integration, this assortment of ambitious projects is expected to ease access to natural resource exploitation and to facilitate the transport of goods more rapidly and more cheaply to regional and international markets, whereby private foreign investors would be attracted to the region (Oxford Analytica, Daily Brief Service, May 26, 2009; Shacinda, 2009).

If the North-South Corridor Model initiative comes to fruition under the current reality, the entropic devastation of tropical forest resources and natural water systems could be beyond comprehension. The DRC's 200 million ha of fabulous virgin forest resources, in particular, are likely to be vulnerable to heavy logging by multinational timber companies poised to take advantage of the new network of roads.

Moreover, deficient—as it were—in national capital, the littoral countries in East Africa would have to loosen their environmental and labor regulations in order to attract foreign investors. Even if they succeeded in this area, the sorts of industrial plants they can attract are likely to be old-generation plants and, by definition, dirty. Because of a shift to the manufacturing sector using old and dirty technologies, these developing states may simply pollute and contaminate their lakes and other limited water resources and, by implication, reduce the amounts of physical, biological, and freshwater resources actually available to them. Oil spills, chemical contamination, sulfuric and nitric acid emissions from manufacturing and power- generations plants, combustion emissions from vehicles, and urban waste discharges can expedite the state transition of Lake Victoria. Nitric and sulfuric acid, in particular, can have far-reaching implications for the limnology, biodiversity, and productivity of the lake. Since emissions of sulfur and nitrogen eventually end up in the lake via dry deposition from the atmosphere or rain, cascading reactions are generated in water as oxidants to form sulfuric and nitric acids (Cox, 1993).

It is instructive to note that oxidized nitrogen results from burning organic matter (the union of nitrogen and oxygen under high pressures and temperatures of car engines) and from the conversion of nitrogen fertilizer to nitrogen oxide by certain groups of soil bacteria (Cox, 1993). There are cautionary lessons to be drawn from the collective experiences of even rich capitalist countries. For example, sulfur oxide emissions in the U.S. rose from 77.5 million tons in 1955 to 113 million tons in 1978, while nitrogen oxides emissions increased from 18 million tons to 64 million tons over the same period. These sulfur and nitrogen oxides are byproducts of manufacturing, heating, electricity generation, transportation, and industrial farming (Cox, 1993).

Given the history of capitalism, we cannot be oblivious to the fact that depletion of natural resources and pollution of the environment are the inevitable consequences of unfettered industrial production. The greater the rates of metabolic transformation of low-entropy throughput, the greater the amounts of high-entropy wastes inserted into the ecology. For example, the amount of organic petrochemicals produced worldwide has risen astronomically from 75 billion pounds a year in the 1970s to 350 billion pounds a year today, grossly undermining our food and water security and consequently, our health (Merchant, 2005). Here, those concerned about the effect of dirty industrialization on Lake Victoria's ecosystem health can learn a good deal from the Chinese model of dirty industrialization.

Lured by the neoclassical axiom that if China diverts the 1,000 metric tonnes of water used in agriculture to produce just one metric tonne of wheat worth $200 to industrial production of finished goods instead, it could produce goods worth $14,000, which China could export and use the revenue to pay for imported food (E. Wilson, 2002). This logic underpins China's strategic choice to make the manufacturing sector pivotal to its economic modernization drive based on discarded Japanese and Western machinery and equipment. Of course, this flawed strategy is based on a fallacious assumption that there are countries with surplus ecological footprints, which can be used to grow food for export to China.

More important, due to a combination of rapid industrialization and rising living standards, the industrial/commercial and residential water use in China is actually expected to grow from 52 billion metric tonnes a year in 2000 to 169 billion metric tonnes by 2025 -- in which case there will be no hydrological savings to brag about (Barlow and Clark 2002). Moreover, the environmental and hydrological cost of rapid Chinese industrialization has already become incomprehensible. China's new industries, chemical agriculture, and domestic users today discharge 25 billion metric tonnes of wastewater into the Yangtze River alone, 80 percent of it untreated. There are over 500 pumping stations along the banks of the Yangtze supplying drinking water to 186 cities. The river's delta is the powerhouse of China's economy, accounting for 50 percent of national GDP and supporting 400 million people (Xinhua General News Service May 30, 2006). In the past, many people had assumed that the Yangtze River (the world's third longest river) could not be polluted because the river's annual discharge of 900 billion metric tonnes of water would dilute and flush out all toxins to the sea. However, in May 2006, Chinese scientists sounded the alarm that the river is in the process of dying. Seventy-four of the 126 aquatic species, which were still flourishing in the river in the 1980s, have already become extinct.

Even though there are 170 wastewater treatment plants along the banks, they are no match for the amount of pollution that has overwhelmed the waste absorption and self-cleaning capacity of the river. The levels of toxic chemicals in the drinking water supplies in some cities are 190 times higher than the safety threshold set by the World Health Organization. Other Chinese rivers are facing similar problems. The Yellow River is said to be so degraded that its water can no longer be used either for human consumption or for irrigation. Eighty percent of China's 50,000 km^2 rivers are so polluted that they can no longer support fish species (Barlow and Clark 2002; E. Wilson, 2002). Furthermore, China today has the worst air pollution in the world, which kills over 400,000 Chinese every year. A third of China's land surface also suffers from acute acid rain and dry acid deposition, mainly resulting from the heavy use of low-grade coal (Fishman 2005).

China's air pollution travels far beyond its borders. Up to 40 percent of air pollution in Japan, for example, is attributed to Chinese dirty industrialization -- something that prompted Japan to dole out $3.1 billion in aid and loans to China between 1997 and 2002, specifically earmarked for environmental and pollution control projects (Fishman 2005). The human cost of China's dirty industrialization is equally staggering. Over 200 million of the 700 million Chinese workers are exposed to hazardous levels of carcinogens such as lead, mercury, cadmium, benzene, nickel, toluene, and other toxic chemicals on a routine basis. These commonplace elements in China are all banned in the United States. As a result of this exposure, there are 4.4 million Chinese who suffer from an acute lung disease known as silicosis (from working unprotected in factories), while millions more suffer from other debilitating illnesses. Millions more have lost their limbs. In 2005 alone, 390,000 Chinese workers died from work-related illnesses (Tofani 2007). The illnesses, amputations, and deaths of millions of workers are the price for the $287.8 billion trade surplus China enjoyed in 2006 with the United States alone. To keep the working class in line with productionism, China has created a 400,000-strong armed police force to deal with social protests and strikes (Cline 2007).

The enduring lesson from the Chinese experience for the East African riparian societies is that a potential shift from subsistence farming and limited irrigation to full-scale dirty industrialization under deregulatory market conditions (shepherded by the pirate state bourgeoisie) can have apocalyptic consequences for the integrity of the lacustrine ecology and for lakeshore communities as a whole. To clarify a point, lest the reader may draw a wrong conclusion from the observations ventilated in this section, let me claim that the question is not that Lake Victoria's littoral countries should not be industrialized. The tall order here is how they should industrialize, under whose control, and for whose benefit.

CONCLUSION

This chapter has shown how human biomanipulation of Lake Victoria's species composition through species introduction, sedimentation resulting from change in land cover and deforestation, and eutrophication from pollution, in combination with climate change, have induced state transition in Lake Victoria, favoring the growth of cyanobacteria, hypolimnotic deoxygenation of substantial portions of the lake. The radical alterations in the hydrobiology, geochemistry, and ecology of the lake are perhaps permanent and irreversible. So long as further changes in land use, exponential increase in agricultural activity in the basin, further human population growth, dumping of urban waste and effluent into the lake, and unsustainable human predation of fish continue, the biolimnological trajectory of Lake Victoria appears to be set toward irretrievable loss. Cyanobacteria dominance of the phytoplankton community (which is both undercutting the productivity of the lake and posing serious health risk for lakeshore communities), the extirpation of a vast number of native species, and the sharp decline in fish harvest are probably the early biological manifestation of the impending biolimnological breakdown of the lake. The new drive for dirty industrialization may simply serve to make entropic degradation of the lake total. As expressed in the opening epigraph to this chapter, this is exactly what Fryer and Iles feared would happen when they wrote their book 40 years ago.

5 | The Structural Conditions in Lake Victoria

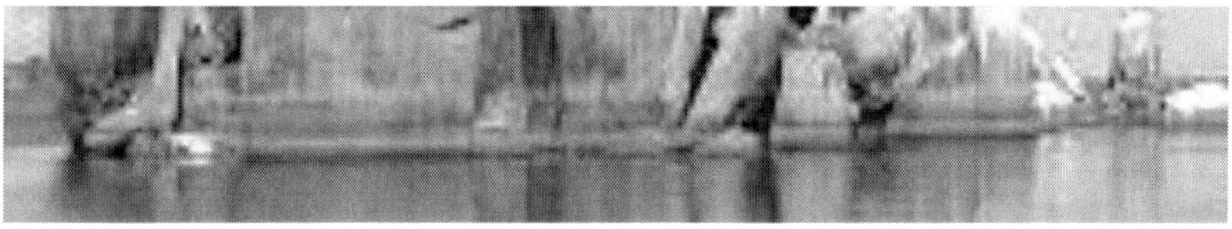

THE STATE AND THE FISHERY MODE OF PRODUCTION

Previous chapters analyzed the state transition of Lake Victoria from the general recognition of how the interplay of natural processes and human-induced changes in the lake's total ecology affect the fishery mode of production. In this chapter, I shall provide contextualization to the dynamic relationship between the social transformation of nature and the uneven distributional impact of this transformation on social classes. The pedagogical orientation informing the analysis here is dedicated to the proposition that the hydrosocialization process, as well as the production and distribution of lacustrine resources, are never politically neutral (Bryant and Bailey 1997; Swyngedow 2004; Robbins 2004; Merchant, 2005). Both the metabolic transformation of Lake Victoria's total ecology (inclusive of lacustrine, urban, and terrestrial) and the hydrosocialization process are embedded in social power, in terms of who would have privileged access to the lake's resources, who would bear the financial burden of producing the resources, and the negative externalities associated with the metabolic transformation of the lake's ecology. Social power enables those who wield it to control the production apparatus and the distribution spigot by establishing domination over both nature and labor. Unequal power relations determine who would control ecological resources and who would bear the consequences of ecological degradation (Swyngedow 2004; Robbins 2004; Bryant and Bailey 1997; Merchant, 2005).

As in all aspects of the capitalist processes of production, accumulation, and appropriation, the mediating role of the state in the fishery mode of production is central. After all, as Karl Marx observed well over a century and half ago, all economic decisions are, in the final analysis, political decisions made or mediated by the state. It thus goes without saying that poverty is not an economic condition but, rather, a condition of politics. In any political environment where social forces clash over access to and control over resources, the presence of the state in the fishery mode of production and marketing is crucial to the outcome of the struggle. In situations where the powerful rule, the state defines the contours of the struggle and ultimately predetermines the outcomes in favor of accumulators. As far as the fishery mode of production in Victoria's riparian countries is concerned, those who control the state have a stake in the operation of the fisheries sector for two reasons.

First, the fishery sector is one of the vital sources of tax revenues and foreign exchange earnings. Given this reality, it is in the interest of the pirate state bourgeoisie to define and regulate the parameters of fish production in Lake Victoria in the name of sustainable fishing. In practice, this excludes the ordinary and powerless fishers from access to lake resources.

Second, in addition to providing subsistence to the poor, the lake performs an important legitimation function in the interest of powerholders. The pirate state bourgeoisie define lake resources as national wealth held by the state in trust on behalf of present and future generations. This formally legitimizes the role of the state in imposing its control over the production and distribution of fish products. In theory, state policies and actions are cast in a broader framework to improve the living conditions of the poor. For example, the Kenyan government explicitly enunciates four objectives as integral to its fisheries policy, generating jobs and incomes, raising the protein intake of the populace by boosting production and supply of low-cost fish, improving living conditions of fishers and their dependents, and making the fisheries sector integral to Kenya's ability to generate hard currency by exporting fish (Abila, 2003; Jansen, 2007).

What has happened since the enunciation of this policy in the early 1980s, however, tells a different story. The government's preoccupation has been with the export aspect of the policy. The magnitude of tax revenues and export earnings from fish has been the yardstick by which success in the fisheries sector has been measured. Under an externally imposed liberal trade regime, the fish factories have been handsomely benefitting from the government's export promotion program. The benefits involve exemption from duties on imports of equipment, low- interest loans, credits, and tax exemptions (Abila, 2003). Yet, these same benefits are denied to artisanal fishers and local fish traders. This reality makes the class struggle between owners of capital and those denied access to the benefit of lake resources unavoidable. In fact, this provides perverse social incentives to lakeshore communities to ratchet up the struggle by chasing immature fish because survival becomes central to them. The veritable consequence of this social reality is serial depletion of species. In the context of debilitating conditions of poverty and the ravages of HIV-AIDS among lakeshore communities, the overly orchestrated slogan of "sustainable" fisheries development rings hollow in the ears of the poor.

The intensification of national social struggles over who should have access to lacustrine resources is eventually bound up with transboundary competition for fish production, simultaneously reinforcing the processes of entropic degradation of lake resources and emmiseration of lakeshore communities. Motivated by the imperative to generate export revenue, the three riparian states have intensified competition between themselves for lake resources. For example, in 2007 Kenya licensed over 80 new foreign companies to scour Lake Victoria and its coastal waters in order to raise its fish export to 230,000 metric tonnes per year (*Africa News*, June 4, 2007).

Likewise, Tanzania has actively boosted its fisheries sector, largely motivated by the need for hard currency. Tanzania's pirate state bourgeoisie has not been disappointed in their effort since, between 1990 and 2003, the country's earnings from fish export sharply rose from just $1 million to $154 million, making fish export the second biggest foreign exchange earner after gold (*BBC Monitoring Africa*, June 18, 2008).

The domestic and transboundary competition for lacustrine resources and the consequent social conflicts have further served to strengthen the mediating role of the state in the processes of fish production and accumulation. Highly synchronized national and regional measures of cooptation, exclusion, and coercion have become crucial instruments to which the riparian states resort in order to contain the class struggle, on the one hand, and regulate the competition among themselves, on the other. The social containment strategy operates at the local, national, and regional levels. A paradigm case of the strategy operating at the local level is the recent innovation of BMUs as building blocks of comanagement under the guise of promoting participatory approach to sustainable utilization and conservation of lacustrine resources.

In theory, BMUs would bring together government agencies, fishing communities of boat owners, boat builders, net makers and menders, fish processors and traders, and other stakeholders to share the responsibility and authority to manage the fishery mode of production for the benefit of all stakeholders. By the end of 2006, there were 1,087 BMUs around Lake Victoria, formed with financial support from the European Union. Each BMU is formally headed by a trained fisheries officer. These local units are expected to come together in the next state at the subcounty or district level, which in turn would become linked to fisheries agencies at the national level (Scullion, 2007). The whole exercise involves limiting the number of fishers with access to the lake, in effect quasiprivatizing the lake by excluding the poor masses from access to lake resources or to markets for fish caught without authorization by state agencies. Now, the social antagonism is redirected toward one between those who have privileged access to lake resources, in collaboration with state agencies, and those who are denied access to the same resources. This is the full application of horizontal coercion. Even international operators are now complicitous in this exercise of social containment strategy in the name of promoting sustainable harvesting. For example, Naturaand of Germany announced in April 2009 that it was going to purchase Nile perch products only from eight certified sites involving 1,000 fish harvesters in Tanzania; the perch products would carry the company's logo (Josupeit, 2009).

However, rather than containing the social contradictions, BMUs have actually intensified the class struggle at the local level, since fishery officers and local beach managers have formed a new stratum of exploiters. Using their newly acquired positions to create wealth for themselves, members of this stratum arbitrarily impound the fishing gear of subsistence fishers, confiscate the fish they catch, and carry out sweeping arrests on the water bodies supposedly to deter "illegal" fishers. The impounded gear and confiscated fish are then resold to the highest bidder, only to line the pockets of the new local petit bourgeoisie. It has even become commonplace among these new local bosses to extort money from subsistence fishers in exchange for their freedom. In fact, the social antagonism between fishers and BMU heads became so sharp, that Museveni issued a presidential directive in 2010 prohibiting security operations on Uganda's water bodies by district fishery officers and BMUs. Henceforth, only the National Ministry of Fisheries, supported by the police, would be authorized to enforce fishing regulations and security operations (Bagala, 2010).

On the national level, the fisheries departments and related agencies have their functions expanded and strengthened; they are the sole authority to register and issue fishing licenses for those who seek to operate in the respective national waters of Lake Victoria. Operational expression of this authority is to be provided through strengthening the coercive capacity of the states. This entails increasing the number of modern patrol vessels by each littoral state, fully equipped with updated communication gear to survey jurisdictional areas, spot illegal fishers, and relay the information in real time, in order to apprehend the fishers and subject them to harsh penalties for retribution and deterrence. For example, Uganda launched ""'Operation Clean" in 2003, which is a national campaign to crackdown on "illegal" fishers and their destructive fishing gear. As the operation succeeded in confiscating what authorities considered illegal gear, many fishers saw their livelihoods destroyed (Grellier et al., 2004). Moreover, in the spring of 2005, Uganda deployed its patrol vessels to its portion of Lake Victoria purportedly to fight piracy, illegal fishing, and cross-border smuggling (LVFO, 2005).

At the regional level, the littoral states of Lake Victoria have developed a series of plans, strategies, and organizations, beginning with the formation of the Lake Victoria Fisheries Organization (LVFO) in 1994 as the primary enforcer of the containment strategy. The frameworks agreed to by the three states include the LVFO Strategic Vision for 1999-2015, the Vision and Strategic Framework for the Management and Development of Lake Victoria Basin, and the Lake Victoria Management Plan. In May 2004, the highest organ (Council of Ministers) of the LVFO also adopted the "LVFO Regional Plan of Action to Prevent, Deter and Eliminate Illegal, Unreported and Unregulated Fishing on Lake Victoria and its Basin." As the title of the agreement suggests, the states would be deprived of a stream of foreign exchange earnings unless collective actions were taken to insure the prevention, deterrence, and elimination of what the signatories define as illegal fishing on Lake Victoria.

Coordination and harmonization of action plans by the three riparian states is thus considered critical to "sustainable" fisheries development. Under the agreement, no fishing boats, no fish collection and transportation vessels, and no fisherfolk would operate on Lake Victoria without prior registration with the respective fisheries departments of the riparian states. In order to obtain a fishing license, a boat owner must demonstrate his or her financial capacity to operate on the lake. The agreement also obliges the signatories to use their market power to deny illegal fishers or illegal fish products access to their markets. Moreover, the riparian states would not permit any fisher who is not member of a BMU to operate on the lake. In order to insure implementation of the regional plan of action, the LVFO would be strengthened in capacity to patrol, monitor, and control all fishing activities on the lake.

Finally, the riparian states have agreed to pool their coercive resources at the regional level with a view to fight piracy, smuggling, and illegal crossborder fishing. In 2002, with funding of 29 million euros from the European Union, the riparian states of Lake Victoria set up a project to take the crucial first steps toward strengthening their coercive ability to protect the lake against the presumed illegal activities. This involves carrying out joint patrol of the lake and using personnel in uniform, supported by 34 rapid patrol boats to combat piracy, smuggling, and harassment of duly licensed fishers (*Africa Analysis*, Oct. 4, 2002).

THE INTERNATIONAL DIMENSION OF STRUCTURAL CONDITIONS IN LAKE VICTORIA'S FISHERY MODE OF PRODUCTION

Among historically known modes of production, capitalism is unique in the sense that the manifestations of its characteristics are contradictory. Capitalism is at once retarding and transforming, displacing and replacing, excluding and including, and marginalizing and empowering. These features of the system are the source of economic exclusion, social injustice, and power inequality. For example, the pre-Nile perch fisheries regime in Lake Victoria was dominated by traditional and artisanal fishers. Boat owners were directly involved in the operation, management, control, and marketing of fish. Most people owned their fishing gear. Harvested by some 50,000 fishers, operating 12,000 fishing boats from the three littoral states, yearly fish production was relatively stable at 100,000 metric tonnes on Lake Victoria in the 1960s and 1970s. Almost 80 percent of the fishers earned their entire income from fishing. Moreover, since fishing activity was a family affair in the pre-Nile perch period, fishers did not own more canoes and gear than they could use. Furthermore, by division of labor, women were active beneficiaries of the primitive mode of production by specializing in fish-processing and marketing (Abila and Jansen, 1997; Jansen, 2007).

However, the dominance of the Nile perch in the lake and the consequent industrialization of fish production with injection of outside capital—supplied by international development banks and donor agencies—radically transformed the social structure of the fisheries regime. Displaced by capital, 83 percent of fishers today own neither canoes nor gear; they have become proletarianized in the service of absentee investors in Victoria's fisheries (Jansen, 2007). Absentee investors have no difficulty in finding investible resources from multilateral or bilateral financial agencies. For example, big owners of four fish factories on the shores of Lake Victoria received $42 million from the African Development Bank, donor agencies, and local financial institutions to capitalize their fish production shops (Abila and Jansen, 1997). By 1995, the establishment of 35 large fish-processing factories on Lake Victoria's shores from the three countries represented total proletarianization of labor, as well as the commodification and globalization of Lake Victoria. This inevitably resulted in a thorouGHGoing restructuring of the division of labor. At the top of the supply chain are foreign companies and/or big domestic operators, involved in acquiring, cleaning, filleting, freezing, packing, and exporting the fish.

Fifteen fish factories were on the Kenyan side of the lake by 1995, each averaging 380 metric tonnes of Nile perch processing capacity per day, of which eight were operated by men of Asian origin and another by an Israeli firm. Asians, Swiss, Israeli, and Russian nationals occupied all top management positions (Abila and Jansen, 1997).

By 1994, there were 11 fish-processing firms on the Tanzanian portion of Lake Victoria, with annual fish-processing capacity of 138,000 metric tonnes. In effect, these firms controlled 87 percent of the maximum sustainable annual fish yield on Tanzania's portion of the lake, even though the annual supply of raw Nile perch then stood at 53,000 metric tonnes. To encourage investment in the export sector, the filleting plants were exempted from taxes (D. Wilson et al., 1999). In 2009, there were 15 fish-processing factories on the Ugandan side of Lake Victoria (Beuving, 2010).

Supported by loans or contracts with large global corporations, and by dominating the value-added downstream activities of the supply chain, these fish corporations have all the advantages of externalizing the cost of fish catches. Since they control all downstream operations of the supply chain, they have an unencumbered ability to impose their quality controls and prices on the upstream operations of the sector. Hazardous conditions and postharvest losses associated with the actual fish production are externalized to local fishers. The prevailing information asymmetry between the owners of fish-processing plants and local fishers about global market conditions also allows the processers to determine the price for fresh fish. The highly convoluted structure of the division of labor in the fishery mode of production has further strengthened the asymmetry of information flow in favor of investors. Either foreign fish- processing firms rely on contractual relationships with fishers, or they rely on intermediation of local agents. In the first case, the processers directly extend credit to fishers in the form of equipment, gear, boats, trucks, and collection boats. In return, the fishers are placed under unconditional obligations to sell their fish to the creditor at prices determined by the processers. Alternatively, the firms use local intermediaries or resident agents who are contractually attached to a particular company or independent agents. These agents either buy directly and transport the fish from the beaches to the processing plants, or delegate such activities to subagents. In this intricate social relationship, monopsony is at work for a number of reasons.

The first is information asymmetry. The actual producers have no idea about the ongoing price for fish on a particular day. The fish corporations are the price setters, while the buying agents serve as price transmitters, and the actual producers are price takers.

Second, fish by their very nature are vulnerable to rapid deterioration. Thus, in order to avoid or minimize postharvest loss, fishers are compelled by circumstances to accept the price set by the buying agents.

Third, the fishers are tied to the agents by credit and loans. The buying agents provide the fishers with credit and loans for fishing gear and supplies, on the condition that the fishers would sell their catches only to their contracting agents. The conditions rob the fishers of their negotiating power.

Fourth, because of the Nile perch-based vertical expansion of fisheries, the capital- intensive nature of fishing operations have had the effect of excluding nationals from participation in downstream operations. Only those persons who have access to international finance capital can afford to purchase specialized collection vessels, transport trucks, ice machines or freezers, and processing factories (Beuving, 2010; Charles, 2001).

In sum, the pirate state bourgeoisie of Lake Victoria's littoral countries, the owners of the fish factories, international market forces, and bilateral and multilateral financial agencies have effectively outflanked domestic fish producers and fish traders (Jansen, 2007).

THE EPIDEMIOLOGICAL DIMENSIONS OF LAKE VICTORIA'S GENDERED FISHERY MODE OF PRODUCTION.

Historically, women have played an active role in Lake Victoria's fishery mode of production. In particular, women have dominated the processing and marketing of fish and fishery products. However, with respect to the scale of overall activities in actual production, women have been limited to near-shore fishing and onshore activities because of gender-defined roles. Women have been regarded as too weak to venture offshore into the lake to do dangerous work. This self-serving pity for women was conveniently connected to the evil spirit; if women went fishing further out on offshore areas, they might anger the spirit, resulting in bad luck upon fishermen (Thado, 1998; Westaway, 2007). The role of women has thus been limited to receiving the fish from the fishermen on the beaches. They would then process the fish by drying or smoking it for local and regional distribution. This gender-specific division of labor, sanctioned by tradition and superstition, readily lent itself to making women historically dominant in the postharvest sectors such as fish-processing, smoking, drying, frying, marketing, and other ancillary services related to making pots, baskets, traps, hooks, and nets (FAO, 2008).

As though the exclusionary effect of tradition was not bad enough, globalization of Lake Victoria has further worsened the status of women. Prior to the internationalization of the fisheries sector, women around Lake

Victoria were at least able to receive most of the catches from boats arriving on communal beaches, which they then smoked, sun-dried, or deep-fried for distribution to lakeside communities. Even in the early years of fishery commercialization, women were still visible in processing Nile perch and tilapia.

However, the completion of the commodification and globalization of the fishery mode of production has totalized the social marginalization of women, as they are now pushed out from the processing and distribution network. Today, bound by contractual obligations, fishers bypass women and hand over their catches to intermediaries waiting on the beaches. The intermediaries wait with their refrigerated trucks, for direct delivery to the fish factories. Today, most women spend their days roving the beaches in search of fish that are unwanted by the fish factories, or squatting by fish factories to collect Nile perch frames (such as head, tail, internal organs and skeleton thrown out by the fish factories) to process and deep-fry for local consumption. Women wait for hours by fish-processing factories until the Nile perch is filleted and the skeleton and other organs of the fish are discarded for collection. In the 1990s, the number of women hunting for perch skeleton at Homa Bay on the Kenyan portion of Victoria became so high that women spontaneously organized themselves into four groups of 200 each. By prior arrangement, the three groups would collect the skeletons twice a week, while the fourth group collected the skeletons once a week (Jansen, 2006; Rabi, 1996; Awange and Ong'ang'a, 2006). Oddly, even the collection of Nile perch frames by women has come under threat from fishmeal plants that process the fish frames for animal feed. The impetus for this change came from the European Union, which recommended that all Nile perch skeletons be sent directly to fishmeal factories in order to maintain hygiene and sanitation in and around the factories. As a result, by 1997, 70 percent of *Rastrineobola argentea* and two-thirds of Nile perch skeletons were sent directly to fishmeals factories, bypassing women processers altogether and further marginalizing them. Until the late 1980s, all Nile perch skeletons, juvenile perch, and *Rastrineobola argentea* were consumed by humans (Njiru et al., 2005; Abila, 2003; Jansen, 2006). The gender demarcation in social stratification in the fisheries sector is very much evidenced in all landing beaches around Lake Victoria, although minor variations may be noticed from beach to beach.

In addition to social and economic marginalization, the scourge of HIV-AIDS has disproportionately overwhelmed fishery-dependent communities, in general, and women, in particular. Of course, this epidemiological reality cannot be divorced from the social context of the fishery mode of production. In their highly acclaimed work on the subject of HIV-AIDS, Barnett and Whiteside (2006) draw a conceptual and operational distinction between susceptibility and vulnerability. The former pertains to existing conditions that make the spread of a particular epidemic likely, while the latter refers to the adverse effect that the disease can have on the inflicted persons or communities. The adverse effects are expressed not only in terms of high morbidity and mortality, but also in terms of economic exclusion and social stigmatization. This distinction is particularly relevant to understanding and analyzing the history, economics, culture, social situation, and human ecology of lakeshore communities. As Barnett and Whiteside (2006) note, such infectious diseases as HIV-AIDS are not random events. They happen because of multiple factors that make people susceptible to the events. Poverty, unemployment, low education, breakdown or absence of social support systems, frequent geographic mobility in search of income, marketization of human relationships, transactional sex, and customs that reward sexual prowess are a crucial cluster of factors that contextualize the determination of livelihood choices and the susceptibility of persons to risk conditions (Barnett and Whiteside, 2006). The spread of HIV-AIDS among Victoria's lakeshore communities is a microcosm of the national and continental disease dynamics. In 2004, 25.4 million people (60 percent of total world) were living with HIV-AIDS in sub-Saharan Africa. More than 57 percent of adult females in the continent were carrying the virus. For every ten men infected with the disease, there were 36 young women living with the disease. Of the 43.4 million orphans in sub-Saharan Africa, the orphaning of 14 million children was due to HIV-AIDS (Barnett and Whiteside, 2006). By 2000, 940,000 Tanzanians had already died from AIDS and left 730,000 orphans, and there were 1 million HIV-AIDS cases (Appleton, 2000).

The geographic dimension of the fishery mode of production is another important contributor to the spread of HIV-AIDS. Most fishing communities around Lake Victoria are dominated by migrants, characterized by extremely high geographic mobility. Not only do these fishers leave their families, friends, and communities behind, and detach from their social anchors, but they also travel from beach to beach in search of income. In this way, they lose important social mores and become principal agents of disease transmission. Furthermore, recreational facilities are virtually nonexistent around lakeshore communities. In most cases, the only recreational outlet is drinking. Alcoholism in combination with the youth of fishers serves to erode social inhibitions regarding proper sexual relationships. The fact that fishers are paid daily rather than weekly or biweekly makes immediate access to drinking that much more tempting. Under present reality, the labor force in Lake Victoria remains largely comprised of socially uprooted migrants, moving from beach to beach and following the seasonal pattern of fish harvests. If this continues, the temporary separation of fishers from their wives and their sociocultural anchors is likely to make the highly mobile fisher susceptible to risky social relations or sexual transactions, and thus vulnerable to the social and economic consequences of the risk taken.

For example, Tanzanian fishers are five times more likely to die from AIDS than are farmers, due to the differences in objective social conditions. A study in Uganda also found that 57 percent of the causes of death among fisherfolk around Lake Victoria were attributed to HIV-AIDS (Grellier et al., 2004; Westaway, 2007; Bene and Merten, 2008). According to another study, the seroprevalence rate among fisherfolks in Kenya was 30.5 percent, which was 4.5 percent times higher than the rate among the general population. In Uganda, the seroprevalence rate among fisherfolks was 24 percent, which was 5.8 times higher than among the general population. Even when compared to truck drivers, generally considered among the most susceptible persons to HIV-AIDS, the seroprevalence rate among Kenyan fisherfolk was 2.1 times higher than among Kenyan truck drivers and 1.8 times higher among Ugandan fishers than among truck drivers. At the time the study was conducted, there were 44,000 Kenyan fishers infected with the disease compared with 8,000 truck drivers, and 33,000 Ugandan fisherfolk had the disease compared with only 5,000 truck drivers (Gordon, 2005).

Although by no means gender specific, the HIV-AIDS epidemic affects women more than men due to a number of objective social factors. Forced out from primary fish production for lack of access to capital, gear, and boats, women generally engage in local beer brewing, food vending, and transactional sex; this makes them totally dependent on men to consume their beer, food, and transactional sex (Grellier et al., 2004; Westaway, 2007; Bene and Merten, 2008). This social reality increases their chances of becoming susceptible to HIV infections, and thus becoming vulnerable to its impact in terms of not being able to generate income to support their children.

Moreover, the commercial transformation of Lake Victoria from one based on multiple species to one based on three species has strengthened the gendered social stratification in lakeshore communities. This has forced women to resort to commercial sex as a source of livelihood. As a result, what some writers have called transactional sex has become institutionalized in lakeside communities (Bene and Merten, 2008). As Steve Bloomfield (2006) aptly described, "the tilapia and the Nile perch have become a deadly catch" for women. Utterly destitute, women cannot buy tilapia or Nile perch to process for the local market to earn a living, support their children, and pay for their education. Therefore, they offer fishers sex in lieu of money in a transaction known as Jaboya. The man gets immediate gratification from the sexual transaction, and the woman gets fish in hopes of selling the fish and staying afloat. In effect, the female sex worker transforms part of her body into a form of primitive capital without accumulation.

Usually, women enter the fish business to provide for their children when they are widowed or divorced. As far as women are concerned, the fishery mode of production and the Jaboya system have become inseparable, creating conditions of both susceptibility and vulnerability. Women are first stripped of their natural right to access lake resources, forcing them into transactional sex, and then, as a result, they are stricken by the deadly HIV-AIDS. As a Ugandan woman dying from AIDS put the case, "When my husband was alive, I used to pay for the fish and go to the market. After he died, I needed to give something else for the fish because I didn't have the money. I used to do it everyday" (Bloomfield, 2006).

Economic exclusion, social marginalization, and susceptibility and vulnerability of people to the ravages of diseases all have political dimensions. Politically marginalized, fishing communities have little or no voice in how social delivery systems are structured and how political interests are articulated. They are far removed from centers of transportation networks, communication, and public health services. For example, 50 percent of Tanzanian and Ugandan fishing boats in Lake Victoria are on islands, where fishers make temporary residence without access to public services of any kind. Even fishing communities on mainland beaches chiefly rely on boats for travel, since roads are either nonexistent or extremely poor. Moreover, because fishing communities are generally absent from electoral politics, national legislations and policy formations do not reflect their concerns. For example, although the seroprevalence rate of HIV-AIDS is significantly higher among fishing communities than in rural areas in Uganda, the concerns of lakeside communities do not figure in the prevention and impact mitigation strategies of the country's 700 HIV-AIDS support service providers (LVFO, 2006; Grellier et al., 2004; A. Gordon, 2005). The absence of sanitation facilities and services around Lake Victoria makes human habitation around the lakes extremely susceptible to opportunistic infections. In 2004, of the 1,500 landing sites around Lake Victoria, only 9 percent had sheds. Only 4 percent had potable water, less than 1 percent had cold storages, only 3 percent had fish storages, only 20 percent had toilets, and only 27 percent were receiving electricity. The results of such dismal social realities are regular outbreaks of cholera, dysentery, typhoid, malaria, and bilharzia infestations (LVFO, 2004b; Awange and Ong'ang'a, 2006).

In sum, several factors make fishing communities susceptible to HIV infections and vulnerable to its impacts. Among these factors are geographic isolation and mobility, lack of access to clean water and sanitation, social marginalization, political underrepresentation, high-risk lifestyle, the prevalence of many epidemiological factors such as malaria and bilharzia, the social degradation of women, alcohol as the sole source of recreation, and the absence of health care facilities and social support networks. It is no wonder that the prevalence rates of HIV among fishing communities around Lake Victoria are between 10 and 40 percent (LVFO, 2006).

THE GLOBALIZATION OF LAKE VICTORIA'S FISHERY MODE OF PRODUCTION AND THE GROWING FOOD INSECURITY OF RIPARIAN COMMUNITIES

One of the great paradoxes in the fishery mode of production in Lake Victoria is the inverse relationship between increased catch of fish and decreasing local food security. This largely has to do with the globalization of the fishery mode of production. Fish is diverted from local markets to the global north to satisfy the wants of the affluent. In addition, significant amounts of fish that were traditionally used as local food are now being processed as fishmeal, which is exported to the global north to keep the livestock revolution going (Jansen, 2007; Charles, 2001). A few decades ago, the conventional wisdom was that a vibrant private fishery system would grow lakeshore communities out of poverty. However, contrary to the rosy promises of yesteryear, with increased harvest of lake resources, the poverty of indigenous communities around Lake Victoria has risen over the decades inversely to the increased amounts of fish exported. For example, in the preNile perch fishery regime, the average annual catch of fish from Lake Victoria was 100,000 metric tonnes compared with 500,000 metric tonnes during the Nile perch era. However, even though fishery production increased by a factor of five, local fish availability dropped by 50 percent and half of the children around Lake Victoria suffer from chronic malnutrition. Per capita fish consumption in Kenya fell from 6 kg per annum in the early 1980s to 3 kg in the late 1990s (Jansen, 1999). Even though 80 percent of Kenya's fish- processing factories are located in Kisumu, the city has the largest number of people suffering from chronic protein deficiency and absolute food insecurity among the country's urban centers. More than 50 percent of Kenyans living in the rural districts bordering Lake Victoria do not receive the minimum daily protein requirements (Abila and Jansen, 1997).

In addition to converting over 60 percent of Nile perch frames into fishmeal, the fish- processing factories have now targeted the *R. argentea* for fishmeal production. Traditionally, *R. argentea* was exclusively for local human consumption. This was particularly important to the food ecology of lakeshore communities, since production of *R. argentea* has been steadily rising in the context of both Nile perch and *O. niloticus* having long become beyond the reach of riparian communities. In 1995, for example, Kenyan fishers harvested 77,000 metric tonnes of *R. argentea*. Between 1989 and 1996, *R. argentea* accounted for 38-45 percent of total fish production on Kenya's portion of Lake Victoria, becoming a crucial component of the local food ecology. However, lured by the potential profit that could be had from the export of fishmeal, fish factories migrated to *R. argentea* to convert it into fish feed for export. By 1995, there were six major fishmeal firms in Kenya, primarily targeting the *R. argentea*. Two of these firms controlled 75 percent of the fishmeal production. In 1995, the fishmeal firms processed 53,000 metric tonnes or 70 percent of the *R. argentea* landed that year into fishmeal, which meant that only 24,000 metric tonnes were left for local people to consume (Abila and Jansen, 1997).

The migration of fish-processing factories to what had generally been considered local reserves served to further erode the food ecology of riparian communities, compounding the growing food insecurity of lakeshore communities. In fact, when news about starving children in the Suba District in Kenya's Nyanza Province became an international story in 2006, a New Zealand millionaire offered to ship dog food for the children, a suggestion that bruised the pride of many Kenyans. With 1,005 km^2 in area and inhabited by 160,000 people, Suba District is Kenya's fishing center on Lake Victoria. In 2005, 20,000 fishers operated 6,000 fishing canoes and dugouts and harvested 42,000 metric tonnes of fish. Of this total, 22,000 metric tonnes was Nile perch, which raked in $29 million in foreign currency for the pirate state bourgeoisie (Abila, 2006).

As the preceding observations suggest, the state of fisheries in Lake Victoria and the growing destitution of riparian communities cannot be understood in isolation from the general state of global fisheries. Indeed, if put in proper perspective, the impact of Lake Victoria's state of commodification and internationalized fish supply on domestic food security should not come as a surprise to anyone. Against the backdrop of global demographic expansion and the growing appetite of the affluent global northerners, fish have become the single-most-traded animal species. According to FAO figures (2008b), global fish production grew from 133.6 million metric tonnes in 2002 to 143.6 million metric tonnes in 2006. In monetary terms, world exports of fish and fishery products grew from $78 billion in 2005 to $91.5 billion in 2006. Developing nations accounted for 50 percent of total exports and developed countries accounted for 80 percent of the total imports in fish and fishery products. The amounts of imports of fish and fishery products in 2006 by the EU, Japan, and the U.S. were $38 billion, $14 billion, and $13.3 billion, respectively.

Although there are no data about non-motorized boats and canoes that fish in marine inshore and inland waters, in 2006 there were 2.1 million motorized fishing vessels worldwide, scavenging for fish in the world's inland waters, seas, and oceans. This number could be even higher if one takes into account the underregistration of vessels in reporting countries. The same year, over 170 million people were employed in the primary and secondary sectors of fisheries. On the assumption that each person engaged in the fisheries industry had at least four dependents, the FAO reaches the conclusion that global fisheries provide livelihoods to 520 million people. These raw FAO data do not actually include the hundreds of millions more who occasionally or seasonally fish in rivers, creeks, ponds, and floodplains to supplement their nutrition (Bene et al., 2007).

One veritable consequence of such massive deployment of fishing vessels and fishers to the world's fishing grounds is entropic degradation of these fishing grounds and overexploitation or depletion of fish species. According to the same FAO report, by 2007 80 percent of the world's fish stocks were overexploited, depleted, or fully exploited, without room for further expansion. Only 20 percent of the fish stocks were either moderately exploited or underexploited. In addition to the overexploitation of fish resources, the exploitation of labor and the maldistribution of fish food are equally glaring. Almost 3 billion people worldwide depend mostly on fish food for their animal protein intake, and if placed in the context of the fact that 840 million people globally were classified as chronically malnourished and hungry and another 2 billion were surviving on less than $2 a day in 2007, the global maldistribution of fish food takes on even greater significance (Bene et al., 2007).

The maldistribution of fish resources takes even greater salience with respect to the poor, since fish is the single most important source of animal protein and macronutrients for 200 million poor Africans, in some cases constituting up to 70 percent of protein intake for the poor (FAO,2008b; Harsch, 2005; Thieme et al., 2005). Even though 80 percent of global fishery production occurs in developing nations, the chief beneficiaries of market integration and globalization of fish production are those with the greatest purchasing power. In 2007, per capita fish consumption in Europe stood at 19.9 kg, at 29 kg in China (rising from 5 kg in the 1970s), at18.6 kg in North and Central America, and South America and Africa were at the lower rungs of the consumption ladder with 8.7 kg and 8 kg, respectively. If we particularize the per capita fish consumption to the countries covered in this volume, the difference takes on even more glaring dimension. The annual per capita fish consumption in Kenya progressively declined over the past 30 years to 6.2 kg, to 8.5 kg in Tanzania, and to 12.1 kg in Uganda (Thieme et al., 2005).

Of course, the global fishery mode of production is fraught with contradictions in terms of resource depletion and distribution. In the 1980s and 1990s, for example, sub-Saharan Africa harvested 7.3 million metric tonnes per year from marine and inland fisheries, which stagnated thereafter around 6.8 million metric tonnes by 2002. Paradoxically, while national per capita consumption declined, sub-Saharan African countries exported $2.7 billion worth of fish and fishery products in 2001. To narrow the growing fish deficit, sub-Saharan African states are advised to exploit their aquaculture potential, which means investing $200 million per year in fish farms to produce more than 3 million metric tonnes of fish per year (Harsch, 2006).

The impact of globalized fish food on African nations manifests itself at both the production and consumption levels. On the production level, the fishery sector is highly penetrated by giant commercial vessels from the developed countries. Because of their mechanized nature, these giant vessels are faster and more efficient in the collection, processing, and transportation of fish than their African counterparts are. In addition to outcompeting small fishers into extinction, they leave behind biological deserts by fishing out important species that used to be the mainstay of livelihoods for millions of people in the continent. It is instructive to note that over 90 percent of the people officially classified as fishers worldwide fall under small-scale category, who are naturally vulnerable to outcompetition by highly mechanized industrial vessels (Bene et al., 2007).

On the consumption level, by globalizing the fish food chain, giant fish corporations have decisively tilted fish food distribution in favor of those with the greatest purchasing power, who happen to be the affluent residents in the global north. Because of high international demand for fish, most of the nutritious fish is exported to the global north. What is left for local consumption are juveniles or fish rejected by the fishery industrial complex, and even these rejected left-overs have become increasingly scarce for the poor. Over half of the people who live in the Lake Victoria basin, for instance, do not receive the required minimum daily calorie intake, while the three riparian states together export over $300 million worth of fish each year from Lake Victoria alone, in order to feed the affluent global northerners in exchange for hard currency. However, as George Monbiot (2008) recently reminded us, this is too familiar to the historian. In the 1870s, when drought resulting from the effects of El Nino destroyed crops in the Deccan plateau of India, the British Viceroy to the colony, Lord Lytton, arranged a record export of 6.4 million hundredweight of wheat to feed the well-to-do in England. In the meantime, between 12 million and 29 million Indians starved to death.

Just as domestic exhaustion of raw materials and internal market saturation precipitated the development of procolonial orientation in the 19th century, today's depletion of fisheries on the home front has led developed countries to embark on neocolonial penetration of African fishing grounds. Members of the European Union have led the way in this regard, with Asian countries following suit. Given the rising demand for fish and fishery products in EU countries, fishers kept modernizing their trawlers and imposing an unbearable toll on Europe's fishing grounds. By 1995, 101,680 modern giant fishing vessels from EU countries were mercilessly ravaging Europe's fishing grounds (Charles, 2001). In its April 2009 consultative green paper on the Union's Common Fisheries Policy, the EU Commission spared no words to blast fishers and their supporting governments for the sorry state of European fisheries. According to the paper, 88 percent of fish stocks in the EU are overfished compared with the global average of 25 percent. Furthermore, 30 percent of the fish stocks are actually harvested beyond their biological limits, in which case they cannot reproduce at the normal rate because the parenting populations are so depleted. The green paper highlighted the fact that overcapacity is largely to blame for the depletion of fish, resulting from the mechanization of fishing vessels. Even though the number of vessels in the EU has recently decreased to 85,000, the absolute decline in the number of boats is overcompensated by the modernization of the vessels. These modern vessels are two to three times more efficient than older vessels in their ability to capture, process, and transport fish.

Beginning in 1979, the EU tried to contain the crisis of accumulation in the fishery mode of production by placing a high priority on African fish production and fish supply. This meant that, having depleted their own fishing grounds, the EU vessels had to go elsewhere in search of fish to feed the predatory consumers in the EU. The easy targets for penetration were African marine resources. The EU had the money and the cash-strapped African kleptocrats were desperate for foreign currency to sustain their power, hence the fish deals between the EU and the African leaders then seemed made in heaven. Over the past 30 years, the EU has signed numerous bilateral agreements with African fish-producing nations—extending from Mauritania in the northwest to Mozambique in the east—for access rights to their sovereign waters; on behalf of its member nations' large fleets of industrial trawlers, the EU pays these African states $400 million per year in access rights to their fishing regions (Rosenthal, 2008; Lafraniere, 2008; Miller, 2007).

In order to justify the penetration and domination of African fisheries, the member states of the EU couched their strategy in the antiquated notion of comparative advantage. On the one hand, European states enjoy comparative advantages in capital, technology, and a skilled fishing force. On the other hand, African states enjoy comparative advantages in having abundant fish resources that go untapped for lack of skilled personnel, technology, and capital (Iheduru, 1995). It follows from this logic that, if the European and African states partner in the exploitation of the untapped fish resources, both sides would equally benefit. European consumers would enjoy an expanded basket of fish, enriching their preferences, while African states would gain handsomely from the transfer of technology, managerial skills, and licensing fees. The African states could plow back the revenue obtained from access fees into the economy to spur further development. Thus, the Euro-African fishing agreements were cast in altruistic terms, mainly to help coastal African states to effectively, optimally, and sustainably utilize their marine resources. In reality, the European states were motivated by the need to perpetuate their geoeconomic exploitation of the continent's marine resources and simultaneously to keep the Soviet Union out of Africa. After all, the Soviet Union and its satellite East European states were ruthlessly fishing on African coastal waters. In 1982, for example, two-thirds of total fish production in West African waters were accounted for by East European vessels; Soviet fishing vessels alone harvested 1.844 million metric tonnes (Iheduru, 1995).

African waters are not limited to European trawlers; a flotilla of industrial vessels from China, Russia, the Ukraine, and South Korea are equally guilty of ransacking Africa's marine resources and marine ecology. For example, as of July 2007 in Mauritania alone, over 344 foreign fishing trawlers were licensed to scavenge for fish in the nation's sovereign waters. Of these, 99 vessels were from China. In Mauritania, octopus is the most highly prized species, accounting for one-half of the country's $140 million in fish exports. In July 2006, Mauritania and the EU struck a six-year fish deal worth $700 million, providing over one-third of the regime's annual budget. It is not uncommon for foreign powers to exchange military equipment for fish, as indicated in 2005 by China's supply of two fighter jets to Mauritania in lieu of money for the fish its vessels collected (Miller, 2007).

These foreign vessels are giant and well equipped to stay in the oceans for weeks at a time, scouring the ocean bed, filleting, and freezing their catches before heading home. For example, the Irish fishing vessel the *Atlantic Dawn* is omnipresent in African waters. This supertrawler is 450 feet long and weighs 50 million pounds, can catch and process more than 7,000 metric tonnes in a single outing, which is more than a Mauritanian fishing village can catch in a year (McCarthy, 2002). The *Segundo San Rafael* fishing vessel, owned by a Spanish company, also scours northwest African waters using vast dragging nylon nets; this single vessel catches 260,000 pounds of octopus in just one 45-day outing, then processes the octopus in the area and exports it to Europe, Japan, and the United States (Miller, 2007). The self-evident reality of this new penetration into African fisheries has been that fishing activity has become increasingly centralized. Fish harvesting, fish-processing, and exporting have all come under foreign controls.

Since the neocolonial penetration into African fisheries began, fish exploitation in West African waters has increased by 60 percent; in fact, 80-90 percent of the fish caught in West African waters are caught by non-African vessels (UN IRIN, Nov. 2, 2007, and Feb. 26, 2008). For example, China National Fisheries Corporation is one of the latest plunderers of African fisheries, with half of its foreign offices located all along the western coast of Africa. Thanks to its dominance in West African fisheries, this Chinese corporation is one of the largest global suppliers of fish to affluent Europeans. The corporation even has a large fish-processing facility in the Canary Islands, where significant amounts of illegally caught fish are handled before exporting to the European market. It is believed that 360,000 tons of fish arrive in the Canary Islands each year, most illegally caught in West African waters (Rosenthal, 2008). It is a small wonder that, between 1992 and 1995, China added 29,745 new modern motorized fishing vessels to its already large fleet, in order to enable its fishers to exploit fishery resources in foreign waters (Charles, 2001).

Today, the EU has become the largest market for fish, which is worth $22 billion per year. Fifty percent of EU fish imports are from developing nations. Although the distinction between legal and illegal fish has dubiously become difficult to quantify, significant amounts of EU fish imports enter the fish market via subterranean channels; by some estimates, more than half of the fish sold in the EU market is considered contraband (Rosenthal, 2008). By some conservative estimates, Africa as a whole loses $1 billion worth of fish annually to illegal harvesters. In the mackerel and sardine fishery sectors alone, Angola loses $49 million per year, which represents 21 percent of its fish export. Mozambique also loses $38 million per year to illegal fishing in the shrimp and tuna industries. In a single neocolonial penetration into Tanzanian sovereign waters in 2001, Taiwan's trawlers scooped $20 million worth of fish. In South African sovereign waters, despite the fact that the legally allowable catch was set at 45,000 metric tonnes per year, neocolonial penetrators actually harvested 320,000 metric tonnes of Patagonian toothfish over a two-year period (UN IRIN, Nov. 2, 2007).

This neocolonization of African fisheries has three long-range implications for the economies and marine ecologies of the countries involved.

First, because the mechanized and highly subsidized non-African industrial trawlers have the ability to sweep an entire fishing ground, the breeding grounds are trampled upon and fish eggs are destroyed. Juvenile fish are unnecessarily caught and then discarded. It is instructive to note that each year, over 30 million tons of juvenile fish are dumped back into the ocean worldwide because since they are considered too small to be sold in the market. The result has been widespread entropic degradation of fish ecology and depletion of African fish stocks. The populations of bottom-dwelling fish in West African coastal waters are believed to have declined by 75 percent since the early 1980s. Lobsters had long been fished out from Mauritanian waters; the country's octopus resources have declined by a third since foreign vessels joined the scramble for fish, and the number of traditional octopus harvesters fell from 5,000 in 1996 to 1,800 by 2002; and a third of Mauritania's trawlers have been driven out of business by foreign trawlers (McCarthy, 2002).

In 2002, the EU Commission's own study revealed that the stocks of important commercial fish off Senegal's coast had declined by 75 percent and other fish species either had already become extinct or were on the brink of complete collapse. However, the Dakar government and the EU Commission proceeded to sign a four-year agreement allowing EU trawlers to fish for tuna and bottom-dwelling fish in exchange for $16 million in annual access fees. Following Senegal's example, in 2006, Mauritania also agreed to allow 43 European industrial trawlers to scavenge its waters in exchange for $146 million per year from the EU, on which the government depended to cushion its annual budget (Lafraniere, 2008).

The state of the *Hoplostethus atlanticus* (orange roughy) in the southeast Atlantic and western Indian Ocean is another illustrative case of the effect of renting out African fishing grounds. The orange roughy (so named because of its bright orange coloration and rough scale) is the latest addition to endangered species. Because the fish live in deep oceans, generally around sea mounds, and travel in aggregation for spawning and feeding purposes, they were late to be discovered. Thanks to technological innovation, powerful fishing gear, and sophisticated modern boats, some Australian fishers located the orange roughy some 30 years ago. Because the orange roughy travel in large formations when spawning and feeding, they are perhaps the easiest target to be scooped up in one swoop. Over a period of 25 years, more than 1 million metric tonnes of orange roughy have been harvested. The danger of a harvest of this magnitude is that orange roughy are known for their low fecundity, slow maturity, and big size, which make them vulnerable to overexploitation and potential depletion. The sexual maturity range of an orange roughy is between 23 and 40 years. In the southeast Atlantic, Namibia dominates the orange roughy fishery and the United States is the prime destination of the fish. Namibian exports of orange roughy to the U.S. market jumped from 95 metric tonnes in 1995 to 3,646 metric tonnes in 1997, before it sharply fell to 200 metric tonnes in

2002, reflecting the rate of overexploitation (Lack et al., 2003). In the case of the western Indian Ocean fishery, orange roughy was discovered around the Madagascar ridge in 1999. In a matter of two years, the fishery was almost emptied by foreign vessels. Even though hard figures are wanting, estimates are that between 12,218 and 50,000 metric tonnes of orange roughy may have been taken from the Madagascar ridge, as the number of vessels rose from just seven to 40 between 1999 and 2001 (Lack et al., 2003).

Second, dislocation, unemployment, and impoverishment of native fishers have become the veritable outcomes of neocolonized fisheries. Prior to the neocolonial penetration of West African sovereign waters, approximately 570,000 full-time artisanal fishers used to harvest 1 million tons of marine fish per annum. Another 1.8 million persons were engaged in fish-processing and marketing. The number of part-time and casual fishers was even greater (Dugan, 2003). However, the livelihoods of these fishers were placed under threat with the arrival of long-distance foreign vessels, since native fishers were no match for the highly mechanized and well-subsidized foreign vessels. The rich countries do subsidize their fishing vessels to the tune of $30 billion per year; the EU provides more than $7 billion per year to its vessels, while China provides $2 billion per year to its own trawlers (Miller, 2007). Unable to compete with foreign industrial vessels, many native fishers have been limited to catching juvenile species in inshore waters or have been driven out of business altogether.

The state of fisheries and the predicament of fishers in Senegal provide a paradigm case of the impact of neocolonized fisheries. Not only does fish provide 70 percent of animal protein to the population, but also one out of every six Senegalese of working age traditionally were involved in fishing activity. The sector used to employ 600,000 Senegalese directly. However, since the government struck a fish deal with the EU, native fishers and the fish itself began to lose ground quickly. Limited by the character of their fishing technology and totally dependent on pirogues, native fishers were confined to inshore waters. The entire offshore fishing regions were left to European and Asian trawlers and fish pirates. Here a double jeopardy was in operation while native fishers overexploited the inshore waters, neocolonial fishers began to deplete offshore fishing areas. In consequence, Senegal's annual fish catch dropped from 95,000 metric tonnes in 1994 to 45,000 metric tonnes in 2005; pirogues fell from over 10,000 in 1997 to 5,615 in 2005; and as the country's fish exports fell by 32 percent, fish-processing plants shed their labor by between 50 and 60 percent (Inter Press Service, Aug. 11, 2008). Ironically, dislocated fishers have become a large group of migrants who try to enter Europe illegally. In 2007, over 36,000 Africans reportedly tried to reach Europe in over 900 wooden boats and pirogues, of whom over 6,000 are believed to have perished in the sea (Lafraniere, 2008).

Third, with the neocolonization of their fishing regions, African states have virtually failed to exercise effective sovereignty over their exclusive economic zones, 200 nautical miles out into the ocean. By a twist of irony, the EU Commission has turned the blame game and accuses these states of being responsible for the depletion of important commercial fish species, claiming they do not enforce the provisions of domestic and international laws. This ignores the reality that the African states do not have the capacity to monitor a vast maritime space to insure illegal fishing does not occur in their sovereign waters. For example, Mozambique's fisheries minister acknowledged in May 2008 that he was severely handicapped by resource limitation to police adequately the country's 2,500 km-long coastline. His claim was because he had just one modern patrol vessel at his disposal, rented at $2 million per year and paid for by Norway. According to the minister, Mozambique should have at least four modern patrol vessels to keep an eye on its 200 nautical miles of exclusive economic zone, which means an investment of $100 million into the acquisition of such vessels; this in part explains why Mozambique chose, in 2007, to rent out its marine resources to 89 trawlers from the EU to collect 10,000 metric tonnes of tuna per year in exchange for $1.3 million from the EU Commission, rather than have its sovereign waters ravaged by fish pirates (*Africa News*, May 31, 2008).

In practice, this means that African waters are open to European and Asian industrial trawlers, whether licensed or not. Industrial fishers are well equipped to elude sovereign authorities. For example, licensed vessels frequently transport their loads of fish under the cover of darkness to other vessels docked offshore, so that they can harvest more the next day. This way, they prolong their stay in the waters and exceed their quotas by a wide margin. Alternatively, they evade quota limitations by registering their vessels in the host country, entering into joint ventures with a native actor, or simply by buying up the fish caught by local fishers and then transporting it to their vessels docked off the coast. In this way, they elude fish restrictions and their financial obligation to the host authorities. Aerial surveys carried out between 1997 and 2001 over Guinea's territorial waters discovered that 60 percent of 2,313 fishing vessels monitored were engaged in illegal activity; 24 percent of the 947 vessels spotted in Sierra Leone's waters and 24 percent of the 926 vessels monitored in Guinea Bissau's waters were involved in illegal operations (Inter Press Service, Aug. 11, 2008). Furthermore, in 2006, more than half of the 104 trawlers fishing in Guinea's coast were found to have been fishing illegally or to have been involved in harmful fishing practices. In neighboring Guinea Bissau, over 40 percent of fish are caught by vessels in open violation of regulations and without licenses. This is a country where 12 fishery ministers resigned in rapid succession over eight years, each disgusted over the plunder of the nation's marine resources by foreign marauders. In the meantime, the EU Commission signed an agreement with the Bissau government in 2007, permitting EU trawlers to harvest tuna, octopus, and shrimp, in exchange for $42 million over a four year period (Rosenthal, 2008; Lafraniere, 2008).

THE RISE OF THE AGROFUEL INDUSTRIAL COMPLEX AND THE SECURITIZATION OF FOOD AND BIOENERGY

As though the existing structural conditions of poverty, resource depletion, and environmental degradation were not harmful enough, the latest threat to Lake Victoria and to the lakeshore communities now comes from the growing popularity of agrofuel production and control of food sources as integral to national security. Arab petrostates, rapidly industrializing countries, and Western corporations are currently rushing to control Africa's land resources on unprecedented scales. Even though the threat from the shift to agrofuel production to Lake Victoria's littoral countries may not appear immediate now, there are several ramifications that could be catastrophic in the long run for local, regional, and national food and environmental security.

The fact that Lake Victoria's watershed extends over five countries suggests its expansive ecological terrain will be vulnerable to substantial modifications as a result of converting huge land masses to agrocommercial plantations and diverting massive quantities of water resources to feedstock production and agrofuel generation. This may modify the structures and functions of the basin's ecosystems. In addition, water pollution, organic and inorganic residues, and effluents associated with biofuel production will unavoidably enter the natural water channels for insertion into the lake.

National food security is the sum total of all food components originating in all ecosystems, including terrestrial, riverine, lacustrine, and marine systems. A change in any part of these systems inevitably generates reverberating changes in other components of the total food ecology. Conversion of land from food production to agrofuel production inevitably represents substantial reductions in the overall food amounts available for human consumption. Since lakeshore communities are already poor, the price hikes for foodstuffs only deepen the poverty of those communities.

The shift from agrarian food production to agrofuel production, by definition, necessitates scale economy. This change concentrates land and freshwater resources in ever fewer and fewer hands, and is reliant on mechanization, massive use of water, and heavy applications of agrochemical fertilizers, herbicides, and pesticides, all of which will have harmful impacts on the natural water systems in terms of quantity and quality. In quantitative terms, the diversion of virtually inordinate amounts of water resources to support the large-scale agrocommercial plantations, dedicated to the production, processing, and distribution of agrofuels, will unavoidably result in a substantial reduction in the flow of water to natural water systems and, by extension, to Lake Victoria. The qualitative aspect of agrofuel production can be even more harmful to the lacustrine and riverine systems. The insertions of nutrient-rich organic and inorganic matter (such as reactive nitrogen, phosphorus, pesticides, and herbicides) from the agrofuel plantations and sludges from agrofuel-processing plants will affect aquatic life and accelerate the eutrophication process.

The dispossession and destitution of millions of people resulting from the alienation of land, for purposes of agrofuel production, can surely deepen poverty and trigger unprecedented migration of rural people to lakesides or urban centers. This follows the classic trajectory of land alienation for the production of primary commodities, such as coffee, tea, and cotton, just to mention a few. The increased migration of dispossessed people to the lakeshores, which are already overcrowded, can exacerbate the conditions of destitution, feminization of poverty, social tensions, and wetland destruction.

Most well-watered, fertile land has long been alienated by national elites and transnational corporations for cash crops. New acquisitions of land for biofuel production can only come from further destruction of forests. Not only will this expansion undermine the capacity of terrestrial vegetation to capture solar energy to make chlorophyll during the photosynthetic process, producing oxygen and sequestering carbon dioxide, but it will also undercut the evapotranspiration and precipitation regimes. The local microclimates will be grossly changed and cycles of drought may take on a stable feature.

As discussed in Chapter 3, wetlands are the last line of defense of the lake studied here, absorbing or neutralizing nitrogen, phosphorus, and other metallic substances before their insertion into the lake. Agrofuel plantations are very likely to intensify the encroachment on wetlands and riparian vegetation, since some of the feedstocks (such as sugarcane) require wet and moist soils to grow.

Finally, in addition to being impacted by the events and activities that take place in its immediate ecology and its macroregion, Lake Victoria is equally vulnerable to consequences of global agrofuels production because of the biogeochemical cycles and biospheric processes. For example, the deforestation of the Amazon and Congo rainforests, and the conversion of Southeast Asian peatlands to oil palm plantations to boost agrofuels production, will have unavoidable impacts on global climate, from which the Lake Victoria basin cannot escape. In addition, the conversion of food crops to biofuels has the potential to radically transform the global price structure of key food sources, from which the people in the East African region cannot escape being adversely affected.

As capitalism enters its latest phase (perhaps its final one), the natural systems and the livelihoods of millions of people who depend on aquatic and terrestrial resources in Lake Victoria's riparian countries have now come under a new wave of assaults from corporate land grabbers in overt collaboration with the pirate state bourgeoisie. The riparian countries are now required to rapidly transition to carbohydrate economies where agricultural crops, forests, bushes, shrubs, and grasses would be converted to agrofuels to fuel the reproduction of capitalism on an unprecedented scale.

Kenya's national elites have determined to offer large tracts of land to foreign land grabbers. Lured by the prospect for quick personal enrichment, maintenance of the entrenched patronage system, and containment of the legitimation crisis, the pirate state bourgeoisie have offered a large area at Homa Bay on the shores of Lake Victoria to foreign concessionaires. This scheme known as the Ngima Project would be owned by a European consortium led by the Brussels-based HG Consulting Firm, responsible for raising the 330 million euros required to jump-start the project. HG Consulting lists on its website (www.hgconsulting.eu/documents/ HGConsulting-Ngima-Press-Kip.pdf) the benefits that would accrue to Kenya and the local communities. Ngima Project, as envisioned, is an integrated sugarcane production system with dual purposes: to annually produce 100,000 metric tonnes of sugar for domestic consumption and 259 million liters of bioethanol for export. Initially, outgrowers would supply the sugarcane grown on 42,000 ha. In the name of emission reductions by virtue of producing green energy, the Ngima Project would be certified to receive a carbon credit equivalent to 217,000 metric tonnes, which the consortium can sell to global polluters. In order to mollify local opposition, the consortium would give the unions of outgrowers a token 5 percent stake in the project.

Contrary to the highly inflated promise of environmental and social benefits, the project as packaged is bound to produce enormous costs in terms of eroding ecological integrity and human security. In addition to wetland degradation and habitat destruction, the agrochemicals used in the cultivation, harvesting, and production of sugarcane and the other chemicals used in processing the cane into ethanol as well as the organic residues and sludge from the ethanol plant will be inserted into Lake Victoria. This will accelerate the modification of biotic and abiotic conditions for aquatic vertebrate and invertebrate species and will worsen the eutrophication of the lake. Vinasse, for example, if inserted into the lake in significant amounts, can result in widespread fish kill. Vinasse is known to have high organic matter and potassium content, and it is widely used in sugarcane plantations to improve soil structure, water retention, and to increase other nutrients in the agro-ecology (Berthelot, 2009).

Second, the cost of shifting from producing food crops to growing, harvesting, and supplying the sugarcane by outgrowers and the quantity of labor invested in the production chain would not be properly accounted. Only the volume and quality of the sugarcane involved in the project would be counted. Moreover, in addition to converting the buffering wetland of Lake Victoria to sugarcane production, massive amounts of water resources would be diverted to the agrofuel production system. The Ngima Project would be crushing almost 18,000 metric tonnes of sugarcane per day for nine months of the year, producing huge amounts of sludge, which is very likely to be inserted into Lake Victoria.

Furthermore, the Arab petro-state of Qatar has acquired 40,000 ha in the Tana River Delta in exchange for $2.3 billion intended for port rehabilitation in Kenya. In total disregard of communal interests and long range hydrobiological implications for humans, livestock keeping, and wildlife, the National Environmental Management Authority in Kenya approved 20,000 ha for agrofuel production in this delicate area in the southeast region of the country in June 2008. This is just one installment of a 140 km^2 Tana River Delta scheduled to be converted to sugarcane plantation for ethanol production (Rice, 2008). The dominant Mumias Sugar Company, which put up almost $300 million for the project, has promised thousands of jobs and infrastructure development in the area. However, since the project would potentially consume one-third of all available freshwater, local herders, recession farmers, fishers, wildlife, and riparian vegetation would be cut off from their critical life-supporting resources. In addition to the fact that pastoralists bring up to 60,000 livestock at any particular time to graze on the wetlands, a vast number of other animals including elephants, hippos, lions, primates, reptiles, sharks, and more than 345 species of waterbirds make their home in the delta's pristine mangroves, forest savanna, swamps, and beaches (Rice, 2008). Since the Tana River crisscrosses semiarid regions, this area along the Kenyan coast is known for experiencing food shortages on almost a regular basis. Therefore, the drive toward industrial sugarcane plantations for purposes of agrofuel production is bound to worsen the food ecology of many communities.

Another controversial agrofuel project under way as of this writing involves the Dakatcha Woodlands in the coastal plains of Kenya. The woodlands are a composite of forests, woods, and thickets held in trust by a local council on behalf of the 20,000 indigenous people. In addition to being a global hotspot of biodiversity, because of the rare plant and animal species found there, this pristine ecosystem stores water, protects the soil, and provides livelihood to the people. The indigenous people use the Dakatcha

Woodlands to grow maize, cassava, and pineapple, to graze their herds, to draw drinking water, and to gather medicines and firewood. In total disregard of indigenous interests, Kenya's pirate state bourgeoisie conceded 50,000 ha of land in this global area to a Milan-based Italian company to produce biodiesels from Jatropha trees for export to the European markets. This company has already leased 1 million ha of land for agrofuel production in other African countries. Not only would this project destroy the livelihoods of local communities, but it would also violate EU requirements that savings from agrofuel production must be 35 percent higher than conventional sources of energy, and that sources are not to be grown on lands known to have high carbon stock or high conservation value (Action Aid, 2011).

In fact, a U.K.-based leading consultancy in life cycle analysis of agrofuels (North Energy) calculated that the GHG emissions that will result from destroying the Dakatcha Woodlands to plant, grow, harvest, and process Jatropha would exceed the savings from using agrofuels by a wide margin. Furthermore, they calculated that transporting the diesel to markets in Europe would spell disaster, not only for local communities, but also for the quality of air. North Energy found emissions from Jatropha production in Dakatcha to be 2.5 to 6 times higher than fossil fuel equivalents, resulting from the destruction of woodland and scrubland that would be required to grow Jatropha. This means that the Jatropha project would openly violate existing European sustainability criteria of achieving 35 percent savings from renewable sources of energy (Action Aid, 2011).

With respect to agrofuel potential, Tanzania is ranked among Africa's top agrofuels producers. The Tanzanian elites, donor countries, and agrofuel transnational corporations view the country as having three crucial advantages.

First, only 6 percent of Tanzania's 88 million ha of land is at present under cultivation, which suggests that most of the supposedly idle land can easily be put under biological feedstock production. In fact, one of the sweeping claims made is that Tanzania can become self-sufficient in terms of its energy needs by converting just 300,000 ha of its 4.6 million ha currently under cultivation to agrofuel production. Even if Tanzania chooses not to convert any part of existing arable land to agrofuel production, it could easily bring 2 million ha of land from its 88 million "unused" land for sugarcane, Jatropha, and oil palm plantations. This would be enough to produce 16 billion liters of agrofuels per year, worth $7 billion. As the argument goes, in addition to securing energy independence and the badly needed hard currency for Tanzania, this untapped potential would make Tanzania a major supplier of agrofuels to global markets while creating 1 million new jobs (Liversage, 2009; Kearney, 2006a).

Second, the supposed untapped irrigation potential of the country can be harnessed for agrofuel plantations. Not only does Tanzania share three of the ten largest lakes in the world, but also has nine major rivers and a large number of smaller water bodies throughout the country that can be tapped to irrigate vast agrofuel plantations. In 2005, a German government-supported study of Tanzania's agrofuel potential actually identified some 20 areas with high irrigation potential, having almost 2.5 million ha considered suitable for growing biological feedstocks. The Tanzanian government then followed up the study by setting up a national agrofuel task force, drawn from eight ministries, to give high priority to agrofuel production. The task force was mandated to solicit funding from the G-8 Global Bioenergy Partnership, the EU, the FAO International Bioenergy program, and other bilateral and multilateral sources.

The World Bank and the U.S. government, through USAID and the Millennium Challenge Corporation, had already begun providing financial and technical assistance to outgrower programs in Jatropha and oil palm growing areas. In order to lure foreign agrofuel agencies and reduce the transaction costs of investment in the bioenergy sector, the Tanzania Investment Center promptly established a database of suitable growing regions with one-stop shopping facilities for land acquisition and registration (Kearney, 2006). By then, three international agrofuel companies had sugarcane operations under way and were producing bioethanol; three groups from the U.K., the U.S., and Malaysia were exploring the potential for more than 100,000 ha oil palm plantations (T. Kearney, 2006a).

Third, as an associate member of the European Union, Tanzania enjoys preferential access to the EU market, giving the country a competitive advantage over Brazilian bioethanol (Beattie, 2008; De Keyser and Hongo, 2005). Banking on these presumed advantages, a large slew of contenders have been vying for Tanzanian land, water, and biotic resources.

SEKAB of Sweden led this platoon of agrofuel transnationals in the Tanzanian bioenergy sector. According to its website, SEKAB promises to make both Tanzania and Mozambique not only energy-independent, but also major exporters of several billion liters of bioethanol. This is because both countries have "unused" land that is four times the size of Sweden, excellent tropical climatic conditions ideal for feedstock production, and abundant water resources. The amount of bioethanol exported from Tanzania alone could easily replace the volume of gasoline used in Swedish and Norwegian private cars combined. As of 2009, SEKAB had either already acquired or was in the midst of negotiating to acquire 400,000 ha of land in Bagamoyo in the Coast region, mainly for sugarcane plantations. Sugarcane is a crop known for its water intensive needs. The land acquisition included 200,000 ha on the banks of the Rufiji and Wami Rivers, on which two nature reserves and subsistence farms depend for regular flow of freshwater (Gordon-Maclean et al., 2009).

The Dutch Bioshape acquired 81,000 ha of land in Kilwa District, known for its high- value biodiversity, in complicity with Tanzanian authorities and without the knowledge of local communities. The company's original aim was to put its land concessions under bioenergy crop production using outgrowers, providing them with seeds, fertilizers, pesticides, and herbicides (Gordon-Maclean et al., 2009). The firm promised thousands of jobs and paid $676,000 for the initial installment of 34,000 ha, of which 60 percent went to district authorities even though they were not entitled to it. Ironically, Bioshape filed for bankruptcy in March 2011 and folded up its operations in Tanzania without creating a single permanent job (Freereporter, 2011).

In 2004, the Dutch Diligent Energy System also started Jatropha cultivation based on an outgrower scheme. The firm intended to expand its 81,000 ha possession to 200,000 ha over a period of 15 years, principally under outgrower schemes. This Dutch firm saw a number of virtues in relying on outgrowers, including potential for unlimited growth, avoidance of land ownership conflicts, insulation from external shocks, lower cost of investment, and the overall better use of local farmers. A German agrofuel firm also began Jatropha production in 2009, as part of its strategic vision to put 200,000 ha under Jatropha production in virtually all Tanzanian districts.

The Belgian-dominated FELISA put 4,358 ha of its projected 10,000 ha project under oil palm cultivation for biodiesel. The consortium established oil palm nurseries with 42,000 seedlings. In 2009, it distributed 10,000 seedlings to 990 outgrowers, with further plans to secure additional supplies of oil palm from small producers under a contractual arrangements grown on some 60,000 ha, which it had identified as suitable for oil palm production (Liversage, 2009).

The Wilma Group of the United States also teamed up with a local Tanzanian group to put 60,000 ha under agrofuel production in the Kagera region, which is an integral part of Lake Victoria's watershed ecology. The British CAMS Group acquired 45,000 ha to grow sweet sorghum, in order to produce bio-ethanol. China secured the first installment of 300 ha to grow rice; and in early 2009, the Saudi petro-state was negotiating with Tanzania to lease 500,000 ha.

In the Kigoma region in the watershed of Lake Tanganyika, a Malaysian-Indonesian firm began oil palm cultivation on 8,000 ha, based on a plantation combined with an outgrower scheme.

Three other Western groups were scouting the Kigoma region with an eye to put 100,000 ha under oil palm cultivation. In 2008, British Sun Biofuels started a Jatropha plantation on 8,000 ha, as its first installment of

40,000 ha in Tanzania's Coast Province. To quash local opposition to the project, the company paid a meager $220,000 to be divided among 152 farmers and promised to spend 5 percent of its earnings on local social amenities such as schools and health clinics. The farmers were told that the project would result in thousands of year-round employment with good pay. However, the deal was shrouded in secrecy and the farmers did not know anything about the type and scale of work promised, or the pay scale of the work. Moreover, the lease is for 99 years and was given to the company without regulation and stipulation, which means that the villagers are powerless if they want to reclaim their land. Since population in the area is expected to rise, there will be no land left for distribution to future generations (Beattie, 2008).

To deflect criticism against the neocolonial vandalization of the country's land and freshwater resources, and to lure local communities into believing that they would benefit from the transition to bioenergy crop production, Tanzania's pirate state bourgeoisie produced a long catalog of benefits that could be had from agrofuels. These supposed benefits include: national energy security; employment; creation of markets for agricultural energy crops; saving of foreign exchange earnings; generation of hard currency; contribution to greener and cleaner environment through reduction of GHG emissions; halting deforestation and desertification through drought-resistant crops such as cassava, sweet sorghum, and Jatropha; facilitating technology transfer; and enhancing national food security (Gordon-Maclean et al., 2009). When subjected to these areas of close scrutiny, however, these claims do not seem to hold water.

First, agents of the state and agrofuel transnationals took advantage of villagers' lack of knowledge about land rights and the nature of the deals involved to grab precious land and freshwater resources. SEKAB, for example, intent on exploiting villagers' lack of information, bypassed official channels and directly went to villagers to negotiate about land acquisition. The company never mentioned how much land was needed and whether the agrofuel was for domestic consumption or export to Europe. This way, the company was able to amass land at nominal prices, which simply worsened the dispossession and poverty of villagers. In some areas, the compensation was as little as $250 per household (Gordon-Maclean et al., 2009).

Second, hopeful of growing out of poverty by shifting to bioenergy crop production, farmers are faced with perverse incentives to devote all their time to agrofuel production, consequently abandoning food production to women and children. This situation, rather than enhancing national food security, actually increases poverty. After all, it defies common sense to argue that Tanzania's national food security would be enhanced by transitioning to agrofuels production, when fertile land and freshwater resources are alienated from actual food producers and are instead given to agrofuel transnationals.

Third, contrary to official claims, the transition to an agrofuel production system promotes deforestation rather than nature preservation, as corporations vie to acquire increasingly more land to expand their bioenergy plantations. For example, the concessions given to Bioshape by the Tanzanian pirate state bourgeoisie included logging rights to between 200,000 and 800,000 cubic meters of valuable hardwood—worth $50-$150 million—to finance, in part, its biofuel project by selling the hardwood. In fact, Bioshape unilaterally cut and sold additional timber without permission in the nearby biodiversity-rich nature reserve (Freereporter, 2011).

Not to be outmatched by its neighbors, Uganda also initiated a number of agrofuel projects covering an array of sugarcane, palm oil, Jatropha, sunflower, and castor production. Most of Uganda's biomass production projects are in Lake Victoria's watershed. In August 2008, the Bush administration extended a $527,000 grant to Uganda to help study the agrofuel market potential there. The euphemism used to justify the grant is the need to balance agrofuel production and food security by providing incentives for cost-effective exploitation of agrofuel resources (U.S. Federal News Service, Aug. 28, 2008). Two particularly controversial projects dramatized Uganda's determination to expand its agrofuel sector aggressively. In 2007, Sugar Company of Uganda was given a green light to excise 7,200 ha from the Mabira rainforest on the edge of Lake Victoria, for purposes of cane plantations for ethanol production. The second project involved BIDCO clearing forest on the islands of Bugala and Kalangala on Lake Victoria for palm oil. Even though public outcry halted the operations of the two companies, BIDCO had already destroyed 6,000 ha of pristine forest on Bugala Island (Smolker et al., 2007).

Uganda's pirate state bourgeoisie see the popular protests as minor bumps on the road to transitioning to agrofuel production on a national scale. Indeed, the regime plans to obtain 16 percent of total national energy requirements from agrofuels by 2017, to which domestic and foreign companies have actively responded by spreading Jatropha plantations, mostly based on outgrower schemes. This new drive for agrofuels has spurred competition with national food production for labor, land, and freshwater resources. As one Ugandan commentator (Kagolo, 2010) observes, the shift to agrofuel production entails double jeopardy. On the one hand, loss of biological diversity and environmental degradation will result from application of synthetic fertilizers, pesticides, and herbicides in order to maintain Jatropha monoculture plantations, and, on the other hand, food insecurity will occur, as farmers would be lured by false promises to revert to more and more Jatropha cultivation. It should be noted here that Jatropha is hardly more productive than other sources of agrofuels. For example, it is calculated that one acre of land in Uganda can possibly host 430 Jatropha trees. The trees could generate 430 liters of biodiesel per year, just enough to run one car. Since Uganda had over 600,000 vehicles as of this writing, millions of ha of arable land, forest reserves, and wetlands must be taken out from food and/or forest and wetlands to make way for Jatropha plantations, all in order to produce millions of gallons of biodiesel every year (Kagolo, 2010).

At the risk of restating what I have already said at the beginning of this section, there are several threats to Lake Victoria that come from the new wave of the scramble for agrofuel production in the lake's basin that warrant highlighting.

First, the diversion of increasingly more land and water resources to agrofuel production will sharply reduce the scope for expanding the general terrestrial food ecology to meet the need of the growing human population. As a result, more and more people are likely to migrate to Lake Victoria to eke out a living as fishers. The resulting horizontal growth of fisherfolk will place the lake under even more severe stress.

Second, the reckless application of reactive agrochemicals to boost the quantity and quality of bioenergy crops throughout the basin is bound to increase the insertions of harmful pollutants into the lake, thereby worsening the state transition of the lake and making its rehabilitation even harder in the future.

Third, forest and woodland resources in Lake Victoria's basin will come under pressure from the growing greed of agrofuel entrepreneurs and the further marginalization of the poor, as they clear more forests and woodlands to make way for plantations and the cultivation of biomass.

Fourth, apart from the extensive penetration of Lake Victoria basin by agrofuel interests, clearance of forests in other ecoregions such as the Congo basin, Mozambique, and Zambia—for purposes of agrofuel production—poses grave danger to the integrity of Lake Victoria, since these ecoregions have also become the premier target for penetration. In October 2007, for example, China and the regime in Kinshasa signed a $1 billion land deal to convert a massive 3 million ha of virgin forest to oil palm plantations (World Rainforest Movement, 2008). A deforestation of this magnitude is particularly disconcerting because this pristine rainforest in the Congo basin is critical to carbon sequestration, climate regulation, precipitation, and evapotranspiration regimes, with far-reaching implications for Lake Victoria. Taking advantage of the country's internal weaknesses and of the unredeeming qualities of the pirate state bourgeoisie, Western and Asian logging companies have long targeted the DRC's rainforest for wanton vandalization. Between 2001 and 2005 alone, for example, the regime in

Kinshasa and foreign logging firms signed 107 separate deals to log 15 million ha of rainforest. If one includes the areas that are logged illegally in the rainforest, the figure could be even higher. As some estimates suggest, if the current rate of deforestation continues, the amount of carbon dioxide emissions from the deforestation in the DRC alone could reach 34 billion metric tonnes over 60 years (Simms, 2009).

If we put the conversion of these huge forested areas to oil palm production within the DRC in a basin-wide context, the threats from agrofuel companies to the rainforest in the region loom even larger. In neighboring Congo Brazaville, Italy's energy company (ENI) has invested $3 billion in forest resources for purposes of oil palm plantations and fossil fuel exploration (World Rainforest Movement, 2008). Likewise, the rainforest of the Central African Republic has drawn the attention of agrofuel companies. The EU Commission, for instance, considered the Central African Republic a promising producer of bioenergy crops, since 45.3 million ha of the country's total land mass of 61.8 million ha was defined as suitable for agriculture to grow cassava, sorghum, sweet potatoes, sugarcane, soy, and a variety of trees, all considered suitable for agrofuels production (Smolker et al., 2007). Therefore, if the new scramble for the Congo basin forests reaches its apogee, the results of deforestation and degradation could be catastrophic, for not only the Congo basin's own biodiversity and climate, but for such other ecoregions as the Lake Victoria basin. After all, the Congo basin forests—second only to the Amazon in size—contain 70 percent of Africa's total vegetation, influencing all rainfall and weather patterns with immense regional and global implications (Smolker et al., 2007).

The threat coming from the increasing popularity of oil palm plantations (particularly in Tanzania and Uganda where the catchment areas of Lakes Victoria and Tanganika are now seen as most suitable for monoculture plantations) involves the heavy costs of dispossession, deforestation, biodiversity loss, and pollution. In virtue of the fact that industrial oil palm plantations are monocultures, between 80 and 100 percent of all animal species will be wiped out when the conversion of forests in the Lake Victoria basin to industrial plantations occurs. The erosion that accompanies the clearance of massive forests will unavoidably lead to the siltation of inflowing rivers and the sedimentation and eutrophication of Lake Victoria. These consequences will lead to further deterioration of drinking water quality and substantial modification of biotic and abiotic conditions for aquatic living things. To compound the problems even further, the toxic air that results from burning forests to make way for industrial plantations will worsen pollution and climate change, as well as the insertion of sludges from biomass processing plants into the lake.

Moreover, since industrial oil palm plantations are dependent on heavy use of petroleum-based fertilizers and other agrochemicals that adversely affect wetlands and natural water systems, the propagation of oil palm plantations in the catchment areas of Lake Victoria and the great lake region of Africa cannot be seen as a welcomed development. This is a fact, no matter how proponents of biofuels advertise it an answer to poverty and underdevelopment.

Disturbingly, in order to cover the harmful consequences of the global shift to agrofuels production, agrofuel oligopolies and the region's pirate state bourgeoisie have forwarded a series of assumptions and conclusions fraught with myths, false promises, and misrepresentations.

The first myth pertains to the skillful conflation of development and poverty reduction with a transition to agrofuels economy. The notion orchestrated here is that Lake Victoria's riparian countries (indeed the whole of Africa) can reach their Millennium Development Goals by converting their vast terrestrial resources to lucrative biofuel production. Supposedly, this process entails creating jobs, raising the living standards of the mass population, and realizing their targets in poverty alleviation in the short run and poverty prevention in the long run. For example, in order to rationalize its loans to oil palm planters in the Ivory Coast, the International Finance Corporation (IFC, the commercial arm of the World Bank) couched its arguments for the loans it offered in terms of employment, poverty alleviation, and environmental sustainability associated with oil palm plantations. As the IFC director stated, "Oil palm plantations will generate more employment and higher living standards, and promote environmentally sensitive agricultural production" (Tauli-Corpuz and Pamang, 2007).

In a nutshell, as the argument goes, this biofuel-based "green revolution" is the long awaited answer to Africa's mass unemployment, food insecurity, energy or financial dependency, and environmental degradation. If Tanzania, for example, puts 300,000 ha under bioenergy crop cultivation, the country can easily create 300,000 jobs; the multiplier employment effect can add millions more jobs as biological feedstock producers, oil extractors, transporters, fertilizer suppliers, soap makers, and traders are added to the roster. The cumulative impact of upstream and downstream activities associated with agrofuel production would annually contribute 4 percent in GDP growth to the Tanzanian economy (De Keyser and Hongo, 2005). This gross assertion, however, runs counter to the historical record and empirical evidence.

The story is very familiar, since we have heard it many times before. The countries covered in this study have been told many times that if they only specialize in what they could do best, they would be in superior positions on the international market to generate surplus revenue. This means specializing in the production and supply of tropical cash crops for global markets such as tea, coffee, cotton, cocoa, sisal, and related primary commodities where their comparative advantage lies. They could then use the surplus revenue from export of primary commodities to pay for their imports, including food, energy, intermediate goods, and advanced technologies, all presented as essential for export-led growth. Nevertheless, the disastrous results are familiar to all of us. A small Zambian farmer, Joasam Ndaabona, put it best:

> This Jatropha reminds me of cotton. Many years ago, when Dunavant came here, they promised that if we grew cotton, we would be paid lots of money. We stopped growing traditional maize to make more money from cotton. But when the time to sell it came, we were paid very little. We went hungry because we had neglected growing our traditional crop maize. (Mundia, 2008)

Contrary to the oversold banner of agrofuels as an answer to Africa's dire poverty, further specialization in agrofuel production can deepen the crisis of food shortages, destitution, and hunger. In fact, the transition to agrofuels production can worsen the landscape of poverty and hunger by dispossessing millions of farmers of their holdings. The expropriation of 50 million ha in southern Brazil, Bolivia, northern Argentina, and Paraguay, to make way for energy biomass, has left hundreds of thousands of people landless. This situation is worsened by making food prices beyond the reach of the poor. A number of experts have suggested that a percentage rise in food prices can drive 16 million more people to poverty. Therefore, if the transition to agrofuels becomes total at the expense of food production, the number of chronically hungry people worldwide would increase by 600 million by 2025 (Holt-Gimenez and Shattuck, 2008).

Moreover, contrary to the myth portraying the market for agrofuels as free of fluctuations, biotic resources used in agrofuel production can be subject to boom and bust, owing to structural oversupply, natural disasters, or monopsony. In Kenya, for example, under the banner of "green politics for poverty reduction," small farmers were goaded into converting their entire landholdings to biological feedstock production. In 2000, the pirate state bourgeoisie mobilized their agents to persuade subsistence farmers to grow Jatropha on the idea that this South American plant grows on marginal land that cannot support food crops. Furthermore, they persuaded the cultivators that Jatropha is drought resistant, requires little water and care, promotes rural development and poverty eradication, enhances energy security, and that Jatropha seeds contain more oil than any other plant, allowing farmers to overcome their grinding poverty or become rich quickly. Moreover, state agents promised farmers that the government would supply certified Jatropha seeds that could grow in arid regions, would provide extension services advising farmers on how to grow Jatropha, and would buy the seeds at $13 per kilo of Jatropha seeds (Wadhams, 2009).

The logic shoved down the throat of poor farmers was that they could grow Jatropha on their poor and eroded land for cash, and then use that cash to buy food and other essential things, or even pay for their children's education. There were some early successes in some parts of Kenya as prices for biological feedstocks generated relatively modest incomes for producers. In early fall 2008, however, the prices for Jatropha, castor, and Croton oil seeds virtually collapsed, and growers were dragged into poverty even deeper than before. In April 2008, for example, the average price for a pound of feedstock raked in $5, and by the following fall, a farmer was lucky enough if he could sell the same amount of oil seeds for a quarter of a dollar (Bwakali, 2008). Thus, contrary to the claim of the new biomasters, the growing demand for agrofuels is bound to aggravate the food situation in Lake Victoria's riparian countries. What the evidence strongly suggests is that food production and agrofuels are competitors, not complementary to each other. As Molony and Smith (2010) correctly note, because of the transition to agrofuels economy, there will be less food to consume in the future as more and more crops that would otherwise be used for human consumption will be converted to agrofuels.

The second myth orchestrates the prospect of bioenergy-based job creation. Of course, this claim runs counter to the history and logic of capitalism. First and foremost, the tendency of capitalist bioenergy crop production is toward centralization of production in order to optimize profit through scale economy. The establishment of large-scale bioenergy plantations is bound to rely on increased mechanization. In light of this inherent tendency to scale economy and mechanization, the bioenergy sector is very unlikely to absorb all the people it displaces, much less to create new jobs.

The Bolivian case is an excellent illustration of the dangers posed by a shift to bioenergy production. Of the 270,000 ha annual natural forest loss that occurred between 2000 and 2005 in Bolivia, 100,000 ha were accounted for by the conversion of the forests to soy production for export under pretenses of promoting national development, creating jobs, and alleviating poverty. Contrary to these claims, however, specialization in the production and export of soybeans has not created the advertised jobs. In fact, 80 percent of the huge soy fields are owned by a few wealthy individuals from outside the soy growing regions (Lewandowski and Faaij, 2006).

To be sure, the place of labor in the growing agrofuel plantation system is disconcerting. In Brazil, the 500,000 sugarcane workers, who cut the cane manually in the sweltering tropical heat, are condemned to live in squalid conditions, exposed to debilitating illnesses, cut off from towns, and with pitiful pay and no recourse for grievances (Smolker et al., 2007; ETC Group, 2010). In Indonesia and Malaysia, the status of plantation workers is no better. Exploitation has always been accompanied by extremely poor pay, dire living conditions, and exposure to hazardous chemicals. In the agrofuel plantations, more than 25 different chemicals are sprayed by laborers with little to no safety procedures. In Malaysia alone, 30,000 women are deployed to spray these chemicals without proper protective gear (Smolker et al., 2007).

Moreover, the manner in which the emerging relationships between the agrofuel giants and indigenous labor are arranged do not bode well for labor. The preferred corporate stratagem is to develop agrofuel plantations based on outgrower programs, where companies supply seed, fertilizers, pesticides, and equipment on loans to growers at prices predetermined by the companies. Upon harvest, the growers are under contractual obligations to deliver the feedstocks to the same companies, again, at prices determined by the companies.

In Zambia, for instance, Marli Investments established a 30-year contract with biological feedstock providers, which would extend credit to outgrowers to buy seedlings, fertilizers, and pesticides. Upon harvest, the growers would pay the money back along with several deductions from their cash for licensing services, management fees, and membership dues. In addition, the growers are under obligations to replace, at their own expense, any trees that are damaged or die. Growers can neither sell their feedstocks to a competitor, nor can any member of their household grow feedstocks for anyone else (Mundia, 2008). Thus, there is an insidious aspect to these outgrower schemes. In addition to the reality that agrofuels companies are the ones that determine the quality, quantity, and price of feedstocks, outgrowers are paid for the quantity they supply, not for labor invested in the quantity of feedstocks supplied. The typical outgrower would naturally work his feedstock field day and night to make sure that quality meets the buyer's specification. However, the number of hours spent tending to feedstocks is not counted. He may use the labor of his children and wife, which does not enter the bookkeeping. There is also the question of uncertainty. Due to floods and droughts, there may be harvest failures. The outgrower bears all risks associated with such extreme events while the buyer internalizes the benefits of feedstocks.

The third myth rests on the spurious assertion that the abundant "unproductive" or "unused marginal land" in Victoria's littoral countries and other tropical African countries being sought is suitable only for biofuel production, but not good for growing food crops. The agrofuel industrial complex and the pirate state bourgeoisie find the orchestration of the myth of "idle and unproductive land" as the best selling point. There are wild estimates about the availability of this so-called idle land for biofuel production. Field et al. (2008) suggest that there are 386 million ha of abandoned land worldwide that easily can be harnessed for bioenergy production without impinging on food production or biodiversity protection, while others say that there are as many as 2.1 billion ha of land worldwide that can potentially be converted to agrofuels production (Searchinger et al., 2008). A report from the Global Environment Facility (GEF 2006) also identifies more than 1 billion ha of land worldwide that is considered totally degraded, which could not be cultivated for food crops, but could be used to grow biological feedstocks to generate at least 100 exajoules of energy annually.

For example, Jatropha could be grown on such supposedly severely degraded land, since this hardy plant is promoted as one that could survive in harsh environments on poor soil, with little water and inputs, and without disrupting food production or biodiversity. One should not lose sight here of the truth that the attraction of the agrofuel industrial complex to this Central American plant is motivated by the consideration of Jatropha as highly lucrative, in terms of its bio-diesel yields. A company can supposedly generate 2,000 barrels of Jatropha-based biodiesel from a square mile of land, compared to 200 barrels from maize and 100 barrels from rice from the same amount of land.

The definition of "degraded" land is highly contested since proponents of agrofuels often distort the presentation of what is "degraded" or abandoned.

For example, Brazil (overly sensitive to the charge of promoting rentierism) readily reclassified 200 million ha as "degraded," deemed suitable only for bioenergy crops. While in truth, however, these lands in the Pantanal (the world's largest freshwater wetlands rich in biodiversity), the woody savanna of the Cerrado, and the forests of Mata Atlantica are where indigenous communities derive their livelihoods (Holt-Gimenez and Shattuck, 2008).

Moreover, the character of communal ownership of land in Africa has recently generated a good deal of confusion, allowing the rentier class of African elites to exploit the egalitarian mode of landholdings to their advantage. The Tanzanian situation is a case in point. In the name of liberalizing the economy, in the 1990s Tanzania's pirate state bourgeoisie began tampering with the received customary laws that governed land tenure in the country. The present land law makes the state the sole owner of land in Tanzania,in which case land cannot be bought and sold or allocated without state intermediation. Communities and individuals have usufructary rights to develop and use the land; they may even rent out the land to developers and investors with the knowledge of the state or its agents. The land law authorizes the state to lease land to foreign investors ostensibly to promote rural development, poverty alleviation, local and national food sufficiency, and energy independence. In practice, however, monopoly control of land resources allows the pirate state bourgeoisie to appropriate land for themselves and/or rent out to foreign investors and oligopolies. When land is rented out to investors in areas where cashew and mango trees are tended by families, nominal compensation is given to the owners of those trees, which means that the land is not "idle" or "unoccupied."

Thus, what the facts on the ground indicate is that, contrary to the deceptive corporate definition of land resources in Africa, there are no idle or "unused" land resources that are not used by indigenous communities. The so-called unused marginal land is actually part of the general ecology, performing critical ecosystem functions, such as capturing solar energy for photosynthetic action, sequestering carbon, and storing water. Therefore, conversion of vast lands, presented by proponents of biofuels as idle, inevitably reduces the complexity of ecosystem structures and functions. Furthermore, by extension, this undermines their ability to provide critical ecosystem services. The so-called unused marginal lands are where the majority of the rural poor in Africa live. The land that is being alienated for purposes of agrofuel production, actually belongs to pastoralists, or poor peasants.

This occurrence is not surprising since it is consistent with the classic trajectory of alienation of nature and dispossession of people under predatory capitalism. It is, perhaps, important to remind ourselves constantly of the historical record. Ever since the beginnings of the colonial era, poor Africans have constantly been pushed off their original well-watered and fertile land to make way for colonial plantations. This pattern of land alienation and accumulation never stopped with the end of colonialism. The postindependence African pirate state bourgeoisie have continued to push the poor farther and farther off to marginal lands. In fact, 73 percent of people in sub-Saharan Africa today live in rural areas, generally considered marginal, and African pastoralists use 40 percent of all land in the continent for grazing their herds, as well as for other purposes (Liversage, 2009). The poor have always used the natural resources on the marginal lands for subsistence by grazing their livestock, picking medicinal plants, gathering firewood, and collecting water. Conversion of these so-called marginal lands to agrofuel production means further depriving the already deprived African majority of their livelihoods (Braun and Meinzen-Dick, 2009; ETC Group, 2010). In effect, when apologist researchers, the new global biomasters, and their host elites speak of "unproductive" and "unused" land, they are actually speaking of land resources not yet used for commodity production under market conditions for purposes of profit generation and accumulation. More important, the interests of the agrofuel industrial complex and the pirate state bourgeoisie are not limited only to what is considered "unused marginal land," as their demand for land resources runs the whole gamut of land types.

For example, the land given to SEKAB in the Wami and Rufiji River basins in Tanzania is fertile with good water resources. Small farmers in these areas use the land and water resources for growing rice and other staples. In fact, for many communities, the Rufiji River and its wetlands are the only sources of drinking water during dry seasons. Yet, thousands of small farmers were evicted to make way for SEKAB's 9,000 ha sugarcane plantation, and thousands more await their potential evictions as of this writing (Knaup, 2008). The reason is simple: the planned sugarcane plantations require fertile soil and abundant water resources since sugarcane sucks up lots of water.

In addition to all the concerns highlighted above, the prevailing power differential between the contracting pirate state bourgeoisie and the agrofuel corporations makes the land deals egregiously lopsided in favor of foreign corporations, and to the detriment of the poor. As the report of the International Food Policy Research Institute states:

> Even though some of the land lease agreements make provisions for investments in rural development, these deals may not be made on equal terms between the investors and local communities. The bargaining power in negotiating these agreements is on the side of the foreign firm, especially when its aspirations are supported by the host state or local elites. Smallholders who are being displaced from their land cannot effectively negotiate terms favorable to them when dealing with such powerful national and international actors, nor can [they] enforce agreements if the foreign investors fail to provide promised jobs or local facilities. Thus unequal power relations with land acquisition deals can put the livelihoods of the poor at risk. This inequality in bargaining power is exacerbated when the smallholders whose land is being acquired for foreign investment projects have no formal title to the land, but have been using it under customary tenure arrangements. Since the state often formally owns the land, the poor run the risk of being pushed off the plot in favor of the investor, without consultation or compensation. (Braun and Meinzen-Dick, 2009)

Moreover, since the agrofuel industrial complex and the host pirate state bourgeoisie control the production and dissemination of information, local communities never know the whole truth, if any at all. Reliant on corrupt indigenous intermediaries, foreign land grabbers can and do manipulate the trusting local communities. In April 2006, for example, British Sun Biofuels reported that it consulted 11 communities in Tanzania and falsely represented that 10 of the 11 communities had given their consent to the land acquisition deal. In reality, several of the communities were not even aware of the content of the land deal. Those that were consulted had actually attached relevant conditions to their approval, such as restricting Jatropha cultivation to only demarcated areas. In fact, one community leader accused Sun Biofuels of clearing restricted forests for Jatropha cultivation without even bothering to contact community elders first (Knaup, 2008). Moreover, Sun Biofuels paid a token compensation of $220,000 for the 9,000 ha it acquired in the area, to be divided among 152 farmers in compensation for the mangoes and cashew nuts they lost on the expropriated land. Since Sun Biofuels had a long-range vision of expanding its holdings to 50,000 ha, the consequences of dispossession of this scale for local communities would not be difficult to imagine (Knaup, 2008). Aside from the unreasonable amounts paid to the farmers, the idea of compensating farmers exposes the insidious nature of the land deal itself. If the land was "unused" or "idle," why does the company pay compensation in the first place? If the compensation is for the mango and cashew trees, does not this tell us that local communities own the land and use it sustainably for subsistence? It is more complex than any of the answers to these questions can satisfactorily explain. Therefore, it is important to note that the communities are compensated only for what they grew on the land, and not for the land itself.

In sum, since there is not enough "unused" or "unproductive" land suitable for agrofuel production, any additional alienation of land resources will have to come from either converting more arable land to agrofuel production or drawing down ecological resources at present contained in grasses and forests. It is worth noting that, of the 13.5 billion ha of total global inventory of land areas, 8.3 billion ha are a composite of grasses and forests and 1.6 billion ha are croplands. If one leaves forests and protected land out of consideration for conversion to agrofuel production, there remains only between 250 million and 800 million ha of lands for potential crop production to feed the expected additional human populations.

The conundrum does not end here. There is the daunting issue of water. By definition, agrofuel production depends on large-scale agrocommercial plantations, which require continuous application and maintenance of massive amounts of water resources. So every ha of land and every gallon of water used to produce agrofuels is inevitably subtracted from resources needed to grow food to feed the poor. The water requirements of agrofuels production are in the order of 70 to 400 percent more than fossil fuels, wind, or solar energy, depending on the agro-ecological conditions where the bioenergy crops grow. For example, in 2007, 45 billion cubic meters of freshwater were dedicated to agrofuels production worldwide, which was six times the total amounts of water resources that the entire global population used for drinking (International Council for Science, April 2008).

For reference, it takes 2,900 gallons of water to grow a bushel of corn in Oklahoma and it takes another 4 gallons of water to generate 1 gallon of ethanol; 3,500 liters of water are required to produce sugarcane for each liter of sugar bioethanol; with soy, it takes 14,000 liters of water to generate one liter of biodiesel; and it takes a whopping 20,000 liters of water to generate one liter of biodiesel from Jatropha (ETC Group, 2010; Smolker et al., 2007; Ernsting and Rughani, 2008; Berthelot, 2009). Given this unsavory reality, it comes to choosing between using massive amounts of water resources to produce agrofuels to meet the demand of the affluent global northerners or using the same amounts of water resources to grow enough food to feed the 1.2 billion poor global southerners (Liversage, 2009). The choice entails diverting morc land that is arable, using scarce water resources for agrofuel production, and running the risk of deepening poverty and mass emmiseration. The alternative entails converting more and more grassland and forested areas to make way for agrofuel plantations and then running the risk of exacerbating the rate of global deforestation and, by extension, global warming.

What the evidence and observations discussed in this section indicate is that, while the agrofuel industrial complex and their host pirate state bourgeoisie loot the biomass in the Lake Victoria basin for purposes of agrofuel production and hard currency generation, the broad masses in the riparian countries are bound to remain mired in poverty, pollution, contamination, and utterly degraded ecology. The looters are putting in enabling environments to establish their grip on the region's biotic resources under the veneer of promoting development, reducing poverty, creating jobs, and insuring energy self-sufficiency. By appropriating more and more biomass in the Lake Victoria basin to produce agrofuels, the agrofuel oligopolies and their host elites are accentuating the biolimnological vulnerability of Lake Victoria to entropic degradation. The transition to agrofuels economy can only represent further deforestation, wetlands destruction, soil erosion, loss of biological diversity, shortages of freshwater, pollution, urban and agricultural runoff, as well as lake sedimentation, eutrophication, and acidification.

CONCLUSION

This chapter has shown how the macrostructural contours of accumulation have accentuated the biolimnological vulnerability of Lake

Victoria to degradation, as well as the dispossession, destitution, and alienation of lakeshore communities. The lake's biolimnological vulnerability and the sorry state of the communities' social existence are rooted in the fact that the pirate state bourgeoisie have handed over both lacustrine and terrestrial natural resources to global corporations and their domestic hangers-on. As if the globalization of Lake Victoria's fish resources were not bad enough, the terrestrial food ecology in Lake Victoria's littoral countries is now severely strained by the new wave of multifront aggression on biotic resources for purposes of bioenergy production. This new aggression on biotic resources has several long-range implications for food security, biodiversity conservation, and climate change.

First, the experience in the fishery mode of production has demonstrated that since bioenergy production requires scale economy, only those individuals and groups with upfront capital, technology, and managerial skills are bound to be the prime movers into the bioenergy sector. This means that local farmers would be pushed off their farm holdings. In addition, since the owners of capital, technology, and managerial skills are foreigners, bioenergy production in the countries covered in this study would be very much in the service of accumulators, as well as consumers in the global north.

Second, the shift to bioenergy production is bound to reduce the amounts of cropland and water resources available for food production. As noted in Chapter 2, the majority of the people in the Lake Victoria basin are already in extreme poverty. If the littoral countries of Lake Victoria are to overcome hunger locally and nationally, more food must be produced, which means more cropland and water resources must be found. After all, the populations of the three East African countries are doubling every 20 years or so, pointing to the urgency to invest several billion dollars annually in agriculture. This is required in order to increase cereal production substantially, if the pirate state bourgeoisie are modestly to contain the crisis of legitimacy. This suggests that the imperative to meet the existential needs of new human arrivals, coupled with the urgency to uplift the people already mired in extreme poverty, would require conversion of additional forested areas and grasslands into cropland. This would bring the necessity of survival into direct collision with the agrofuel-based logic of accumulation over competition for the same terrestrial resources.

Third, as the life-cycle analysis of bioenergy production suggests, the climate benefits of agrofuels (as presented by technocentric cornucopians) are ambiguous at best and misleading at worst. Under conditions of unlimited global market growth, which is the primary objective of capitalist accumulators, the relative gains from a transition from a primarily fossil fuel-based economy to an agrofuel-based one can easily be nullified by the hyperacceleration of production and consumption of biobased goods. More demand for agrofuels means more demand for low-entropy throughputs, which, in turn, means more high-entropy wastes resulting from the cultivation, production, distribution, and consumption of agrofuels. In light of this reality, the notion that we will solve all our climate problems by resorting to agrofuels can be dangerous.

In order to assess the gains and losses from increasing reliance on agrofuels objectively, we need to carefully do our homework. That is, assessing the potential climate gains from agrofuels in relation to the entire supply chain from the conversions of forests, grasslands, and croplands to bioenergy crop production. After all, the use of such inputs as petroleum-derived fertilizer and pesticides for the growing of agrofuel feedstocks, the processing of the feedstocks into fuels, and the burning of the biofuels as end products to run our vehicles and machines, all of this represents the foregone carbon sequestration, net greenhouse gas reductions, and climate change abatement. This condition requires guarding against the fallacy of misplaced concreteness, drawing spurious analogies to cover a whole spectrum of issues.

For example, Brazil's sugarcane-based bioethanol is touted as having a 90 percent GHG reduction. From this simple instance, it is possible to draw a general conclusion that other biotic resources can yield the same result, without examining the underlying conditions under which the Brazilian bioethanol production occurs. In fact, as a report of the UN Global Environment Facility (2006) correctly notes, there is a partial exception to Brazil's success in bioethanol production. Brazil enjoys extraordinary endowment in terms of excellent climate, good soil, diversified plant species, and abundant land and water resources, all of which make the cost of bioethanol production from sugarcane competitive with petroleum-based energy. In contrast, 50 percent of global sugar production takes place in such countries as Kenya and Tanzania, where conditions are not as favorable as those in Brazil. This reflected in the fact that production costs in these countries are three times those in Brazil.

Finally, the transition to agrofuel production is bound to exacerbate the social antagonisms between the pirate state bourgeoisie and the broad masses. Overwhelmed by the logic of necessity of survival, the masses in the littoral countries of Lake Victoria are waking up to the cruel realities of dispossession, destitution, and alienation. These conditions are ushering in a new phase in the struggle for survival and dignity. Emblematic of this new social reality is the April 2007 episode, when Ugandans in Kampala went to the street in protest over the president's decision to excise 7,000 ha from the 32,000 ha Mabira forest reserves in the watershed of Lake Victoria for sugarcane plantation. Because the recipient was a company owned by Asians, the class struggle took on a racial character as the protesters raided Asian-owned stores. In the ensuing skirmishes an Asian was killed, to which the police reacted by killing two protesters and wounding a number of others. The incident exposed Museveni's long held contempt for the idea of national resources conservation. Thinking that he was uttering words of wisdom, he actually disdainfully commented on the issue by saying, "It is more difficult for a backward country to guard against environmental degradation than for a camel to go through the eye of a needle" (Pearce, 2007). Reinforcing the president's disdain, his environment minster blamed the protesters for having the temerity to raise the issue of protecting Mabira forest reserves, since poor Ugandans were already destroying forests throughout the country eight times the size of Mabira for cultivation and firewood (Pearce, 2007). According to the minister's logic, since Mabira forest reserves would be encroached upon by the poor anyway, it was in the interest of the pirate state bourgeoisie to convert the forest to sugarcane plantations for agroethanol production before the poor took over the reserves. Such are the men at the apex of power with whom the dispossessed masses have to reckon.

6| A Way to the Future

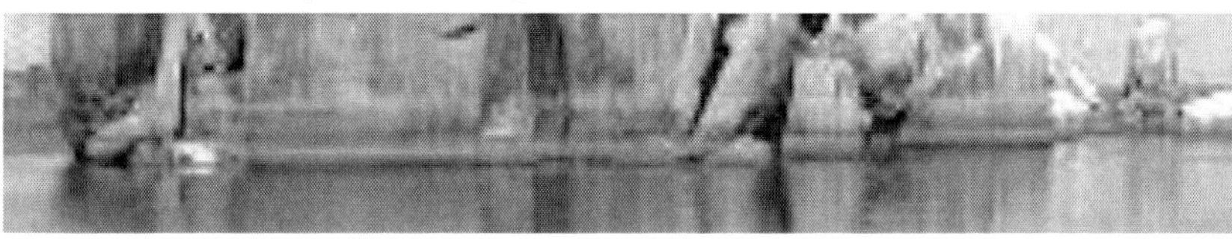

We shall never cease from exploration and the end of all our exploring will be to arrive where we started and know the place for the first time.
T. S. Eliot, *Four Quartets*, 1943

In October 2007, the journal *Aquatic Ecosystem Health and Management* dedicated a special issue to Lake Victoria, carrying 11 professional articles, all dramatizing and bemoaning the impending hydroecological collapse of the lake. Just as they were unanimous in their lamentations over the fate of Lake Victoria, the authors were equally unanimous in their recommendations. Each harped on the transformation of managerialism into the religion of the fishery mode of production as the solitary answer to the state transition of the lake. The authors' collective recommendations include comanagement of the fishery mode of production, stringent control of entry into fishery, reduction of fishing effort, establishment of maximum boat ownership, imposition of a lakewide regime of monitoring and surveillance, reduction of the number of gill nets, increasing mesh sizes, eradication of illegal beach seining, establishment of closed areas and seasons, promotion of harmonized interriparian fishery policies and regulations, restoration of wetland ecology, and creation of protected reserves for endangered species as insurance against depletion and/or extinction of native species.

Other writers have echoed similar themes, assigning primacy to the imperative to establish urgent comanagement for the entire Lake Victoria fisheries, reflecting symbiotic partnership between the states and fishing communities (Brown et al., 2005; Schindler et al., 1998). In line with the rights and duties defining this partnership, fisherfolk will determine who will fish, when they will fish, and how much they will fish in order to conserve lake resources. Both as the final arbiter and an interested party because of its need for hard-currency generation, the state would also play a pivotal role in this exercise of lake resource conservation.

For example, if fishers overstep their right to sustainably harvest, the state would step in to facilitate regulation of mesh sizes in order to alter food web interactions and harvest rates, to reduce fishing effort, to monitor predator-prey relationships, and to promote entry and exit strategies for harvesters to insure sustainability or recovery of the threatened fish species. Moreover, the state would furnish external or internal compensation for those fishers who voluntarily exit from the fishery sector in the interest of species conservation. The state here becomes central to the preservation and restoration of the fishery mode of production in Lake Victoria. It does so by using its levers of power to regulate fishing patterns, to reduce predation conditions, to use food web interactions as a means of conservation by altering the lake's species composition or the total biomass of specific species, and to control the sources of anthropogenic eutrophication of Lake Victoria.

The analyses and recommendations of the authors who contributed to the journal remind me of the attribution made to Karl Marx, that philosophers are adept at interpreting the world in manifold ways, but the tall order is how to change it. On the surface, the authors' collective approaches and proposals appear sound. A close examination of them, however, would reveal fundamental contradictions. All the authors condemn the state for not doing enough to conserve lacustrine resources and protect the fishery mode of production. Yet, they make a turnaround face and ask of the same state to enforce the recommendations they put forward.

Moreover, the authors view the open-access character of Lake Victoria as the single most important variable that primarily accounts for the overfishing and degradation of the lake. From the authors' standpoint, the remedy would be a state-run commercial enclosure of the lake. This shows how epistemological limitations and class orientation have combined to prevent researchers from comprehending the logic of necessity to survive, on the part of the poor fishers, and the logic of accumulation, on the part of big operators as the principal drivers of serial depletion. Furthermore, the authors cannot see the simple fact that big operators work in partnership with the pirate state bourgeoisie, in effect, making the state central to the problem of overharvesting.

Moreover, contrary to the authors' assertions, what attracts poor fishers to Lake Victoria is not the open-access character of the lake, but rather desperation of the dispossessed masses, since the pirate state bourgeoisie has utterly failed to provide alternative means of survival to them outside the fishery mode of production. The authors of the various works cited in this volume have also miserably failed to grasp the simple fact that the social character of the pirate state bourgeoisie is incompatible with the imperative to sustainably use and judiciously conserve lake resources for the betterment of present generation of fishers and for meeting the needs of future generations. Actually, asking the pirate bourgeois state to enforce all sound recommendations vital to the health of Lake Victoria is akin to making the fox the keeper of a chicken coop, and then expecting him not to eat the chickens so that other non-fox parties can have a fair share of the chickens.

A PROLIFERATION OF STUDIES TO SOLVE THE PROBLEMS

The general observations I have made above are not limited to the authors of the articles I have referred to in the opening paragraph of this chapter. Indeed, my criticism covers an entire industry of intellectual entrepreneurship in the fishery mode of production. In a seemingly endless search for answers to the prevailing challenges obtaining in Lake Victoria, national, regional, and international fishery managers and researchers have long spoken of the urgency to promote sustainable fisheries development without grasping the sociopolitical context within which such development can occur. Accompanying the proliferation of microsegmented studies, pilot projects, and programs, the reproduction of technical papers has become a teeming industry. Expatriate commando experts, from time to time, are deployed to the riparian states to advise on how to conduct biological counterinsurgency operations to halt and then reverse the depletion of fish resources and the entropic degradation of the lacustrine ecology.

Surfing the websites of riparian organizations and a review of the literature they have produced on how to promote sustainable fisheries reveal remarkable redundancies in orientations, approaches, and missions. However, the fundamental assumption remains unaltered doing more studies of the lakes' potential for purposes of creating the requisite conditions for the biomanipulation of the processes of aquatic ecology remains the sine qua non of the research agenda. As if linguistic substitutes would do the job, the studies and documents are replete with such high-sounding and mellifluent terms as stakeholder participation, participatory management, comanagement, regional integration, regional lakewide management, multidisciplinary management approach, integrated sustainable ecosystem approach, sustainable livelihood approach, and so on.

On paper, Lake Victoria's littoral countries have also signed protocols and conventions on how to achieve a proper balance between a sustainable resource development and aquatic resource conservation in the lake. To arrest, supposedly, the further ecological deterioration of Lake Victoria, the nations concerned, aided by external donors, have launched the Lake Victoria Fisheries Organization, the Lake Victoria Environment Management Project, and the Lake Victoria Basin Commission. The international sponsors of these organizations and researchers untiringly repeat that the root causes of lacustrine degradation are runaway populations, overfishing, changes in landscape in the basin, and pollution. Consequently, the organizations are asked to make comprehensive studies of fish biology, hydrodynamics, fish genetics, fish statistics, remote sensing, socioeconomics, and legal and institutional studies.

Long before the recommended solutions are implemented, the same organizations are again asked to do more of the same studies if there are still external funders. The investigations will supposedly yield the means for optimal management of lake resources, in ways that serve present and future generations. On the one hand, the particular national fisheries organizations are expected to collaborate on the whole gamut of issues and work in coordination with other projects. On the other hand, the basin-wide organizations are supposed to provide an overarching framework for sustainable harvesting by addressing basin-wide issues of biodiversity conservation, assessment of the fish stocks and their biology, promotion of integrated fisheries management, pollution control, and forest preservation or restoration. These are in order to make lakewide optimal management durable. This so-called optimal management is also supposed to solve problems of postharvest losses arising from poor infrastructure, lack of marketing opportunity, the use of destructive fishing gear and techniques, and the absence of complementary community education and participatory management.

In addition, the means to achieving this objective remain the same continuing the export-led growth in order to generate, ostensibly, resources needed to fund the various fisheries development projects and programs. For example, in 2007, researchers at Uganda's elite university generated a lengthy report advising the government on how to reach its goal of earning $1.2 billion in hard currency from the export of fish and fish products. They blamed Uganda's wanton inability to exploit Europe's ready-made market access on its fragmented businesses, poorly managed fisheries, and subsistence farming. The solution, they said, was government promotion of public/private partnerships in the interest of scale economy. In their formula, every dollar the government invested in such partnerships would translate into net returns of $13. Consequently, by 2011, Uganda could earn $1.24 billion from the export of fish and honey alone to the European Union. To achieve this, Uganda must upgrade its fishing facilities to European sanitation and health specifications, since 40 percent of the annual fish catch is rejected because of poor handling. The reference point for the researchers was the 1990-2006 period, during which Uganda's export revenue grew by 80 percent (from $178 million to $877 million); this was a period when fish export also increased by 96 percent from $5.5 million in 1991 to $143.6 million in 2005 (*Africa News*, Oct. 11, 2007).

What is missing from the analyses of these researchers is whether the increases in foreign exchange earnings have made a difference in the lives of Uganda's toiling fishers. Lacking in proper training or blinded by class interest, these researchers are unable to see the obvious fact that the so-called export-led growth is a means to perpetuating the predatory character of the state. As Sauper's documentary film amply demonstrates, despite lip service paid to the need to lift fishery-dependent communities out of poverty by exporting fish, the supposed beneficiaries have yet to receive any tangible dividends from this extroverted growth strategy. For example, the government in Kampala has an official policy on the books that states that 65 percent of all taxes collected from the fishery mode of production would be plowed back to the counties where the actual activities take place. Twenty-five percent of the taxes would go directly to the local communities. However, there is no evidence that this is actually happening. It is instructive to note that people in Uganda's fisheries sector pay eight different types of taxes, compared to only two types of taxes in the case of farming communities (Grellier et al., 2004).

In many crucial respects, Uganda's university researchers and the contributors to the aforementioned journal are public purveyors of what E.Wilson (2002) calls ontological reductionism, scientism, and simplism. However, as said before, this "original sin" is a product of both intellectual training and class self-interest. Indeed, writing about fishery-dependent communities and the fishery mode of production in Lake Victoria (as a proxy for the general epistemological debate over structural poverty of people, accumulation of private wealth, and entropic degradation of living aquatic and terrestrial resources) cannot be understood in isolation from the politics or class position and academic training of the researchers. Thus, epistemologically speaking, the challenge among those of us preoccupied with the condition of Lake Victoria today is not only how to examine the state transition of the lake, but also how to examine the causes of social pollution entailing the reproduction and dissemination of manufactured knowledge. This says that, without first tackling the social pollution of knowledge, no progress can be achieved in our understanding of the causes and consequences of the state transition of Lake Victoria. Furthermore, the recommendations we put forward will have no meaning or significance.

CHANGES NEEDED TO SOLVE THE PROBLEMS

Confronting the crisis of knowledge must begin with a frank recognition that the crisis in Lake Victoria is systemic. We have long reached a tipping point in the state transition of the lake, where a particular focus on the lake has become an exercise in futility. The tall order for the riparian societies, and those who are concerned for the welfare of lakeshore communities and the fate of the lake itself, is how to save the whole food ecology and the entire environment in which the system of production takes place. This is to say that since the state transition of Lake Victoria stems from the prevailing neocolonial system of production and the hegemonic ascendance of the pirate bourgeois state, our research program ought to be how to develop a counternarrative of contestation against the real source of the problem. In other words, the social obligation of those of us engaged in professional writing ought to be how to find ways of cleaning up the social pollution of knowledge.

With this in mind, I am about to violate the convention of reproducing recommendations on how to solve the crisis in Lake Victoria. I believe that the lacustrine crisis is not going to be solved by simply offering a long catalog of proposals, no matter how sophisticated our methodology or how powerful the theoretical arsenal at our disposal. Systemic change must occur. Therefore, in the rest of this chapter, I will broaden the discourse to a system-level analysis, in order to situate the fishery mode of production within the general food ecology.

As I outlined in Chapter 1, we are all influenced and shaped by, or indoctrinated into, competing schools of thought. For the sake of simplicity, I will narrow these schools into economic dogmatism, economic distributionism, ecological economics, and historical materialism. Today, these four perspectives inform the general debate over the human condition, by offering competing historical descriptions, theoretical contentions, and presumably practical propositions relative to the causes and consequences of structural poverty and ecological degradation. Thus, if we are to come to grips with the crisis in the fishery mode of production in Lake Victoria, the first order of business ought to be how to resolve the crisis of knowledge manifested in the differences among writers who belong to different schools of thought. Consistent with which the above observation is made, I turn in the rest of this chapter to untangling the juxtapositions of the aforementioned intellectual orientations.

As far as the condition in Lake Victoria goes, one of the enduring problems today is that state piracy and corporate intellectualism are joined at the hip, complicating analyses of the causes and consequences of the lake's state transition. Operating from the citadel of neoclassical economics and scientism, corporate intellectuals place the greatest value on the sophistication of their empiricist research designs, and on how they market them to corporate funders, public institutions, and the pirate state bourgeoisie. Researchers from the global north regularly fly in and out of Lake Victoria to study different aquatic species or different chemical properties, only to tell us that their new findings simply confirm the old ones. In the same way, national fishery researchers stretch their hands out to funders in the global north for research crumbs to carry out research on supposedly new species or physical properties. However, they also tell us that their findings confirm previous research findings. Then, both foreign and national researchers present an elaborate catalog of recommendations to their funders and the pirate state bourgeoisie on how to rehabilitate the lake or how to restock the species therein.

What is missing from the elaborate research designs and proposed solutions is the extent to which the prevailing mode of production has contributed to the present state of species depletion and lake degradation. Also missing is whether the recommendations put forth could be carried out within the same mode of production, which has been the primary cause of species depletion and lake degradation in the first place. Expatriate researchers are totally oblivious to the capitalist socioeconomic system within which the laws of thermodynamics and evolutionary biology operate, and they are reinforced by self-interest to make all sorts of guesswork about how aquatic resources must be exploited. Thus, expatriate researchers and national fishery managers often end up peddling commercial scholarship with the full knowledge that their careers and their future research projects are contingent upon recycling old findings in new forms. In order to purportedly give credulity to their work products, they situate their assumptions, analyses, and conclusions in conjured-up models and rehearsed axioms. All the while, in truth, most of the work is simply reprocessed knowledge or private conjectures. It is probably with this in mind that Shackle (1972: 8) insightfully observes:

> A private judgment of value is, of its nature, essentially and inescapably a conjecture of what is the valued object or system will be able to do, a conjecture of its potentialities, of its future. Valuation is expectation. What is vital is that expectations are conjectures; let us say figments, resting on elusive fragmentary and confusing evidence whose interpretation and suggestion can change from moment to moment with no visible cause. Valuation is expectation and expectation is imagination.

Given their class position, it is no coincidence that expatriate researchers and national fishery managers (baptized consciously or spontaneously into the orthodox dogma of neoclassical economics) situate the fishery mode of production within the context of market fundamentalism. Since they theoretically contend that capitalism is a self-driven and self-regulating system that eventually finds its natural equilibrium condition, unless externally generated perturbations interfere with its otherwise smooth operations, they cannot see that there are structural flaws inherent in the prevailing capitalist system of production. As a result, they cannot perceive the need for structural change in this system. Because they are socialized into the presupposition that capitalism (as a self-generating and self-regulating system) is naturally equipped to spontaneously and endogenously surge forward through a series of negations and affirmations, a contention captured in Joseph Schumpeter's aphorism "creative destruction," they cannot see that there is an inexorable relationship between the serial depletion of species and lacustrine degradation, and the capitalist mode of production within which resource depletion and ecological disintegration take place. Given this a priori position, if the performance of the fishery mode of production does not match the essential conditions of the market, then it must be owing to exogenous shocks or disturbances. These have to do with cumbersome regulations or the incompetence of fishery managers, not the prevailing system of production.

According to this logic, politicians, in their zeal to promote equity through redistribution of income, or simply for reasons of corrupt practices, have the propensity to meddle in the fishery mode of production. In so doing, they create conditions of price distortions, rent-seeking behavior, mismanagement, and corruption. In the end, they achieve neither social equity nor market efficiency in the fishery sector.

Given this theoretical position, the proposed neoclassical solution is to leave the market alone, which in this case is the fishery mode of production. Through efficient allocation of resources and optimal production of goods and services, the market for fish and fishery products will generate growth, the benefits of which are expected to trickle down to all segments of society. In the end, poverty alleviation through growth will result from competitions and transactions in the marketplace under conditions of freedom. Thus, "poverty alleviation through growth" becomes the favorite mantra of neoclassical theory builders and official growth ideologues.

The chief enforcers of the theory and ideology of growth are corporate intellectuals and fishery managers, backed by financial resources from foreign public institutions, philanthrocapitalist foundations, and global financial institutions such as the World Bank and the UN Food and Agriculture Organization.

Within this broad theoretical framework, the role of the state is specified as providing only enabling services to operators in the fishery sector. It was under this growth ideology that big operators (backed by international finance capital) set up 35 fish-processing factories on the shores of Lake Victoria. Their purpose was not to feed lakeshore communities but to serve the affluent global northerners. There was no consideration of whether the lake and the aquatic resources therein could support such number of capitalist operations; everything was left to the market to sort it out. This is the cause and catalyst for the serial depletion of fishery resources and for the destruction of the livelihoods and food security of local people. This lesson is still unlearned.

For example as Uganda's university researchers (who are chief promoters of the neoclassical growth ideology and oblivious to the pirate bourgeois character of the state) still counsel the state to build modern infrastructure that will enable private fish operators and exporters to meet European sanitation and quality specifications. This is so that they would be in a position to minimize postharvest losses, key to maximizing the flow of foreign exchange earnings. This is a clear case of how researchers can be indoctrinated into believing that the prevailing social system is beyond scrutiny since capitalism is a natural given. The net effect of such epistemological position is that the neocolonial capitalist mode of production becomes immune to examination and contestation.

Satisfied with their collective recommendations about how the fishery mode of production in Lake Victoria could be protected under the stewardship of the pirate state bourgeoisie, fishery managers and expatriate researchers have hitherto avoided the much-needed soul searching on whether the necessity of survival and the logic of accumulation can proceed indefinitely within the present structure of control and domination without accelerating the entropic breakdown of Lake Victoria. Given this reality, the greatest challenge for public intellectuals today is how to mount an effective scholarly counteroffensive against the peddlers in commercial scholarship and their international funders that provide legitimacy to the otherwise illegitimate pirate bourgeois state. The vital question today is no longer about economics or natural science, but rather, about politics.

Economic distributionism hosts a large number of writers, including Keynesian globalists, welfare economists, and humanist philosophers. All economic distributionists accept the fundamentals of a market system. However, they tend to modify this belief in the market by highlighting the importance of politics and the impact of power relations on the capacity of people to generate, control, and use wealth. The belief is embedded in the contention that the market alone does not and cannot generate equity, and the purpose of politics is to restrain the excesses of income inequality. Within this general rubric of distributionism, different accents are placed on the different aspects of poverty. Keynesian global distributionists situate the fishery mode of production (at least by indirection) within the urgency to modernize and expand the general food ecology in Africa, with the help of financial resources from donor countries. According to some global distributionists, the price of all the projects needed to lift 800 million Africans out of poverty would not be huge if donor countries live up to the modest development assistance they pledge.

In 2004, a group of Keynesian globalists, led by Jeffrey Sachs of Columbia University, wrote what amounted to be a 100-page development manifesto entitled "Ending Africa's Poverty Trap." The authors of this manifesto start out by positing that poverty in Africa is but one manifestation of lack of development. The latter, in turn, results from lack of capital, an uneducated workforce, overpopulation, the feminization of poverty, tropical disease ecology, and poor infrastructure. Given these structural constraints, Africa remains mired in underdevelopment and vicious poverty. Since net capital formation is the difference between what is produced and what is consumed nationally, the continent cannot feed itself. More so, it cannot save toward national capital formation, which is the necessary yeast to leaven the dough for virtuous growth. After all, the 800 million combined population of the 33 sub-Saharan African countries had a per capita income of $271 as of 2004, hardly enough to push them out of poverty (Sachs et al., 2004). It stands to reason that countries that do not have enough to save cannot have adequate capital to invest in modern agriculture, utilitarian education, public health, and modern infrastructure, and thus cannot begin the process of development.

Presentation of the case in this way suggests that the initial big push should come from overseas, with massive infusion of capital and technology. The inflow of capital would pay for replenishing soil macronutrients (nitrogen, phosphorus, and potassium) with synthetic fertilizers, pesticides, and herbicides. At the same time, the inflow of capital would pay for improved or genetically modified seeds, educating Africans and promoting their health, providing agricultural extension services, and for building roads, ports, and railways. These steps would allow African farmers to produce more than what they can consume, and then take their left-overs to markets to sell. Since the proceeds from the sale of surplus agricultural goods are beyond the needs of the producers, the money would enter the circuit of national capital formation.

However, in the authors' view, Africa's structural poverty is so deep that the conditions for surplus generation are not in prospect without foreign assistance. More important, since the structural factors of the continent's underdevelopment are widespread and well entrenched, incremental flows of foreign aid would not work; it requires a kind of shock therapy as an "exit strategy from the poverty trap" (Sachs et al., 2004).

For example, if Tanzania were to reach its Millennium Development Goals by 2015, it would require an investment of $4.2 billion per year in various development projects between 2004 and 2015 to combat the various elements of poverty and underdevelopment. Likewise, Uganda should invest $3 billion per year to achieve its Millennium Development Goals.

In sum, if the sub-Saharan states are to attain their Millennium Development Goals, they need constant flows of foreign aid in the range of 20 to 30 percent of their GDP, which adds up to $50 billion per year for a decade or more (Sachs et al., 2004). In Sachs' (2008) computation, the 33 sub-Saharan African states could graduate from foreign aid by 2025, if the donor countries had kept their pledges to double aid to Africa from $25 billion in 2004 to $50 billion per year until 2015 and then to $120 billion per year between 2015 and 2025. In Sachs' view, this sum is not huge if spread around the 800 million sub-Saharan Africans, since it amounts to just $150 per African per annum. In other words, in order for Africa to overcome the poverty trap, the income of the continent must rise from the 2007 level of $1,400 at purchasing power parity to $4,000 per person per annum by 2015, tripling foreign aid for at least 16 years to comprehensively combat the root causes of hunger and become developmentally self-sustaining (Sachs, 2008).

In 2009, a task force co-chaired by Catherine Bertini and Dan Glickman (formerly head of the UN World Food Program and Secretary of Agriculture during the Clinton Administration, respectively) produced a lengthy document under the auspices of the Chicago Council on Global Affairs, entitled The Chicago Initiative on Global Agricultural Development. Despite some divergences in emphasis, the authors of this manifesto essentially echo the Keynesian global distributionist vision, already forwarded by the Sachs group. The focus of the Chicago task force is on the need to reassert American leadership in the fight against hunger. The target is rural Africa. In the authors view, with the yearly injection of $30 billion into agricultural development in the global south, the world can solve global hunger for good.

For example, with the investment of just $2.9 billion per year, the world can help reduce the number of hungry and malnourished Africans by 144 million by 2020 (Bertini and Glickman, 2009). According to this blueprint, donor assistance in Africa can go toward the provision of agricultural extension services, promotion of agricultural research, development of rural agricultural infrastructure, and improvement of development assistance deliveries to farmers. Cumulatively, these efforts will spark a kind of "second green revolution" in Africa (Bertini and Glickman, 2009).

The macrodescriptions of the human condition in Africa, meticulously detailed in the two manifestos reviewed here, are indisputably powerful and moving. The focus on the terrestrial food ecology as the key to solving Africa's hunger, poverty, and underdevelopment is equally laudable. By indirection, these Keynesian globalists are making a strong case for the sustainable livelihoods approach to the fishery mode of production, discussed in Chapter 1. The reasoning is very sound. Rehabilitation of Lake Victoria and the conservation of aquatic resources therein cannot be imagined without making a revolutionary alteration in the terrestrial food ecology of the riparian countries. Nevertheless, however noble the Keynesian globalist vision is, the prescriptions are fanciful for several reasons.

The contention that Africa's underdevelopment is due to lack of capital starts out from a faulty empirical presentation is the first reason. Contrary to conventional position, the leakage of too many resources out of Africa, not shortage of foreign aid, is responsible for the continent's underdevelopment and structural poverty. In fact, according to a study of 30 African countries, the magnitude of financial resources that exited from those countries between 1970 and 1996 was equal to 145 percent of what those countries owed in external debt. In 2007, the cumulative external debt of sub-Saharan African countries was $195 billion. For comparison, at the 2005 G-8 Summit, donors pledged $25 billion in aid for sub-Saharan Africa, while the region registered $50 billion in capital flight that year (I. Taylor, 2010). As Ian Taylor correctly notes, the problem in Africa is not lack of resources, but rather the stashing away of those resources in the advanced capitalist countries. In September 2004, for example, it was reported that there were 100,000 millionaires on the continent, worth $600 billion, probably most of it held in Western banks. According to the 2004 World Bank report, as much as 40 percent of private wealth in Africa was circulating in the global north (I. Taylor, 2010; Yohannes, 2008).

Additionally, there is the perennial problem of leakage. Most of the so-called foreign assistance is leaked back to where it originates, in the form of exorbitant remuneration of the expatriate experts who claim to know tropical Africa and payment for purchases of expensive inputs and equipment. After deducting the kickbacks given to the African pirate state bourgeoisie and the expenses of dependency on foreign aid receipts, the net amount left is too ridiculously small to even make a dent in the continent's structural poverty.

Furthermore, the continent's historical condemnation to specialization in primary commodities remains to be one of the root causes of the problem. Instead of growing food crops to feed their people, African states devote their vital water and land resources to grow cash crops. These crops are neither consumed domestically nor benefiting the toiling farmers who are left with empty bags after harvesting the crops. These cash crops are for the enjoyment and benefit of affluent northerners, multinational corporations, and the domestic African pirate state bourgeoisie. While global corporations pocket hefty profits from the sale of fish and fishery products, cash crops, and cut flowers produced in Africa, end users in the advanced countries enjoy the consumption of these products by paying affordable prices. However, African producers are always paid way far below the cost of production.

A second reason that the advanced countries can and should funnel tens of billions of dollars each year to uplift Africans out of poverty betrays understanding of the nature of capitalism. Even if there is goodwill on the part of politicians, it is inconceivable that they would invest enormous political capital to raise billions of dollars in taxes from their anxiety-driven and restless citizens in order to help poor Africans at a time when the crisis of accumulation is threatening the general living conditions in donor countries. After all, the advanced countries have entered a new phase in the secular evolution of capitalism, in which the Keynesian capitalist welfare state is being dismantled left and right. In fact, the serious shortfall between what is pledged by the donors and what is actually delivered to the recipient countries betrays the fallacy entailed by the notion of foreign assistance as engine of growth.

For example, despite the customary ritual pledges of 0.7 percent of their GNP in Official Development Aid for countries of the global south, developed countries delivered $100 billion, which was a mere 0.2 percent of their combined gross national income (GNI) of $35 trillion in 2007, leaving a huge deficit of $145 billion (Sachs, 2008). In other words, the donor countries have a poor record of keeping their pledges. In 2001, for example, 20 developed countries pledged to funnel $410 million annually, or a total of $2.9 billion over seven years into UN-controlled accounts to help poor countries cope with the effects of climate change. These pledges were subject to further review by 2008. However, over that seven-year period, a total of only $260 million was contributed to the UN accounts by the donor countries, something that prompted Secretary General Ban Ki-moon to accuse the donor countries of backtracking on their pledges (BBC News, Nov. 25, 2009).

The record of donor assistance specifically for agriculture in developing countries is even poorer. At a time when the UN Food and Agriculture Organization was urging donor countries to transfer $30 billion annually in additional global agricultural assistance to enable developing countries to feed themselves, donor countries actually began to drastically slash existing agricultural assistance programs, much less to furnish fresh resources. Between 1980 and 2003, the advanced countries progressively slashed their aid transfer for agriculture in developing countries by 64 percent, from $5.3 billion to $1.9 billion. In Africa alone, U.S. assistance to agriculture declined by 85 percent, from $400 million in 1984 to $60 million in 2006 (Bertini and Glickman, 2009; Thurow, 2010). Despite this poor record, Keynesian global distributionists, like Jeffrey Sachs, still inadvertently nurture the illusion that Africa could graduate from poverty, hunger, and underdevelopment with steady nourishment coming from the advanced capitalist countries. In his 2008 book, Jeffrey Sachs untiringly made the case that if donor countries annually funnel $300 billion to poor countries from their combined GNI of $45 trillion by 2015, out of which $120 billion would ideally go to Africa annually, Africa would be on its own feet, ready to wean itself from official development aid.

There is a third empirical problem associated with the Keynesian distributionist vision involving the potential impact of $75 billion or even $120 billion transferred to Africa annually. Such conjured-up figures actually tend to underestimate the cost of jumping off African economies because, if the World Bank's computation is correct, overcoming Africa's huge infrastructure deficit alone would require the injection of $80 billion per year in external assistance, equally divided between investing in new modern infrastructure and maintenance of existing ones. In its report card, the Bank reminds us that, in 2002, Africa received just $4 billion in foreign assistance for infrastructure. In fact, Africa's need to develop its power-generation capacity entails half of the $80 billion the World Bank recommended in fresh foreign aid flow, while 30 percent of the funding would be used to refurbish the continent's dilapidated infrastructure (World Bank, February 2009).

Finally, there is a danger implied by the push for the so-called second green revolution, as the heaven-sent solution for Africa's poverty and underdevelopment. This second green revolution is supposed to proceed along two complementary lines. On the one hand is the classic green revolution driven by improved conventional seeds and the use of reactive nitrogen fertilizers and pesticides. On the other hand is the introduction of genetically engineered plant and animal species into African agriculture. What is currently in vogue among biotech oligopolies and corporate intellectuals, is the notion that without the use of green revolution technologies or a rapid shift to the gene revolution, the number of food insecure Africans will climb to 500 million by 2020. In the view of these commentators, this should not happen because the continent can avert such prospects by importing existing "green" and "gene" technologies already available in the advanced capitalist countries.

Because sub-Saharan Africa still commands half of the earth's unused arable land, land resources would not be a constraint on the imagined agrarian revolution if only African governments embraced the protection of corporate intellectual property rights and relaxed regulations that hinder commercial release of biotech crops. As we are told, by transitioning to the second green revolution, the annual output of African agriculture presently valued at $250 billion can easily reach $880 billion per annum by 2030.

Moreover, the transition of African agriculture to the second green revolution would spur a boom in postharvest processing activities, as well as in demand for reactive fertilizers, pesticides, herbicides, and specialized seeds, generating additional $275 billion per year in global revenue (Thurow, 2011).

For these analysts, India is a textbook case for the early success in the green revolution. In the 1960s, Western experts and corporate interests thought that India's food crisis was an opportunity to test the idea of a green revolution. American commando experts were dispatched to the Punjab region in northern India and the Rockefeller Foundation provided the financing for the revolutionary transformation of Punjab's food ecology. Promised a rose garden, Punjabi farmers stopped growing beans, vegetables, and related greens. Instead, they turned to planting high-yield wheat, rice, and cotton. To sweeten the pot, the Indian government supplied the farmers with subsidized chemical fertilizer and seeds. Farmers blanketed their fields with over 20 million wells to irrigate their new crops with groundwater. The agrarian miracle was real and tangible in the short run as annual wheat production rose from 11 million metric tonnes in 1961 to 73 million metric tonnes in the mid-1970s while rice production also rose from 53 million metric tonnes to 70 million metric tonnes over the same period. India not only weaned itself from dependency on imported food, but also became a net wheat and rice exporter.

Forty years later, however, India's green revolution proved to be a mirage. The high-yield crops have become too chemically dependent and too water thirsty. The radical shift from rain-fed farming to irrigation-based agriculture meant depleting the region's aquifers. At the beginning, underground water was plentiful; it never seemed depletable. In order to harvest three or four crops a year, farmers endlessly pumped water to fill the irrigation channels. Soon, however, the farmers pumped so much water that the water table began to sink drastically by three feet or more a year. Today, water companies are drilling Punjab's fields to bore the wells deeper and deeper (up to 200 feet) to find water. In turn, farmers are forced to install powerful pumps to force the water up. Today, 175 million Indians are fed with grain produced by over-pumped water. Moreover, instead of the precious fresh water, farmers are getting more and more saline brackish water, which is damaging the roots of the crops. Because these powerful pumps are very expensive, the farmers must borrow money from local usurers at twice the conventional bank rates because banks turn farmers away if they have no demonstrable or tangible collateral.

The shift to chemical dependency also meant increasing the application of fertilizer. Yet, the more farmers applied fertilizer, the more the soil became deficient of natural macronutrients. Today, Punjabi farmers are said to use three times as much fertilizer as they did 30 years ago to harvest the same amount of crops. Thirty years ago, spraying the fields with pesticides twice a year was enough to keep insects at bay. Today, Punjabi farmers must spray their fields up to twelve times a year, since harmful insects have increasingly become resistant to many types of insecticides (Zwerdling, 2009; Brown, 2011).

There are several cautionary lessons from the Indian experience with the green revolution with pertinent implications for Lake Victoria's littoral states.

First, in the context of the prevailing social order, the transition to a green or gene revolution is likely to favor agrarian accumulation in the hands of the few with political access and capital. In the end, this strengthens the symbiotic partnership between the pirate state bourgeoisie, domestic big operators of the plantation system, and global corporations. As we saw in Chapter 5, the widening and consolidation of this symbiosis has already found practical expression in the new land grab for agrofuel and for more cash crop production.

Second, the transition to a green or gene revolution can only deepen the disjunction between the principles of ecological sustainability and the logic of accumulation with the introduction of antibiodiversity monocrop plantations that are dependent on synthetic chemicals and continuous extraction of vast quantities of water resources. Moreover, the transition to monoculture plantations and industrial agriculture will only strengthen the stranglehold of global corporations as suppliers of synthetic agrochemicals and genetically modified seeds on Lake Victoria's riparian countries.

Third, increases in grain and feedstock production because of the shift to a green or gene revolution does not necessarily guarantee national food security and eradication of poverty. On the contrary, food availability to the poor may worsen as a result of the expenses incurred by corporate producers to pay for genetically engineered seeds, synthetic fertilizers, pesticides, and herbicides, as well as from the diversion of more and more grain to agrofuels production. In India, for example, despite consecutive bumper grain harvests following the introduction of green technologies, some 400 million Indians still remain below the poverty line, that is, one in every four of the global poor is an Indian today. This empirical observation belies the presentation of food shortages as a cause of poverty, hunger and malnutrition.

The crucible of the problem hinges not on production but, rather, on who controls the global distribution channels and prices of food. A UN commission recently reported that food losses and wastes are at the root of global poverty, hunger, greenhouse gas emissions, and climate change because, in the final analysis, food losses and wastes represent the squandering of scarce freshwater resources, land, energy, labor and capital. According to the report, of the 2.3 billion metric tonnes of cereal crops produced annually, 1.3 billion metric tons, worth $1 trillion, are lost or wasted at the various points of production, distribution and consumption (*Africa News*, Dec. 3, 2013). Consumers in affluent countries waste 222 million metric tonnes of food every year; for reference, the annual net food production in sub-Saharan Africa is 230 million metric tonnes. While in developing countries, 40 percent of food is lost at postharvest and processing phases due to poor infrastructure, inadequate storage facilities, and low levels of technical capacity, over 40 percent of the food in rich countries is wasted during the distribution and consumption phases as people simply throw away food because the appearance of the items does not look good (FAO Media Center, May 11, 2011). Even though 38 million Americans depend day after day on local and regional food banks for survival, between 160 billion and 290 billion pounds of the 500 billion pounds of food produced in the United States annually are wasted along the different points of production, distribution, and consumption (Bloom, 2010). This empirical evidence suggests that, short of a radical reordering of the prevailing mode of production and distribution, the transition to a green or gene revolution in Africa can only serve to perpetuate the monopoly domination and control over national resources by the pirate state bourgeoisie and their international sponsors.

It is against the backdrop of the above empirical observation that some distributionists view the problem of poverty as inhering not in production or availability of food in the market but, rather, in the distribution of food. Amartya Sen (1981), for example, in his celebrated work, *Poverty and Famines: An Essay on Entitlement and Deprivation*, argues that droughts, crop failures, food shortages, famines, and other natural phenomena do not and cannot enjoy causal priority in explaining hunger and structural poverty, even though these natural events may have contributory implications for the circulation of food. Determinants of structural poverty, for Sen, inhere in a society's legal constructs and social systems that inhibit an equitable distribution of food or purchasing power. Unless people have full command over their own food or have enough purchasing power, the mere abundance and availability of food in the market does not and cannot insure food security.

The policy prescription that logically follows from this epistemological orientation is for nations to dismantle existing legal and social frameworks and put in their place new constructs that are pro-poor and reasonably equitable. Within this general epistemological framework, liberal cosmopolitans make similar arguments with respect to the causes and effects of structural poverty. Two salient points are particularly highlighted in liberal cosmopolitan discourse. The first refers to the presentation of structural poverty, not as a function of uneven distribution of natural resources, but rather as unjust exploitation of and radically unequal access to natural resources. Second, the mass populations of the world (for no fault of their own) are born into unfortunate preexisting conditions of poverty (Pogge, 2002; Held, 2002).

The proposed solution for the prevailing radical inequality in the distribution of wealth is the introduction of regulatory ethos and redistributive conception of justice into the market system. To give practical expression to this belief, liberal cosmopolitans present ideational frameworks in which all human beings are viewed as having natural entitlement to equal moral standing, political citizenship, and impartial recognition and treatment. Recognition of this natural equality of humans places the burden on the rich countries to transfer some of their unjustly acquired wealth to poor countries. Thomas Pogge (2002), for example, proposes the creation of a Global Resources Dividend in which taxes can be collected from natural resources users for distribution to the poor countries.

Even though welfare economists and liberal cosmopolitans offer an important corrective to Keynesian global distributionism, they seem to lack a clear understanding of how differential power relationships generate structural inequalities in wealth that bedeviled many nations. Unequal distribution of power in a society entitles some individuals or groups access to resources while denying others equal access to the same resources; social marginalization and political underrepresentation make the powerless poor utterly invisible to decision makers and therefore irrelevant to the exercise of political power as it pertains to the creation, accumulation, and appropriation of wealth (Chambers, 1983). From an ideational perspective, taking a distributionist position appears plausible. After all, who would quarrel about the central distributionist thesis that all humans have natural entitlements to the necessities of life?

The focus on distribution alone, however, inadvertently projects the notion that the prevailing capitalist mode of production is sound and can be made responsive to ethical considerations, concerns, and appeals. In the final analysis, the distributionist epistemology tends to sustain the illusion that the deep crisis of poverty and entropic devastation of ecological resources can be solved by pushing more crumbs downward from the rich minority to the poor majority. Advocates of welfare economics and liberal cosmopolitanism gloss over the fact that power structures, social systems, and legal constructs are precisely the methods by which unequal distribution of wealth under capitalism is maintained and perpetuated. The enduring lesson of history suggests that, under capitalism, productive and distributive activities are never socially or legally neutral; the primary function of social power and law in society is to differentiate among individuals and groups by determining who will have privileged access to natural resources and produced goods and services (Swyngedow, 2004; Robbins, 2004; Yohannes, 2009). Even Adam Smith (the grandfather of the free-market system) long acknowledged the unavoidable coexistence of opulence of the few and the indigence of the many as the essential, and yet unfortunate, condition of wealth creation and accumulation under capitalism. Indeed, perpetuation of the opulence of the few and increasing marginalization of the indigent majority is the mother of all contradictions of capitalism. This observation was well elaborated by Karl Marx and other writers.

In his materialist conception of history, Marx (1977) identified the logic of accumulation as determining the trajectory of history and associated relationships. The search for wealth motivates capitalists to invent new technologies and to invest more and more in the means of production. This tendency to innovate and invest has two major implications for social relations of production and nature. First, new inventions and methods of production accelerate the rate of metabolism at which natural resources are transformed into goods with exchange value, resulting in ever rising entropic degradation of ecosystems. Second, with new inventions and the consequent efficiency gains, the capitalist mode of production becomes increasingly less dependent on labor; the greater the sophistication of machines, the higher the productivity of labor, and therefore the fewer the workers that are needed to run the production process. What emerges from the rapid modernization of the means of production is thus the crisis of ecological degradation stemming from the hyperacceleration of resource extraction and the crisis of poverty.

In the context of these two types of crises, owners of capital inevitably experience a crisis of accumulation. In order to escape this deepening crisis, owners of capital then increase the speed with which they extract ever larger quantities of natural resources through the application of modern science and technology, establishing giant dams, modern hydropower plants, fish-processing facilities, modern logging companies, and so on and so forth. The logic of accumulation thus results in unemployed "surplus" population, on the one hand, and in the erosion of the regenerative and waste absorptive capacity of the ecology to which the poor are disproportionately vulnerable, on the other. If the ecological crisis and the accumulation crisis are the twin manifestations of the modernization of the means of production, labor redundancy and the consequent social marginalization of the mass population are the twin results.

The tautological point stated here is that the accumulation crisis manifests itself in mass poverty and entropic degradation of the ecology simultaneously. Precisely because accumulators internalize the gains from modernization to themselves while simultaneously externalizing the harmful consequences of modernization to the socially disempowered masses, the accumulators are bound to experience a legitimation crisis that can be contained only through the use of cooptation of the intellectuals and coercion and violence against the toiling populace. To do this, capital accumulators depend on the state, in which case the state becomes not a neutral actor but the predominant instrument of domination and control (Yohannes, 2009).

It is only in the context of the materialist conception of history that we can understand the fishery crisis and the entropic degradation of the lacustrine ecology in East Africa. Demand and supply conditions do not occur in a vacuum; they have powerful economic, social, and institutional contexts. The contextualization occurs when structural conditions of production and distribution are put in the service of accumulation, which inevitably leads to owners of capital and laborers clashing over ownership of the means of production or the distribution of benefits. The class struggle manifests itself in various ways, including "illegal" fishing, piracy, theft of fishing gear and equipment, and damage to property. In early 2008, for example, a group of what the Kenyan government labeled as a roving band looted five trucks loaded with Nile perch for export from Lake Victoria and destroyed another two trailers carrying Nile perch. Since Nile perch accounts for 84 percent of Kenya's total fish exports, the serial depletion of harvestable fish and the consequent class struggle have been having a significant impact on the operation of Kenya's 17 fish-processing plants (*Africa News*, Feb. 25, 2008).

To come full circle to where we started, my purpose in devoting such large space to the contending approaches to the food ecology is not to make the reader forget about the central thesis of this work, but rather, to suggest that the crux of the problem in Lake Victoria today is largely a result of the pollution of knowledge that has hitherto served the pirate state bourgeoisie well. Cleaning up this social pollution of knowledge is the necessary prerequisite to squarely confront the technical and practical problems, not only in Lake Victoria, but also in the entire region as well.

The challenge for theory builders today is how to situate the state transition of Lake Victoria within the general food ecology; and this requires theoretical clarity in order to objectively and comprehensively address the factors that have induced serial fish depletion and entropic lake disintegration. I am hopeful that the insights gathered from ecological economics and historical materialism will go a long way in charting a new theoretical path. This can happen only if the crisis in the fishery mode of production in Lake Victoria is understood to be just a small part of the general crisis of primary accumulation. Therefore, the state transition of Lake Victoria may not be halted and reversed until and unless the riparian societies and countries wean themselves from total dependency on natural capital stocks; continued dependence on natural resources alone is likely to lead to the complete collapse of their aquatic and terrestrial ecosystems. As the World Bank (2005) correctly notes, the most critical component of national wealth is intangible capital, which consists of human skills, rule of law, transparent institutions, and effective governance.

The reason the rich countries in the world are rich is not because they have inexhaustible natural capital stocks, but because they have higher proportions of intangible capital in relation to their natural and produced capital. The 30 OECD countries, for example, have average total per capital wealth of $440,000, consisting of $10,000 in natural capital, $76,000 in produced capital, and $354,000 in intangible capital. While Switzerland (with the highest total per capita wealth in the world) could enjoy $648,000, the average total per capita wealth for Burundi, the DRC, Malawi, Mozambique, Kenya, Tanzania, Uganda, and Zambia was $3,335. This suggests the primacy of authentically authoritative knowledge and people-centered politics.

To return to the central question proposed in the title of this chapter and as the various chapters of this work cumulatively have shown, the only way forward in Lake Victoria's riparian countries is to vigorously challenge the pirate state bourgeoisie; without doing so, any plausible change in the neocolonial capitalist mode of production, in general, and the predatory fishery mode of production, in particular, cannot be imagined. In order to vigorously contest the prevailing social order and to search for an alternative to it, we need the insights of both historical materialism and ecological economics.

The imperative to articulate an alternative to the neocolonial capitalist order requires us to move away from formal deductions and toward feasible and realistic objectives. This requires of us the urgency to abandon the exploitative system at present driven by the logic of destructive creation in favor of a system that is at once pronature and prolabor, which amounts to "communitarian economics". This term does certainly evoke in the minds of defenders of unfettered capitalism an image of a system controlled and directed by the state. However, communitarian economics is the total repudiation of both the naturalization of capitalism and the religionization of Eurasian pseudocommunism. It is a system that seeks to empower producers to control the total fruits of their labor in ways that establish an organic balance between the natural world and the social world in both of which they live.

Conceptually, the construction of a communitarian economic order involves the simultaneous adoption of top-down and bottom-up approaches. The first approach is primarily political, while the second is principally economic and social. As I have abundantly made clear throughout this book, so long as the hegemony of the pirate state bourgeoisie remains unchallenged, no national progress can take place in any of Lake Victoria's littoral countries. The endless reproduction of research products and recommendations on how to save Lake Victoria has served no purpose thus far other than creating the illusion that something is being done. Therefore, the first order of business ought to be how to launch a vigorous contestation against the presence of the predatory state in the fishery mode of production; this is a top-down approach to the crisis in the fishery mode of production.

There are three fundamental reasons why the pirate state bourgeoisie will remain an obstacle to progress of any sort in addressing the entropic degradation of all aquatic or terrestrial resources.

First, by now it is clear that the major driving forces in the state transition of Lake Victoria have been the commodification and globalization of the lake through the intermediation of the state, mainly in the interest of tax collection and foreign exchange generation. Given this reality, the pirate bourgeois state cannot give up its dependency on the lake for streams of financial resources. This suggests that the pirate state bourgeoisie cannot take any meaningful initiative to reverse the state transition of Lake Victoria any time soon without the diminution of taxes and foreign exchange earnings. Moreover, given the parasitic and authoritarian character of the state, the endless talk about promoting comanagement as the answer to the crisis in the fishery mode of production is utterly illusory. Only unequivocal expulsion of the state from the fishery mode of production may contain and hopefully reverse the state transition of Lake Victoria. This mandates a throughgoing communitarianization of the fishery mode of production. This is hardly a novel idea. Tribal communities in the precapitalist era had used traditions, customs, and rituals to sustainably use the lake by carefully monitoring and observing seasons of fish reproduction and maturity. Fishing grounds were allocated not to individuals but, rather to tribal groups who managed their areas by regulating entry and exit seasons. Contemporary fishing cooperatives can be modeled after this valuable tradition.

Second, as indicated in the core chapters of this volume, the widespread presence of agrochemical-dependent commercial plantations and the growing human biomass in the Lake Victoria basin are among the major contributors to state transition of the lake. Out of necessity, the dispossessed people for the most part depend on liquidating their natural capital stocks. When they cut the forest for cultivation, firewood, construction materials, and when they overgraze their grasslands, they create permissive conditions for soil erosion that affect Lake Victoria in terms of sedimentation. Furthermore, when unable to survive on degraded terrestrial resources, the dispossessed poor migrate to the lake in the hope of making a living as fishers, thereby overpopulating lakeshore communities.

On the other side of the ledger, agrocommercial plantations not only divert precious land and water resources away from food crop production but also generate high-entropy waste in the form of toxic residues of reactive fertilizers, pesticides, and herbicides that end up in Lake Victoria. The pirate state bourgeoisie actively provide enabling services to the expansion of commercial plantations because they depend on receipts from the sector in the form of taxes and hard currency.

The point here is that, without radical transformation of the terrestrial production system, the entropic consequences of terrestrial resource appropriation by the indigenous masses, the opulent owners of the plantation system, and the latest land grabbers cannot be addressed. This is something that requires central direction and coordination by a well-organized political entity, which is the state. The greatest conundrum in East Africa today is that the pirate state bourgeoisie have no redeeming qualities to undertake meaningful steps to address the crisis in the terrestrial system of production. And, short of wide ranging reform in the terrestrial system, the future of Lake Victoria will grow even grimmer.

Third, as abundantly indicated in Chapter 4, the rapid transformation of the urban ecology around Lake Victoria is perhaps the single most important source of eutrophication of the lake. It has long been known that urban residents and industrial establishments simply dump their untreated wastes into the lake on a regular basis. Planned development and management of the urban ecology will require robust and enforcible regulations supplied by responsive and responsible state institutions. Obviously, the challenge is to create such institutions in a political environment where the pirate state bourgeoisie are hegemonic. One cannot have clean, efficient, effective, and responsive urban public administration without first having a clean and accountable state. What the above three reasons together suggest is that the future of Lake Victoria cannot be analyzed and understood in isolation from the future character of the state in the lake's littoral countries.

PROTOTYPES OF SUCCESSFUL COMMUNITY ACTIONS

The bottom-up approach is a call for the construction of local ways of life independent of the state. This involves mass mobilization, mass participation, and mass action to build egalitarian economic order from below. This supposes microevolutionary processes, involving communities to take charge of the biosocial environment with complete ownership of the means of production. Obviously, in a situation where the pirate state bourgeoisie are in ascendance, mass mobilization and mass action will require the active participation, direction, and secular leadership of public intellectuals. There are a number of prototypes generated by public intellectuals that have shown great promise.

For example, Kenyan Nobel Prize winner Professor Wangari Maathai single-handedly led Kenyan women to embark on self-emancipatory projects through a grassroots initiative known as the Green Belt Movement. The effort concentrated on planting all purpose trees. The result has been that thousands of women established 6,000 tree nurseries, and planted more than 30 million trees over a period of 30 years. Women have been able to sustainably harvest the trees for nutritiously edible fruits, fuelwood, and construction materials. This way, the women have been able to achieve a degree of economic self-sufficiency and financial autonomy, while, at the same time, performing invaluable conservation services. This is an incipient form of communitarian economics, an essential building block for nationwide humane economic order.

The experience of Kenyan women is being emulated elsewhere by other women in the region. In 2004, Tanzanian women in the Tabora region embarked on planting indigenous trees alongside traditional crops. The project started with 18 conservation women's groups with a total membership of 1,200 and with seed money of $78,000 from a nongovernmental organization. The wild fruits from the trees are processed into a variety of juices, jams, and wines, now being marketed in Tanzanian cities, including Dar Es Salaam, Arusha, and Mwanza (The East African, March 24, 2008). In addition to preserving such important indigenous trees as miombo trees, the women in these groups have raised their incomes by an average of $28.5 a month, which is significant in the context of Tanzanian living standards.

The nongovernmental One-Acre Fund, created by Andrew Youn, has also provided an important prototype of how communitarian economics can be established. The fund started in 2006 with 40 farming families in Kenya, providing them with such services as farm inputs on credit, regular training; market access for their output, insurance coverage for bad weather, and easy repayment plans for the seed money they borrow. The program draws on the importance and experience of cooperatives since farm families are grouped together as a work unit. In just four years, One-Acre Fund went from working with 40 families to 30,000 families in Kenya and Rwanda, where farming families have doubled and/or tripled their harvests. The program has become so successful in promoting self-sufficiency in local food production that the fund plans to reach one million families within a decade (Thurow, 2010).

The World Agroforestry Center has also embarked on a very innovative undertaking in rural Malawi. The aim of the program is to prove that nature-driven preservation and restoration of soil fertility is actually superior to dependency on reactive nitrogen in the long run. According to soil ecologists, frequent use of synthetic fertilizer not only pollutes the general environment but also contributes to soil degradation by making it acidic and dry unless it is used in combination with such measures as planting atmosphere nitrogen-fixing trees and growing legumes. Prompted by the 2002 and 2005 droughts in Malawi, which together made eight million Malawians vulnerable to chronic food insecurity and hunger, in January 2006 the World Agroforestry Center started in a project known as the Agroforestry Food Security Program with $4 million from Irish Aid. By the end of 2010, the program had already reached 200,000 poor farmers, comprising 1.5 million people, with tree seedlings and regular training on how to plant nitrogen-fixing fertilizer trees; fast-growing trees for firewood, construction material, and fodder for livestock; and fruit trees to supplement the nutrition and income of the farmers. Because families in the program now regularly get sufficient fiber, vitamins, minerals, fatty acids, protein, water and energy from regularly consuming avocadoes, peaches, bananas, guavas, and mangoes, obtained from the fruit trees they have planted, the gain in health benefits is enormous. These integrated activities have already made a big difference in the lives of traditionally dispossessed farming families by insuring food security, nutrition, income, shelter, health, energy, restoration of soil fertility, and environmental sustainability.

This pioneering program has amply demonstrated that, through continuous cultivation of maize with nitrogen-fixing Gliricidia trees, Malawian farmers have been able to harvest more than 5 metric tonnes of maize per ha in good years, or an average of 3.7 metric tonnes per ha over a long period without using synthetic fertilizers. Before this innovative practice was introduced, the farmers barely managed to produce 1 metric tonne of maize per ha on average. Here is the exciting rub: since Gliricidia trees can live up to 20 years, continuously capturing nitrogen from the atmosphere and incorporating it into the soil, avoiding the risk of soil acidity and dryness, Malawian farmers can obtain natural fertilizer for that long period of time free of charge. The World Agroforestry Center calculates that, if half a million Malawian farmers with average holding of 0.2 ha planted nitrogen-fixing fertilizer trees, the nitrogen naturally fixed by the fertilizer trees would amount to 250 kg per ha a year. If the farmers were to buy this amount in the form of synthetic fertilizers, it would cost them $5.8 million a year (Pye-Smith, 2008; UN IRIN Africa, April 19, 2011).

The experiences from the Green Belt Movement in Kenya, the women's conservation groups in Tanzania, the One-Acre Fund in Kenya and Rwanda, and the Agroforestry Food Security Program in Malawi clarify a number of important issues. To begin with, the collective experiences amply demonstrate how organic agroforestry and organic farming are not only feasible but also how they can readily expand the total food ecology available without compromising the requirements of ecological sustainability. Sustainable expansion in terrestrial food production also takes the pressure off aquatic resources. With abundant food available around them, rural farmers will stay put rather than being forced by objective material circumstances to migrate to lakeshores to earn a living as fishers. Furthermore, reliance on fast-growing and productive trees by millions of rural people for firewood, construction material, and fodder for their livestock takes the pressure off natural forests and woodlands. This serves to preserve and strengthen the watershed ecology of natural water bodies by reducing the conditions of soil erosion and lake sedimentation. Finally, reliance on village nurseries for fertilizer tree seedlings and seeds allows farming communities to bypass the global market for expensive and environmentally polluting synthetic fertilizers. More important, the system of organic agroforestry and organic farming is self-sustaining. The use of synthetic fertilizers can certainly boost production in the short run. However, the environmental and financial costs can prove to be prohibitive in the long run for such poor countries as Malawi which have to import reactive fertilizers. For example, beginning in 2005 the Malawian government with the financial support from donor countries introduced fertilizer and seed subsidies from which some 1.7 million smallholder farmers reportedly benefited in the short run.

The program became so successful that some Keynesian global distributionists, like Jeffrey Sachs, saw the temporary boost in maize production as affirmation of their advocacy for chemical-dependent industrial agriculture (Denning et al., 2009). Between 2006 and 2010, Malawi enjoyed consecutive maize surplus so much so that not only did the 2007-08 maize surplus save the country $120 million in its food import bill but also allowed the country to export over 320,000 metric tonnes of maize to Zimbabwe in December 2007 (UN IRIN Africa, September 2, 2010). However, the cost to the country was enormous. By 2008, the input subsidies program was siphoning away 16 percent of the national budget, compelling the government to substantially scale down the program. Even after cutting back on fertilizer and seed coupons, the budget for the 2010-2011 program was still a hefty $152.3 million (UN IRIN Africa, April 19, 2011). Yet, the regime learned soon enough that the subsidies program could not meet the minimum fertilizer requirements when smallholder farmers publicly protested the inadequacy of the coupons they received. This shows how dependency on imported agricultural inputs can make farmers vulnerable to the caprices of markets and governments. Moreover, producing more maize using synthetic fertilizer does not necessarily guarantee national food security. For example, despite bumper maize harvests five years in a row, the number of food- insecure Malawians actually increased from 147,492 in 2009 to 1.1 million in 2010 (UN IRIN Africa, September 2, 2010).

The microlevel prototypes in the terrestrial food production system can also be replicated in aquatic systems. In fact, some proponents of local food self-sufficiency have proposed integrated multitrophic aquaculture production at the livelihood level as the potential answer to the growing poverty, hunger, and malnutrition affecting millions of people in the Lake Victoria basin. This does make sense since lakeshore communities are already highly dependent on the region's wetlands for biomass production and cultivation of crops by taking full advantage of the high primary production potential and moist soils of the wetlands. On the one hand, integrated multitrophic cultured farming allows livelihood producers to combine a variety of fish species in keeping with the biological and ecological requirements of the species involved, thereby fostering economic diversification and environmental sustainability. On the other hand, livelihood producers can also take full advantage of the synergy of fish farming and crop cultivation in the same area (FAO, 2008b). For example, the waste products from the crop cultivation system can supply nutritional inputs to the fish production system and vice versa. Such practices could potentially obviate the need to depend on toxic synthetic pesticides and fertilizers in order to boost production and, by extension, protect the soil, air, and water from contamination and the lake from eutrophication. If the livelihood production system is based on the practices of organic aquaculture, the integrity of the wetlands and that of Lake Victoria could be even better protected. Kipkemboi and others (2007), for example have advanced the concept of fingerponds, or small ponds used as fish hatcheries, as viable option to improve the livelihoods and enhance the food security of smallholder farmers at the subsistence level around Lake Victoria by promoting integrated aquaculture-agriculture production systems in which the farmers combine crop cultivation with fingerpond-based fish farms. The idea here is that subsistence farmers can use crop residues, byproducts and manure to fertilize the fingerponds, and use the fingerpond water to grow crops. Their pilot study in Kenyan wetlands by Lake Victoria suggests that farmers can become food secure by promoting aquaculture-agriculture production networks by simply using seasonal flood recession or underground water. This means that they can set up fingerponds alongside their crop plots in the wetlands. The result of the experiment suggests that farmers can produce 200 kg per ha a year from 192 square meters of fingerpond, which translates into annual gross earnings of 752 Euros (Kipkemboi et al., 2007). The longterm value of family-based or communitarian-oriented integrated multitrophic aquaculture production system can expand not only the total food ecology available but also take off the pressure from Lake Victoria, allowing lake species to restock themselves. Obviously, promoting a multitrophic aquaculture production system is not and cannot be a substitute for communitarianizing the lake itself. If the state transition of Lake Victoria is to be contained and reversed, the drive for industrial fishery production system must be combated.

CONCLUSION

In sum, the cases briefly discussed in the preceding paragraphs compel one to sound a note of optimism that not all is lost in Victoria's riparian countries if the right approach is used. The challenge for public intellectuals and social activists is how to strengthen the small points of light and foster the scope for their expansion on national and regional scale. This requires not subjecting Lake Victoria to more studies that are empirical in order to generate more recommendations, but, rather, to contest, persistently and vigorously, the neocolonial mode of production and to challenge the pirate bourgeois state that is solidly entrenched in all the countries covered in this work. After all, Lake Victoria is probably the most studied lake in the world. By now, we know this much: too many exactions on lake resources and too many insertions of high-entropy wastes into the lake explain the lake's state transition. We also know this much: desperation of the toiling masses and the greed of the few wealth accumulators are responsible for the exactions on lake resources and the insertions of wastes into the lake. So long as the conditions of desperation and greed remain entrenched, to think that Lake Victoria's state transition can be halted is tantamount to entertaining perennial illusions. The daunting challenge facing lakeshore communities as well as those who depend on terrestrial resources is how to make a case for systemic change in the fundamental means of livelihoods. Nevertheless, this onerous task belongs not to the masses, but to public intellectuals who have acquired objective understandings and subjective wills to give back to the masses that made their education possible by their sweat. At this point in human history where politicians see mediocrity as an asset and intellect as a liability, mendacity as a virtue and honesty as a vice, only the engaged scholar activists can supply the much-needed secular leadership to social activism.

Bibliography

Abila, Richard. (2006). "Dog food offer to starving children in Kenya's leading fish exporting district." *Lake Victoria Portal*, www.lake-victoria.info/page/97.html accessed July 11, 2010.

Abila, Richard O. (January 2003). "Fish trade and food security: are they reconcilable in Lake Victoria?" FAO Fisheries and Aquaculture Department, Rome.

Abila, Richard, and Jansen, Eirik G. (1997). *Socioeconomics of the Nile Perch Fishery on Lake Victoria, From Local to Global Markets: The Fish Exporting and Fishmeal Industries of Lake Victoria, Structure, Strategies and Socioeconomic Impacts in Kenya.* Nairobi, Kenya: The World Conservation Union.

Achieng, *A. P. (1990).* "The impact of the introduction of the Nile perch, Lates niloticus (L), on the fisheries of Lake Victoria." *Journal of Fish Biology*, 37, Supplement-A, 17-23.

Action Aid. (2011). "Jatropha biofuels in Dakatcha, Kenya, the climate consequences." Found at dakatcha_biofuels_report_march_10_2011_-_final_tr1_and_2_mc.pdf accessed May 6, 2011.

Africa Analysis (London). 2002). "Joint protection force for Lake Victoria." October 4. P. 17.

Africa Analysis (London), (2002) "Coffee era is ending in Buganda". August 23. p. 18.

Africa Analysis (London). (2001). "Kenya coffee faces bean counters." July 27.

Africa News. (2013). "Africa: new report offers menu to solve the global food gap." December 3.

Africa News. (2008). "Tanzania: ADB approves U.S. $156 million budget support." October 30.

Africa News. (2008). "Uganda: country to lose over Sh96 billion over illegal fishing." September 30.

Africa News. (2008). "Uganda: illegal fishing threatens Nile perch exports." August 28.

Africa News. (2008). "Tanzania: EAC's 5 billion US dollars Lake Basin plan." August 8.

Africa News. (2008). "Mozambique: Fisheries Ministry has just one patrol boat." May 31.

Africa News. (2008). "Uganda: coffee earns country Sh44-billion in 2007." May 15.

Africa News. (2008). "Tanzania: dwindling fish stocks alarm President." May 14.

Africa News. (2008). Tanzania: Tabora to raise tobacco output." April 4.

Africa News. (2008). "Uganda: cellphone coverage over Lake Victoria promised." March 17.

Africa News. (2008). "Why fish sector is ailing." February 25.

Africa News. (2007). "Uganda: Country's fish, honey export earnings to hit $1.2 billion by 2011." October 11.

Africa News. (2007). "Kenya: government allows 84 foreign firms to fish its water." June 4.

Africa News. (2007). "Uganda: country could lose more than 90 percent in 30 years." May 8.

Africa News, (2007). "Kenya: 16 million living in extreme poverty." April 27.

Africa News. (2007). "Uganda: tobacco smokes life out of Western Nile peasants." June 10.

Africa News. (2007). "Tanzania: World Bank approves funds for Mkukuta poverty reduction." April 25.

Africa News. (2007). "Uganda: is Kenya overexploiting Lake Victoria?" January 10.

Africa Research Bulletin: Economic, Financial and Technical series. (2007). "Tobacco: Tanzania." vol, 45 Iss. 3, p. 17790A-17790B.

Aguilar-Manjarrez, J., and Nath, S. S. (1998). *A Strategic Reassessment of Fish Market Potential in Africa*. CIFA Technical Paper No. 32. Rome: FAO.

Alexander, Susan E., et al. (1997). "The interaction of climate and life." In Daily, Gretchen C. (ed.) *Nature's Services: Societal Dependence on Natural Ecosystems*, pp. 69-86. Washington D.C.: Island Press.

Allison, Edward, et al. (2009). "Vulnerability of national economies to the impacts of climate change on fisheries." *Fish and* Fisheries, 10 (2), 173-196.

Anderson, David and Throup David. (1985). "Africans and agricultural production in colonial Kenya: the myth of the war as a watershed," *The Journal of African* History, 26 (4), 327-345.

Anderson, Jon Lee, (2009). "Gangland: who controls the streets of Rio de Janeiro?" *The New* Yorker, October 5, pp. 47-50.

Anderson, A. M. (1961). "Further observations concerning the proposed introduction of Nile perch into Lake Victoria." *East African Agriculture and Forestry Journal*, 26 (4), 195-201.

Appleton, Judith. (2000). "'At my age, I should be sitting under the tree': the impact of AIDS on Tanzanian lakeshore communities." *Gender and Development*, 8 (2), 19-27.

Appropriate Technology, (2008). "Bountiful rice harvests for West Africa," September. pp. 12-13.

Appropriate Technology. (2008). "Increasing Africa's rice crop." June. pp. 17-19.

Aquastat, Uganda, FAO's Information System on Water and Agriculture, June 2006.

Aquastat, Kenya, FAO's Information System on Water and Agriculture, May 2006.

Ase, Lars-Erik. (1987). "A note on the water budget of Lake Naivasha, Kenya, especially the role of Salvinia molesta Mitch and Cyperus papyrus L." *Geografiska Annaler. Series A Physical Geography*, 69 (3-4), 415-429.

Attwood, Karen, (2008). "Kenya's economy in Crisis: political violence has caused L1-bn in damage," *The Independent* (London), January 25, p. 56.

Awange, Joseph and Ong'ang'a, Obiero. (2006). *Lake Victoria: Ecology, Resources, Environment.* Berlin, Germany: Springer.

Baffes, John. (2003). *Tanzania's tea sector: constraints and challenges*. Washington D.C.: The World Bank.

Baffes, John. (2002). *Tanzania's cotton sector: constraints and challenges*. Africa Region Working Paper Series No. 42. Washington D.C.: the World Bank.

Bagala, Andrew, (2010). "Police call for order as rival forces were after Lake Kyoga fishermen," *The Monitor*, (Kampala), May 11.

Bagumire, Ananias et al. (2009). "Potential sources of food hazards in emerging commercial aquaculture industry in sub-Saharan Africa: a case study of Uganda." *International Journal of Food Science and Technology*, 44 (9), 1677-1687.

Bagumire, Ananias et al. (2008). "Analysis of environmental and chemical residues in products of emerging aquaculture industry in Uganda as case study for sub-Saharan Africa." *Food Additives and Contaminants*, Part B-Surveillance, 1 (2), 153-160.

Baldauf, Scott, (2011). "Hunger and food security: is Africa selling the farm?" *The Christian Science Monitor*, February 6.

Bales, Carter F. and Richard D. Duke. (Sept.-Oct. 2008). "Containing Climate Change: an opportunity for U.S. leadership." *Foreign Affairs*, pp. 78-89.

Balirwa, John S. (1998). *Lake Victoria: Wetlands and the Ecology of the Nile Tilapia, Oreochromis Niloticus* Linne. Ph.D. dissertation, A. A.Balena/Rotterdam-Brookfield, the Netherlands.

Barel, R. et al. (1985). "Destruction of fisheries in Africa." *Nature*, 315, 19-20.

Barlow, Maude and Clark, Tony. (2002). *Blue Gold: The Fight to Stop the Corporate Theft of the World's Water*. New York: New Press.

Barnett, Tony, and Whiteside, Allan. (2006). *AIDS in the 21st Century: Disease and Globalization* 2nd ed. New York: Palgrave Macmillan.

Barnosky, Anthony D. et al. (March 3, 2011). "Has the earth's sixth mass extinction already arrived" *Nature*, 471: 51-57.

Barrett, Hazel R. et al. (1999). "Globalization and the changing network of food supply: the importation of fresh horticultural produce from Kenya into the UK," *Transactions of the Institute of British Geographers, New* Series, 24 (2), 159-174.

Battisti, David S. and Rosamond L. Naylor. (January 2009). "Historical Warnings of Future Food Insecurity with Unprecedented Seasonal Heat." *Science,* 323 (5911), pp. 240-244.

Bayer. (2010). "Bayer Healthcare." www.viva.vita.bayerhealthcare.com/ scripts/pages/en/press/.../13898/? Accessed November 21, 2010.

BBC News. (2011). "Food prices: World Bank warns millions face poverty." August 14.

BBC News, (2009). "Climate change help for the poor has not materialized." November 25.

BBC Monitoring Africa. (2009). "Kenya government probes says 25 powerful public officials allocated forest land." April 3.

BBC News Monitoring Africa. (2008). "Uganda says losing 90,000 hectares of forest cover." October 13.

BBC News Monitoring Africa. (2008). "U.N. official said Uganda's high population fueling poverty." August 23.

BBC News Monitoring Africa. (2008). "Uganda: World Bank grants Uganda $200 million for poverty reduction." May 8.

BBC News Monitoring Africa. (2007). "Tanzania receives $698 million in aid from U.S.-based corporation." September 19.

Beattie, Alan, (2008). "Tanzania blows hot and cold over biofuels," *Financial Times* (London), May 29, p. 5.

Beeby, Rosslyn, (2007). "U.N. issues warning of critical food shortages," *Canberra* Times (Australia), December 26, p. 25.

Bene, C., Macfadyen G. and Allison, H. (2007). Increasing the contribution of small-scale fisheries to poverty alleviation and food security. FAO Fisheries Technical Paper No. 481. FAO, Rome.

Bene, Chrostophe, and Merten, Sonja. (2008). "Women and fish-for-sex: transactional sex, HIV-AIDS and gender in African fisheries." *World* Development, 36 (5), 875-899.

Berbery, E. H. et al. (2003). "Eta model estimated land surface processes and the hydrologic cycle of the Mississippi basin." *Journal of Geophysical Research*, 108 (D22), 1-52.

Berthelot, Jacques. (2009). "Agribusiness headlong flight to agrofuels and their impact on food security," accessed January 5th, 2010 from: www.landaction.org/spip/IMG/pdf/BerthelotAgrofuels2009.pdf

Bertini, Catherine and Glickman, Dan. (2009). "Renewing American Leadership in the Fight against Global Hunger and Poverty: the Chicago Initiative on Global Agricultural development", The Chicago Council on Global Affairs, available at www.thechicagocouncil.org/globalagdevelopment.

Beuving, J. J. (2010). "Playing pool along Lake Victoria's shores: fishermen, careers and capital accumulation in the Ugandan Nile perch business." *Africa: Journal of the International African institute*, 80.2, 224-248.

Beverton, Raymond J. H. (1959). "A Report on the State of the Lake Victoria Fisheries." Lowestoft: Fisheries Laboratory.

Birdlife International, (undated). "Yala Swamp Complex," available at www.birdlife.org

Bloom, Jonathan. (2010). *American Wasteland: How America Throws Away Nearly Half of* Its *Food (and What we can Do About It).*

Bloomfield, Steve. (2006)."Trading fish for sex: how HIV spread in Western Kenya," *The Independent* (London), December 2006, p. 28.

Boar, Rosalind R., et al. (1999). "Biomass allocation in Cyperus papyrus in a tropical wetland Lake Naivasha, Kenya." *Biotropica*, 31 (3), 411-421.

Boere, G. C., et al (editors). (2006). Waterbirds *Around the World.* Edinburough, U.K.: Stationery Office LTD.

Boggs, Grace Lee, (2007). "U.S. corporation destroys Kenya community," *Michigan* Citizen (Highland Park), March 4-10, A-10.

Boko, Michel et al. (2007). Africa, Climate Change 2007: Impacts, Adaptation and Vulnerability. Retrieved September 17, 2010 from: www.ipcc.ch/pdf/assessment-report/ar4/wg2/ar4-wg2-chapter9.pdf

Bootsma, H, and Hecky R. E. (1993). "Conservation of the African Great Lakes: a limnological perspective." *Conservation Biology*, 7 (3), 644-656.

Bowles, B. D. (1975). "Economic anti-colonialism and British reaction in Uganda, 1936 to 1955," *Canadian Journal of African Studies*, 9 (1), 51-60.

Braun, Joachim von and Meinzen-Dick, Ruth. (2009). "Land grabbing by foreign investors in developing countries: risks and opportunities," Policy Brief No. 13, accessed October 31, 2009 from: www.presentationsistersunion..org/_uploads/fcknw/land_grabbing.doc

Brown, Gardner et al. (2005). "Different property rights regimes in the Lake Victoria multiple species fishery." *Environment and Development Economics,* 10 (1), 53-65.

Brown, Lester R. (2011). "The new geopolitics of food." *Foreign Policy*, May/June, 186: 54-63.

Brox, Ottar (1990). "Epistemological status and analytical utility of common property theory." *Human Organization*, 49 (3), 227-235.

Bryant, Raymond L. and Bailey, Sinead. (1997). Third *World Political* Ecology. New York: Routledge.

Bunker, Steven g. (1985). "Peasant responses to a dependent state, Uganda 1983," *Canadian Journal of African Studies*, 19 (2), 371-386.

Burkholder, J. et al. (2007). "Impacts of waste from concentrated animal feeding operations on water quality." *Environmental Health* Perspectives, 115 (20), 308-312.

Burdette, Marcia. (1984). "The mines, class power, and foreign policy in Zambia." *Journal of Southern African Studies*, 10 (2), 198-218.

Business Daily (Nairobi). (2009). "Kenya: the tragedy that is Lake Naivasha." October 5.

Business Daily (Nairobi), (2007). "Kenya: new fish species causing siltation in lake, say experts." October 16.

Business Time, (Dar Es Salaam). (2004). "Tanzania: cattle lung disease wreaks havoc in Lake Zone." March 12.

Business Wire (New York). (2009). "Democratic Republic of Congo's Mining Industry accepts U.S. $9 billion in investments from China in first half of 2009." Sept. 7.

Butler, Rett. (2008). "Amazon palm oil: palm oil industry moves into the Amazon rainforest," mongabay.com July 9, 2008.

Butler, Rett A. (2006). "Why are biofuels fueling deforestation?" Mongabay.com April 25, 2006.

Butler, Rett. (2005). "Africa tea faces over-production," *Tea and Coffee Trade Journal*, March/April 2005.

Byrnes, Rita M. (ed.) (1990). *Uganda: Country Study*. Washington D.C.: Library of Congress.

Bwakali, John D., (2008). "Kenya: biofuels boom going bust amid falling prices," *Global Information Network* (New York), October 27.

Campbell, Curt M. et al. (2007). The Age of Consequences: the Foreign Policy and National Security Implications of Global Climate Change. Washington D.C.: *Center for Strategic and International Studies*.

Canadian Green Peace. (undated). "Canadian Atlantic Fisheries collapse," accessed December 16, 2009 from: archive.greenpeace.org/comms/cbio/cancod.html

Candia, Steven, (2010). "Lake Victoria choking with sewage," *New Vision* Online, June 8.

Cassidy, Kevin, (2006). "Picking education, not tea, voices of change in Uganda," *World of Work Magazine*, September.

Castillo, Ruth et al. (2007). "The influence of El Nino of 1997-98 on pelagic ostracods in the Humboldt current ecosystem off Peru." *Hydrobiologia*, 585 (1), 29-41.

Cavides, Caesar N. (1984). "El Nino 1982-83." *The Geographical Review*, 74 (3).

Chambers, R. (1983). *Rural Development: putting the last first*. London: Longman.

Changadeya, W. et al. (2003). "Potential of genetics for aquaculture in Africa." *World Fish Center Quarterly*, 26 (3), 31-34.

Chapman, Lauren J. et al. 2004). "Respiratory ecology of macroinvertebrates in a swamp river system of East Africa." Biotropica, 36 (4), 572-585.

Chapman, Lauren et al. (1998). "Witness to mass extinction." *Conservation* biology, 12 (1), 250-257.

Chapman, L. et al. (1996). "Refugia for endangered species from introduced predator in Lake Nabugabo, Uganda." *Conservation Biology*, 10 (2), 554-561.

Chapman, Lauren et al. (1995). "Hypoxia tolerance in twelve species of East African cichlids: potential for low oxygen refugia in Lake Victoria." *Conservation Biology,* 9 (5), 1275-1286.

Charles, Anthony T. (2001). *Sustainable Fishery Systems*. London: Blackwell Publishers LTD.

Charles, Anthony T. (Dec. 1994). "Towards Sustainability: the fishery experience." *Ecological Economics*, 11 (3), pp. 201-211.

Chege, Nancy, (2010). "Lake Victoria: a sick giant," *People and the Planet*, 1995, www.cichlid-forum.com/articles/lake_victoria_sick.php accessed November 13.

Chick, John H. and Pegg, Mark A. (2001). "Invasive carp in the Mississippi River Basin." *Science: New Series*, 292 (5525), 2250-2251.

CIA World Book of Facts, "Kenya," July 2009.

Clark, Colin W. (2006). *The Worldwide Crisis in Fisheries: Economic Models and Human Behavior*. New York: Cambridge University Press.

Cline, Naomi. (2007). *Shock Doctrine: The Rise of Disaster Capitalism*. New York: Metropolitan Books.

Coche, A. G. and Balarin, J. D. (1982). "Kenya: Report of the Preparatory Assistance Mission", Fisheries Development Program in the Lake Basin Region. FAO, Rome, June.

Cocks, Tim, "Global warming could kill Uganda coffee," *Reuters*, February 26, 2007.

Cohen, A., Kaufman, L., and Ogutu-Ohwayo. (1996). "Anthropogenic threats, impacts and conservation strategies in the African great lakes: a review." In Johnson, Thomas C., and Odada, Eric (eds.), *The Limnology, Climatology and Paleoclimatology of the East African Lakes*: 575-622. London: Gordon Breach Publishers.

Cohen, Andrew. (1957). "Uganda Protectorate: annual report," *African Affairs*, 56 (223), 111-122.

Collins, Robert O. (2002). *The Nile*. New Haven: Yale University Press.

Commission of the European Communities. (2009). "Green Paper: Reform of the Common

Fisheries Policy." April. Brussels, Belgium.

Conca, Ken. (2005). *Governing water: contentious transnational politics and global institution building*. Cambridge, Mass.: The MIT Press.

Containers, S. (2010). "Phosphorus: Monitoring and Assessment." U.S. EPA, retrieved September 22, 2010 from: water.epa.gov/type/rsl/monitoring/vms56.cfm

Costanza, Robert et al. (1997). "The value of the world's ecosystem services and natural capital." *Nature*, Vol. 287: 253-260, May 15, 1997.

Coulter, G. W. and Spigel, H. (1991). "Hydrodynamics," In Coulter, G. W. (ed.) *Lake Tanganyika and its Life* History. New York: Oxford UP.

Cox, George W. (1993). *Conservation Ecology: The Biosphere and Bio*-survival. Dubuque, Iowa: William C. Brown Publishers.

Crul, Rudd. (1995). "Limnology and Biology of Lake Victoria." *Studies and Reports in Hydrology*. Paris, France: UNESCO Publishing.

Currie, Kate and Ray, Larry. (1987). "The Kenya state: agribusiness and the peasantry." *Review of African Political Economy*, No. 38, pp. 89-96.

Daily, Gretchen C. (1997). "Ecosystem services supplied by soil." In Daily, Gretchen C. (ed.) *Nature's Services: Societal Dependence on Natural Ecosystems*, pp. 113-130. New York: Routledge.

Daly, Herman E. (1993). "Steady state economics: a new Paradigm." *New Literary History*, 24 (4), 811-816.

Daly, Herman E. (1990). "Sustainable development: from concept and theory to operational principles." *Population and Development Review*, 16, Supplement, 25-40.

Daley, Herman E. and Joshua Farley. (2003). *Ecological Economics: Principles and Applications.* Washington D.C.: Island Press.

Damianopoulos, Richard. (2005). *Market adaptability, industrial divergence and the politics of liberalization in the Kenyan and Ugandan coffee industries* (MA thesis, University of British Columbia).

Danielsen, Finn, et al. 2008). "Biofuel plantations on forested lands: double jeopardy for biodiversity and climate." *Conservation Biology*, 23 (2), 348-358.

Delgado, C. et al. (1999). "Livestock to 2020: The Next Food Revolution." *Food, Agriculture and the Environment Discussion Paper* 28. AFPRI/FAO/ILRI.

Delgado, Christopher et al. (1997). "The impacts of livestock and fisheries on food availability and demand in 2020." *American Journal of Agricultural Economics*, 79 (3), 1471-1475.

De Keyser, Stefan and Hongo, Hamimu. (2005). "Farming for energy for better livelihoods in Southern Africa," paper presented at the Partners for Africa Policy Dialoge Conference, Dar Es Salaam, Tanzania, accessed Nov. 2, 2009 from www.hedon.info/goto.php/598/news.htm

Denning, Glenn et al. (2009). "Input subsidies to improve smallholder maize productivity in Malawi: toward an African green revolution." *Plos Biology*, 7 (1), 23-31.

Deweerdt, Sarah, (2004). "Dark secret of the lake," *New Scientist* (London), February 28, pp. 40-43.

Dinham, Barbara, (2005). "Agrochemical and biotech corporations spur growth of pesticides," *Pesticide News*, No. 68, June, pp. 9-11.

Dodd, Walter K. (2002). *Freshwater Ecology: Concepts and Environmental Applications.* Academic Press.

Donham, K. J. et al. (2007). "Community health and socioeconomic issues surrounding concentrate animal feeding operations." *Environmental Health Perspectives*, 115 (2), 317-320.

Dorsey, Bryan. (April 1999). "Agricultural intensification, diversification and commercial production among smallholder coffee growers in central Kenya," *Economic Geography*, 75 (2), 178-195.

Dugan, P. (2003). "Investing in Africa: the World Fish Center's Africa's strategy in summary." *World Fish Center Quarterly*, 26 (3), 9-12.

Edroma, Eric L. (1974). "Copper pollution in Rwenzori National Park." *Journal of Applied Ecology,* 11 (3), 1043-1056.

East African Business Week. (2008). "Uganda: flowers in thorny year." January 7.

EIU ViewsWire (New York), (2011). "Commodities: EIU's monthly soybeans outlook." September 1.

East African Standard (Nairobi), (2007). "Kenya: farmers earn Sh.3.5 bn from milk sale." July 30.

ETC Group (Action Group on Erosion, Technology and Conservation). (2010). *The New Biomassters: Synthetic Biology and the Next Assault On Biodiversity and* Livelihoods. Communique No. 104, October, 2010.

European Union. (2006). Uganda, available at: www.sfp-acp.eu/EN/Pays/Africa/Uganda/UgandaCountryProfile.pdf. August 2006.

European Union. (2006). Tanzania, available at: www.sfp-acp.eu/EN/Pays/Africa/Tanzania/TanzaniaCountryProfile.pdf. March 2006.

Export Processing Zones Authority, Kenya. Fishery Industry in Kenya, (2005). available at www.epzakenya.com/UserFiles/File/Fishkenya.-pdf.

FAO Media Center, "Cutting food waste to feed the world," May 11, 2011.

FAO. (2009a). *The State of World Fisheries and Aquaculture* 2009. Rome: FAO.

FAO. (2009b). *The State of World Food and Agriculture: Livestock in the* Balance 2009. Rome: FAO.

FAO, (2008a). *Current Situation and Medium-Term Outlook*. FAO Committee on Commodity Problems, Intergovernmental Group on Tea, 18th Meeting, Hangzhou, China. May.

FAO. The State of World Fisheries and Aquaculture 2008b. FAO of the U.N., Rome.

FAO. State of the World's Forests 2007. Rome 2007.

FAO. (2005a). Forest Resources Assessment 2005. FAO Forestry Paper No. 147, Food and Agriculture Organization of the United Nations, Rome.

FAO. (2005b). National Aquaculture Sector, an Overview, Kenya. Rome: FAO.

FAO. (2005c). National Aquaculture Sector, an Overview, United Republic of Tanzania. Rome: FAO.

FAO. (2005d). National Aquaculture Sector, an Overview, Uganda. Rome: FAO.

FAO. (2002). FAO Fisheries and Aquaculture, Peru, Profile.

FAO. (1998). Wetland Characterization and Classification for Sustainable Agricultural Development. Rome.

Ferver, Dan. (2001). "New focus: will black carp be the next zebra?" *Science: New Series*, 292 (5515), 203.

Field, Christopher B. et al. (2008). "Biomass energy: the scale of the potential resource," *Trends in Ecology and Evolution*, 23 (2), 65-72.

Fishman, Ted C. (2005). *China, Inc. How the Rise of the Next Superpower Challenges America and the World.* New York: Scribner.

Fitzhenry, Erin. (July 2002). "The Ogiek Welfare Council," *Cultural Survival Voices*, 1 (3), 11.

Flanary, Rachel and Watt, David. (1999). "The state of corruption: a case of study of Uganda." *Third World Quarterly,* 20 (3), 515-536.

Freeman, Donald. (1988). "The importance of being first: preemption by early adopters of farming innovations in Kenya," *Annals of the Association of American Geographers*, 75 (1), 17-28.

Freeman, Scott. (2002). *Biological Science.* Upper Sattle, N.J.: Prentice Hall.

Freereporter, Stefano Valentino, (2011). "Tanzania: biofuel project's barren promise," *Global Information Network* (New York), March 9.

Fryer, Geoffrey and Iles. (1972). The *Cichlid Fishes of the Great Lakes of Africa: Their Biology and Evolution.* Neptune City, N.J." T. F. H. Publications Inc.

Fryer, Geoffrey. (1960). "Concerning the proposed introduction of Nile perch into Lake Victoria." *East African Agriculture and Forestry* Journal, 25 (4), 267-270.

Gartreli, Beverly. (1983). "British administrators, colonial chiefs and the comfort of tradition: an example from Uganda," *African Studies Review*, 26 (1), 1-24.

Georgakakos, A. et al. (2005). *Study on Water Management of Lake Victoria, Technical Paper 10: Climate Change Impact Assessment.* Uganda Ministry of Energy and Mineral Development, Kampala, Uganda.

Georgescu-Roegen, Nicholas. (1975). "Energy and economic myth." *Southern Economic Journal*, 41 (3), 378-381.

Georgescu-Roegen, Nicholas. (1971). *The Entropy Law and the Economic Process.* Cambridge, Mass.: Harvard UP.

Gershenberg, Irving. (1972). "Slouching towards socialism: Abote's Uganda," *African Studies Review*, 15 (1), 79-95.

Gereta, Emmanuel et al. (2009). "Ecohydrology as a tool for the survival of the threatened Serengeti ecosystem." *Ecohydrology and Hydrobiology*, 9 (1),

Gibbs, H. K. et al. (2010). "Tropical forests were primary sources of new agricultural land in the 1980s and 1990s." *Proceedings of National Sciences*, www.pnas.org/cgi/boi/10.1073/pnas.0910275107 accessed September 17, 2010.

Giddings, Stephen w. (2009). "The land market in Kampala, Uganda and its effects on settlement patterns," *International Housing Coalition*, Washington DC, accessed August 9, 2009 from intlef.org/docs/giddings-kampala.pdf

Gibbon, Peter. (1997). "Of saviours and punks: the political economy of the Nile perch marketing chain in Tanzania." CDR Working Paper 97.3, Danish Institute for International Studies, Copenhagen, June 1997.

Global Environment Facility (GEF), (2006). Report of the GEF-Staff Workshop on Liquid Biofuels, November 27, 2006, accessed December 2, 2009 from: www.gefweb.org/COUNCIL/GEF_C30/C.30.Inf.09.Rev.1.pdf

Global Exchange. (2008). "Child labor." June 1.

Global Response. (2007). "Protect wetlands and communities' rights, Kenya," available at www. Globalresponse.org

Goldschmidt, Tijs. (1994). *Darwin's Dreampond: Drama in Lake* Victoria. Cambridge, MA. MIT Press.

Goldschmidt, Tijs et al. (1993). "Cascading effects of the introduced Nile perch on the detritivorous/phytoplanktivorous species in the sublittoral areas of Lake Victoria." *Conservation* Biology, 7 (3), 686-700.

Gordon, Anne. (2005). "HIV/AIDS in the fisheries sector in Africa. Cairo, Egypt: World Fish Center, retrieved Oct. 19, 2010 from: www.eldis.org/fulltext/hivandaidinthefisherysector.pdf

Gordon, David. (2006). *Naghitutts Gift: Economy, Society and Environment in Central Africa*. Madison: Wisconsin University Press.

Gordon, David. (2005). "Growth without capital: a renaiscent fishery in Zambia and Katanga, 1960s to recent times," *Journal of Southern African Studies,* 31 (3), 495-511.

Gordon-Maclean, A. et al. (2009). *Biofuel Industry Study, Tanzania. Tanzania and Sweden: World Wildlife Fund for Nature.*

Gottingen, Cuvillier. (2006). "Analysis of macrophyte biomass productivity, utilization and its impact on various eco-types of Yala Swamp, Lake Victoria basin, Kenya." *Ecology and Development Series*, No. 48.

Graham, Michael. (1929). *The Victoria-Nyanza and its Fisheries: a Report on the Fishing Survey of Lake* Victoria *and Appendices*, 1927-1928. London: Crown Agents for the Colonies.

Greboval, D., and Mannini, P. (1992). "The fisheries of Lake Victoria: review of basic data." UNDP/FAO Regional Project for Inland Fisheries Planning, RAF/87/099-WP16/92. Bujumbura, Burundi.

Greenpeace Canada, (2009). "The Congo Basin Rainforest," accessed January 6, 2010 from: www.greenpeace.org/canada/.../the-congo-basin-rainforest

Grellier, Rachel et al. (2004). *The impact of HIV/AIDS on fishing communities: situation analysis*. Jinja, Uganda: LVFO.

Gronewold, Nathanial, (2008). "Draining wetlands for development releases large amounts of greenhouse gases," *Climate Wire*, July 21.

Haan, C., Steinfeld, H., and Blackburn, H. (1997). *Livestock and the Environment: Finding a Balance.* Brussels, Belgium: European Commission Directorate-General for Development.

Hails, Sandra. (2000). "The Ramsar Convention on Wetlands, background papers on wetland values and functions", available at www.ramsar.org/info/values_intro_E.htm

Hall, S. R., and Mills, E. L. (2000). "Exotic species in large lakes of the world." *Aquatic Ecosystem Health and Management*, 3 (1), 105-135.

Hammond, Ross. (1997). "The tobacco industry impacts on Tanzania," CORPWATCH, San Francisco's Forum on Global Tobacco Control Policies.

Hanlon, Joseph. (2004). "Do donors promote corruption?: the case of Mozambique." *Third World Quarterly*, 25 (4), 747-763.

Hanson, J. et al. (2008). "Target atmospheric CO2: where should humanity aim?" *Open Atmospheric Science Journal*, 2: 217-231.

Harbeson, John W. (August 1971). "Land reforms and politics in Kenya: 1954-1970," *The Journal of Modern African Studies*, 9 (2), 231-251.

Harrison, Graham. (1999). "Corruption as boundary polity: the state, democratization and Mozambique's unstable liberalization." *Third World Quarterly*, 20 (3), 537-550.

Harsch, Ernest. (2006). "Africa starts a fishing revolution: NEPAD plan proposes poverty reduction, food security and environmental sustainability," *Africa Renewal.* (UN magazine), 20 (10, April.

Hecky, Robert E., et al. (2010). "Multiple stressors cause rapid ecosystem change in Lake Victoria." *Freshwater Biology*, 55 (Supplement 1), 19-42.

Hecky, Robert E. et al. (2006). "African lake management initiatives: the global connection," *Lakes and Reservoirs: Research and Management*, 11 (4), 203-213.

Hecky, R. E. et al. (1994). "Deoxygenation of the deep water of Lake Victoria, East Africa." *Limnology and Oceanography*, 39 (6), 1476-1481.

Heederik, D. et al. (2007). "Health effects of airborne exposures from concentrate animal feeding operations." *Environmental Health Perspectives*, 115 (2), 298-302.

Held, David. (2002). "Cosmopolitanism: ideals, realities and deficits," in Held, David and McGrew, Anthony (eds." *Governing globalization: power, authority and global* governance (pp. 305-324). Cambridge, U.K.: Polity Press.

Hellman, Joel S. et al. (2000). *Measuring Governance, Corruption and State Capture: How Firms and Bureaucrats Shape the Business Environment in Transition Economies*, World Bank Policy Research Paper, http://129.3.20.41/eps/dev/papers/0308/0308004.pdf accessed August 5, 2010.

Hentschel, Thomas et al. (2002). *Global Report on Artisanal and Small-Scale Mining.* London: World Business Council for Sustainable Development.

Hickley, Phil et al. (2004). "Management and ecological note: discovery carp, Cyprinus carpio, in the already stressed fishery of Lake Naivasha, Kenya." *Fisheries Management and Ecology*, 11 (2), 139-142.

Himbury, W. H. (1918). "Empire cotton," *The Journal of the Royal African Society*, 17 (38), 262-275.

Holt-Gimenez, Eric and Shattuck, Annie. (2008). "Agrofuels and food sovereignty: another agrarian transition is possible," Institute for Food and Development Policy, Food First, paper presented to a workshop, retrieved January 19, 2010 from: www.foodfirst.org/files/pdf/Agrofuels%20&%20Food%20Sovereignty%20-%20Canada%202008.pdf

Holttoman, J., and Lehman, J. (1998). "The eutrophication of Lake Victoria." In Lehman, John T. Ied.) *Environmental Changes Response in East African Lakes*, pp. 89-98. London: Kluwer Academic Publishers.

Hughes, Alex. (2001). "Global commodity networks: ethical trade and governability, organizing business responsibility in the Kenyan cut flower industry," *Transactions of the Institute of British Geographers,* New Series, 26 (4), 390-406.

Hughes, R., and Hughes, J. S. (1992). *A Directory of African Wetlands.* Geneva, Switzerland: The World Conservation Union.

Iheduru, Oichukwa. (1995). "The political economy of Euro-African fishing Agreements," *Journal of Developing Areas*, 30 (1), 63-90.

Ihucha, Adam, (2009). "160,000 bales of cotton pile up as prices slump on the world market," *Guardian* (Dar es Salaam), March 13.

Ikiara, Moses, M., and Odink, Joop G. (2000). "Fishermen resistance to exit fisheries." *Marine Resource Economics*, 14: 199-213.

Intergovernmental Panel on Climate Change. (2007). *Contribution of Working Group II to the Fourth Assessment Report of the Intergovernmental Panel on Climate Change,* 2007. New York: Cambridge University press.

International Council for Science, (2009). SCOPE Biofuel Project. *Biofuels: Environmental Consequences and Interactions with Changing Land Use*, April 2, accessed December 16, from: http://cip.cornell.edu/biofuels/

International Diabetes Federation. (2010). "Millions unite for diabetes awareness on world diabetes day," November 12, 2010. www.idf.org/millions-unite-diabetes-awareness-world-diabetes-day2010 accessed November 21, 2010.

Inter Press Service. (2008). "Tanzania: poverty reduction slow despite economic growth." December 4.

Inter Press Service. (2008). "Kenya: tea market tumultuous as trading partners change." September 23.

Inter Press Service. (2008). "West Africa: overfishing linked to food crisis, migration." August 11.

IUCN. (2003a). "Nakivubo Swamp, Uganda: managing natural wetlands for their ecosystem services," Valuation No. 7.

IUCN. (2003b). "Tana River, Kenya, Case Studies in Wetland" Valuation No. 6, Accessed September 14, 2009 from: http://cmsdata.iucn.org/downloads/casestudy06tana.pdf

Jamal, Vali. (1978). "Taxation and inequality in Uganda: 1900 to 1964," *The Journal of Economic History*," 38 (2), 418-438.

Jansen, Eirik. (2007). "Socio-economics of the Nile perch fishery on Lake Victoria: rich fisheries and poor fisherfolk." Nairobi, Kenya: IUCN.

Jansen, Eirik. (2006). """The role of women in the fisheries," accessed January 7, 2010 from: www.sum.uio.no/big_fish/eng/employment/pdf/women.pdf

Jansen, Eirik. (1999). "Out of a lake." www.icsf.net/icsf2006/uploads/publications/samudra/pdf/english/issue_23/art02.pdf accessed October 9, 2010.

Jeheb, Kim et al. (2008). "Nile perch and the hungry of Lake Victoria." *Food Policy*, 33 (1), 85-98.

Johnson, Thomas C. et al. (2000). "The Holocene history of Lake Victoria." *Ambio: a Journal of the Human Environment* 29 (1), 2-11.

Jones, Carey N. S. (1965). "The decolonization of white highlands of Kenya," *The Geographical Journal*, 131 (20, 186-201.

Jorgensen, Sven Erik, and Szirezhez, Wuri M. (2004). *Towards the Thermodynamic Theory of Ecological Systems*. London: Elsevier Publishers.

Josupeit, Helga, (2009). "Nile perch market report," *Globefish*, August, retrieved September 21, 2010 from: www.thefishsite..com/.../nile-perch-market-report-august-2009

Jul-Larsen, E. et al (eds). (2003). "Management, Co-Management, or No Management? Major Dilemmas in Southern African Freshwater Fisheries." FAO Fisheries Technical Paper No. 426, FAO, Rome.

Kabukuru, Wanjohi, (2009). "When Kenya sneezes," *New African* (London), May, pp. 64-65.

Kagolo, Francis, (2010). "Hunger looms as biofuels take root in Uganda," *New Vision* Online, January 19.

Kahl, Colin. (2006). *States, Scarcity and Civil Strife in the Developing World*. Princeton, N.J.: Princeteon University Press.

Kamau, John, (2004). "Kenya: nothing is trickling down," *African Business* London), June.

Kansiime, F. and van Bruggen, J. (2001). "Distribution and retention of fecal coliforms in the Nakivubo wetland in Kampala, Uganda," *Water Science and Technology: a journal of the International Association on Water Pollution* Research, 44 (11-12), 199-206.

Karlsson, Heidi et al. (2000). "Persistent chlorinated pesticides in air, water and precipitation from the Lake Malawi area, Southern Africa." *Environmental Science and Technology*, 34, pp. 4490-4495.

Kategile, J. A. and Mubi, S. eds. (1992). *Future of Livestock Industries in Eastern and Southern Africa.* Addis Abeba, Ethiopia: International Livestock Center for Africa.

Kaufman, L., and Ochumba, P. (1993). "Evolutionary and conservation biology of cichlid fishes revealed by faunal remnants in northern Lake Victoria." *Conservation Biology*, 7: 719-730.

Kaufman, L. (1992). "Catastrophic change in species-rich freshwater ecosystem: lessons of Lake Victoria." *Bioscience,* 42: 846-858.

Keane, G. T. (1935). "Uganda Protectorate: annual report," *Journal of the Royal African* Society, 34 (135), 190-194.

Kearney, T. A. (September 2006a). "Biofuel projects," accessed November 12, 2009 from: www.unido-aaitpc.org/unido-aaitpc/new1/tanzania/Biofuel%20PROJECTS1.pdf

Kearney, T. A. (2006b). "Cotton profile," available at www.zjfdi.com/Upload/Files/aaf001200751195735_2.pdf

Kearney, P. M. et al. (2005). "Global burden of hypertension: analysis of worldwide data." *Lancet,* 365: 217-223.

Keddy, Paul A. et al. (2009). "Wet and wonderful: the world's largest wetlands are conservation priority." *Bioscience*, 59 (1), 39-51.

Kenya Wetland Forum. (2006). "Report of rapid assessment of the Yala swamp wetlands," available at www.eawildlife.org/programme_areas/Yala%20report.pdf

Kiishweko, Orton. (2009). "Tanzania: government all out to increase tobacco production," *Tanzania Daily* News (Dar Es Salaam), November 26.

Kimirei, I, A. et al. (2008). "Changes in species composition and abundance of commercially important pelagic fish species in Kigoma area, Lake Tanganyika, Tanzania," *Aquatic Ecosystem Health and Management*, 11 (1), 29-35.

Kinuthia-Njenga, Cecilia. (2008). *Promoting Biodiversity in and Around the Lake Victoria Basin.* Nairobi, Kenya: United Nations Human Settlements Program.

Kipkemboi, J. et al. (2007). "Integration of smallholder wetland aquaculture-agriculture systems (fingerponds) into riparian farming systems on the shores of Lake Victoria, Kenya: socio-economic and livelihoods." *The Geographical Journal,* 173 part 3, 257-272.

Kitchell, James et al. (1997). "The Nile perch in Lake Victoria: interactions between predation and fisheries." *Ecological Application*, 7 (2), 653-664.

Kiwango, Yustina and Wolanski, Erik. (April 2008). "Papyrus wetlands, nutrients balance, fisheries collapse, food security, and Lake Victoria level decline in 2000-2006." *Wetlands Ecology and Management*, 16 (2), 89-96.

Knaap, Ban der et al. (2007). "Great Lake Victoria fishery: changes, sustainability and building blocks for management." *Aquatic Ecosystem Health and Management*, 10 (4), 481-483.

Knaup, Horand, (2008). "Green gold rush: Africa becoming a biofuel battleground," *Der Spiegel*, September 5, accessed October 19, 2009 from: www.spiegel.de/international/world/0,1518,576548,00.html

Kolding, Jeppe. (1995). "Changes in species composition and abundance of fish populations in Lake Turkana, Kenya." In Pitcher, t. J., and Hart, P. J. (eds.) *Impacts of Species Changes in African Lakes*, pp. 336-364. London: Chapman and Hall.

Koneswaran, Gowri and Nierenberg, Danielle. (2008). "Global farm animal production and global warming: impacting and mitigating." *Environmental Health Perspectives*, 116 (5), 578-582.

Kudhongania, A. and Chltamwebwa, D. B. R. (1995). "Impact of environmental change, species introduction and ecological interactions on the fish stocks of Lake Victoria." In Pitcher, T. J. and Hart, P. J. B. (eds.) *The Impacts of Species Changes in African* Lakes, London: Chapman and Hall. pp. 19-29.

Kweyuh, Pamphil M. (1997). "Tobacco farming in Kenya: the profits deception." Paper presented to the International Lung Heal Conference and Annual General Meeting of the International Union against Tuberculosis and Lung Diseases, Paris, France, September 29 to October 5, 1997.

Lackmeyer, Steve, (2007). "Farming Kenya: Dominion Group works to combine humanitarian and business efforts," *McClatchy Tribune* (Washington D.C.), October 22, , 1.

Lack, M. et al. (2003). Managing Risk and Uncertainty in Deep-Sea Fisheries: Lessons from Orange Roughy. *Traffic Oceania and World Wildlife Fund Endangered Seas Program.*

Lake Victoria Fisheries Organization (LVFO), (undated). Resource Monitoring, www.lvfo.org/index.php? Accessed March 7, 2009.

Lafraniere, Sharon, (2008). "Europe's takes Africa's fish, and migrants follow," *The New York Times*, January 14, p. 1.

Lehman, J. T. et al. (1998). "Lake Victoria phytoplankton ecology: mixing, depth and climate-driven control of lake condition." In Lehman, John T. (ed.) *Environmental Changes Response in East African Lakes*, pp. 99-110. London: Kluwer Academic Publishers.

Lehman, J. T., and Franstator, D. K. (1994). "Nutrient dynamics and turnover rates of phosphate and sulfate in Lake Victoria, East Africa." *Limnology and Oceanography*, 39 (2), 227-233.

Lemonick, Michael D. (2008). "Beyond the Tipping Point." *Scientific American*, 18 (4), pp. 60-67.

Line, Les, (1998). "Peru, epicenter of El Nino, fears for its wildlife," *New York Times*, May 19, p. 4.

Lewandowski, I. and Faaij, A. P. C. (2006). "Steps towards the development of a certification system for sustainable bio-energy trade," *Biomass and Bioenergy*, 30 (2), 83-104.

Liversage, Harold. (2009). "Managing the biofuel boom, smallholders' access to land: innovative models in Africa," accessed November 2, 2009 from: www.un.org/esa/dsd/susdevtopics/sdt_pdfs_meetings/hml029/session22_HLiversage.pdf

Lowe-McConnell, Rosemary. (1997). "EFRO and after: a guide to key events affecting fish communities in Lake Victoria (East Africa)." *South African Journal of Science*, 93 (11-12), 570-574.

LVFO. HIV/AIDS in the Fisheries Communities of Lake Victoria: strategic action plan 2006, final draft. Jinja, Uganda, November, 2006.

LVFO. Strategy and Action Plan for Monitoring, Control and Surveillance of Fisheries on Lake Victoria (final draft), May 2005. Jinja, Uganda.

LVFO. LVFO Regional Plan of Action to Prevent, Deter and Eliminate Illegal, Unreported and Unregulated Fishing on Lake Victoria and its Basin. Jinja, Uganda, May, 2004a.

LVFO. Regional Status Report on Lake Victoria: Frame Surveys for 2000, 2002 and 2004b.

LVFO. Strategic Vision for Lake Victoria, 1999-2015. Jinja, Uganda.

Mackay, J., and Mensah, G. (2004). *Atlas of Heart Disease and Stroke*. Geneva: World Health Organization.

Madeley, John, (1998). "A toxic worm in Kenya's buds," *The Independent* (London), February 1.

Mangora, Mwita. (December 2005). "Ecological impact of tobacco farming in miombo woodlands of Urambo District, Tanzania," *African Journal of Ecology*, 43 (4), 385-391.

Marshall, B. E. and Mumbamba, R. (eds). (1993). Papers Presented at the Symposium on Biological Stock Assessment and Exploitation of Small Pelagic Fish Species in the African Great Lakes Region. CIFA Occasional Papers No. 19. FAO, Rome.

Marx, Karl. (1977). *A contribution to the critique of political economy*. Moscow: Progress Publishers.

Merchants, Carolyn. (2005). *Radical Ecology: the Search for a Livable World.* 2nd ed. London: Routledge.

McCarthy, Michael, (2002). "European fishing fleets plundering West African stocks," *The Independent* (London), March 18,.

Mead, Gary, (1998). "Growers flourishing on Kenya's natural blessings," *Financial Times* (London), April, , p. 49.

Meyers, Norman. (1997). *"The world's forests and their ecosystem services." In Daily, Gretchen C. (ed.) Nature's Services: Societal Dependence on Natural Ecosystems*: 213-226. New York: Routledge.

Mgama, Richard, (2003) "No benefits from Geita, say locals," *African Business* (London), June, p. 51.

Milledge, S. H. et al. (2007). *Forestry, Governance and National Development: Lessons Learned From a Logging Boom in Southern Tanzania.* An Overview. Traffic East/Southern Africa Tanzania Development Partners Group/Ministery of Natural Resources of Tourism. Dar Es Salaam, Tanzania.

Millennium Ecosystem Assessment. *(2005). Ecosystem and Human Well-Being, Diodiversity Synthesis.* Washington D.C.: Island Press.

Miller, John W., (2007). "Global fishing trade depletes African waters," *The Wall Street Journa*l, July 23.

Mines and Communities, (2009). "London calling investigates some capital-land grab connection," July 16, accessed November 7, 2009 from: www.minesandcommunities.org/article.php?c=9355.

Mitchell, Andrew W. et al. (2008). *Forests now in the Fight against Climate Change.* Oxford, U.K.: Global Canopy Program.

Mkombo, Oliva C. et al. (2007). "Towards sustainable exploitation of Nile perch consequential to regulating fisheries in Lake Victoria." *Aquatic Ecosystem Health and Management*, 10 (4), 449-457.

Molssa, H. et al. (1999). "Fisheries Research: towards resources management on Lake Tanganyika." In Lindqvist, O. D. et al., eds. *From Limnology to Fisheries: Lake Tanganyika and other large* lakes. Pp. 3-23. London: Kluwer Academic Publishing.

Molony, Thomas, and Smith, James. (2010). "Biofuels, food security, and Africa." *African Affairs*, 109 (436), 489-498.

Monbiot, George, (2008). "Comment and debate: rich countries once used gunboats to seize food, now they use trade deals," *The Guardian* (London), August 26.

Moreau, Jacques. (1995). "Analysis of species changes in Lake Victoria using ecopath: a multispecies trophic model." In Pitcher, T. J., and Hart, P. J. (eds.) *Impacts of Species Changes in African* Lakes, London: Chapman and Hall. pp. 138-161.

Morgan, W. T. W. (June 1963). "The white highlands of Kenya," *The Geographical* Journal, 129, 140-155.

Moser, Michael et al. (1996). A global overview of wetland loss and degradation, available at www.ramsar.org/about/about_wetland_loss.htm

Mugerwa, Yasiin, (2008). "Danish official criticizes Uganda's anti-corruption policies," *BBC Monitoring Africa*, May 19.

Muller, Daniel et al. (2009). "Forecasting hotspots of forest clearing in Kakamega forest, Western Kenya," *Forest Ecology and Management*, 257 (3), 968-977.

Mundia, Matongo. (2008). "Agro-Fuel in Zambia," *African Biodiversity Network*, Accessed Sept. 22, 2009 from: www.landcoalition.org/cpl-blog/wp-content/uploads/case_studies_16.pdf

Munro, Forbes J. (1983). "British rubber companies in East Africa before the first world war," *The Journal of African History*, 24 (3), 369-379.

Mwakatombe, A. and Mlingwa, C. (2005). "The status of Tanzanian honey trade, domestic and international markets." Paper presented at the Bees for Development, Honey Trade workshop, August, Dublin, Ireland.

Nakkazi, Esther, (2007). "Uganda's coffee producers hope to benefit from Vietnam's dismal year's crop yield after climate change diminishes supply," *The Wip Contributor*, July 22.

Napolitano, Giulio, (2010) "New market access rules, economic crisis affecting seafood industry: developing countries struggle with new challenges getting fish to the market," *FAO Media Center*, Rome, April 26.

Nasi, Robert et al. (2002). "Can they pay our way out of deforestation?" Discussion Paper prepared for GEF, accessed December 11, 2009: www.unep.org/dec/docs/Forest_Ecosystem_Service-Executive_Version.pdf

Naylor, Rosamond L. et al. (2001). "Aquaculture a gateway for exotic species." *Science: New Series*, 294 (5547), 1655-1656.

Naylor, Rosamond L., and Ehrlich, Paul R. (1997). "Natural pest control services and agriculture." In Daily, Gretchen C. (ed.) *Nature's Services: Societal Dependence on Natural Ecosystems*,. New York: Routledge. pp. 151-170.

New Vision (Uganda). (2010). "Uganda: Government set revamp fish sector." 2010.

New Vision (Uganda). (2009). "Uganda has 11 million cattle." October 2.

New Vision (Uganda). (2009). "Uganda: when a big population is an asset." September 15.

*New Vision (*Uganda*).* (2009). "Uganda: Sweden restocks L. Victoria." April 6.

New Vision (Uganda). (2009). "Uganda: honey production==country yet to fully realize potential." January 27.

New Vision (Uganda). (2009). "Uganda: government boosts cocoa as world supply dwindles." January 21.

New Vision (Uganda). (2008). "Uganda: fish farming on the rise." November 4.

New Vision (Uganda). (2008). "Uganda: Lake Victoria turns green." March 7.

New Vision (Uganda). (2007). "Uganda: local leaders lack funds to stop Lake Kyoga from drying up." March 21.

New Vision (Uganda). (2005). "Museveni tips on how to reap big from cattle-keeping." November 9.

Niboye, Elliott Phillemons. (2006). "Impacts of changing pastoral strategies on environmental resources and livelihoods in Tanzania's Lake Victoria basin." Publications ossrea.net/index.php?option=com _content…id accessed July 21, 2011.

Niquen, Miguel, and Bouchon, Marilu. (2004). "Impact of El Nino events on pelagic fisheries in Peruvian waters." *Deep-Sea Research Part II, Topical Studies in Oceanography*, 51 (6-9), 563-574.

Njiru, M. et al. (2008). "An overview of the current status of Lake Victoria fisheries: opportunities, challenges and management strategies," *Lakes and Reservoirs: Research and Management*, 13 (1), 1-12.

Njiru, M. et al. (2005). "Exotic introductions to the fishery of Lake Victoria: what are the management options?" *Lakes and* Reservoirs, 10 (3), 147-155.

Nweke, Onyemaechi C. and Sanders, William H. III. (2009). "Modern environmental health hazards: a public health issue of increasing significance in Africa." *Environmental Health Perspective,* 117 (6), 863-870.

Nyakaana, J. B. et al. (2007). "Population, urban development and the environment in Uganda: the case of Kampala city and its environs," accessed September 4, 2009 fromwww.cicred.org/Eng/Seminars/Details/Seminars/PDE2007/Papers/Nyakaana_-paperNairobi2007-2.pdf

Nyannugasira, Warren. (1996). "Structural adjustment, land reform, and the disenfranchisement in Uganda." *Development in Practice*, 6 (4), 347-351.

Obonyo, Godfrey, (2007). "Cash crops boost country's economic growth," *Daily News* (Dar es Salaam), August 7.

Obulussa, George, (2008). "Norway gives Tanzania $100 million for forests," Reuters, April 21.

Ochumba, Peter, and Kibaara, David. (1989). "Observations on the blue-green algae bloom in the open waters of Lake Victoria, Kenya." *African Journal of* Ecology, 27: 23-34.

Odada, Eric et al, (2004). "Mitigation of environmental problems in Lake Victoria, East Africa: causal chain and policy options analyses." *Ambien*, 33 (1-2), 13-24.

Ogutu-Ohwayo, Richard and Balirwa John S. (December 2006). "Management challenges of freshwater fisheries in Africa." *Lakes and* Reservoirs, 11 (4), 215-226.

Ogutu-Ohwayo, Richard. (1990). "The decline of the native fishes of Lakes Victoria and Kyoga, East Africa, and the impact of introduced species, especially the Nile perch, Lates niloticus, and the Nile tilapia, Oreochromis niloticus." *Environmental Biology of Fishes*, 27 (81).

Oil and Gas Journal. (2010), "Tullow plans Ugandan oil production this year." February 8. p. 10.

Ojuok, J. et al. (2008). *African Journal of Ecology*, 46 (3), 445-448.

Ollago, Daniel O. and Odada, Eric O. (2007). "Sediment impacts in Africa's transboundary lake/river basins: case study of East African great lakes." *Aquatic Ecosystem Health and Management*, 10, 23-32.

Olol-Dapash, Meitamei. July 2002). "Mau forest destruction: human and ecological disaster in the making," *Cultural Survival Voices*, 1 (3), 1.

Orach-Meza, F. L. (1990). *Statistical Sampling Method for Improving the Catch Assessment of Lake Fisheries.* FAO-UNDP Project UGA/87/007, Fishing Notes and Records, Occasional Papers No. 2, May, FAO, Rome.

O'Reilly, Catherine et al. (August 14, 2003). "Climate Change Decreases Aquatic ecosystem Productivity in Lake Tanganyika." *Nature*, 424 (6950), pp. 766-768.

Orimbo, Bernard. (September/October 2008), "Of butterflies, birds, and bees," *World Watch*, 21 (5), 30-33.

Owino, Alfred and Ryan, Peter. (January 2006). "Recent papyrus habitat loss and conservation implications in western Kenya," *Wetlands Ecology and Management*, 15, pp. 1-12.

Oxfam International. (2011). *Land and Power: the Growing Scandal Surrounding the New Wave of Investments in Land.* Oxfam Briefing Paper 151, Oxford, September 22.

Oxfam International. (2008). "Another inconvenient truth: how biofuel policies are deepening poverty and accelerating climate change." Oxfam Briefing Paper, Oxford, June 26, 2008.

Oxford Analytica Daily Brief Service. (2009). "Africa: donor support boosts North-South corridor," May 26

Oxford Analytica Daily Brief Service. (2009). "China/Africa: new scramble for agricultural resources." January 29.

Page, S. E. et al. (1999). "Interdependence of peat and vegetation in tropical peat swamps." *Philosophical Transactions: Biological* Sciences, 354 (1391), 1885-1897.

PANOS (Nairobi). (2006). "Kenya: the real costs of cut flowers." November 17.

Pass Limited. (2001). "Investment potential in the tea industry"; "Investment potential in the cotton industry"; "Investment potential in the sisal industry"; available at www.pass.ac.tz/

Pearce, Fred, (2007). "Biofuel plantations fuel strife in Uganda," *New Scientist* (London), April 19,.

Pearce, Fred, (2006). "Uganda pulls plug on Lake Victoria," *New Scientist* (London), February 11-17, p. 12.

Pelletier, Nathan, and Tydmers, Peter. (2010). "Forecasting potential global environmental cost of livestock production 2000-2050." *Proceedings of the National Academy of Sciences of the United States of America*, 107 (43), 18371-18374.

Pepper, David. 1996. *Modern Environmentalism: An Introduction.* London: Routledge.

Perrings, Charles. (1987). *Economy and Environment.* New York: Cambridge University Press.

Pitcher, Tony J. (2001). "Fisheries: manage to rebuild ecosystem? Reconstructing the past to salvage the future." *Ecological Application*, 11 (2), 601-617.

Pitcher, Tony J. (1995). "Thinking the unthinkable: a candidate model for predicting sustainable fish yields of introduced fish species in African lakes." In Pitcher, T. J., and Hart, P. J. (eds.) *The Impacts of Species Changes in African Lakes: 495-516. London: Chapman and Hall.*

Pogge, Thomas W. (2002). *World poverty and human rights: cosmopolitan responsibilities and reforms.* Malden, Mass." Blackwell Publishers.

Population Council. (2005). "Population, environmental degradation and human well-being: report of the Millennium Ecosystem Assessment." *Population and Development Review*, 31 (2), 389-398.

Postel, Sandra, and Carpenter, Stephen. (1997). "Freshwater ecosystem services." In Daily, Gretchen C. (ed.) *Nature's Services: Societal Dependence on Natural Ecosystems*: 195-210. New York: Routledge.

Princen, Thomas, (2005). *The Logic of Sufficiency.* Cambridge, MA: MIT Press.

Pye-Smith, C. (2008). *Farming Trees, Banishing Hunger, How an Agroforestry Programme* Is *Helping Smallholders in Malawi To Grow More Food and Improve Their Livelihoods*, b15589.pdf. Nairobi, Kenya: World Agroforestry Center.

Rabi, Marcela. (1996). "Lake Victoria case study: case No. 388," accessed December 19, 2009 from: www.American.edu/TED/victoria.htm

Reynolds, J. E., and Greboval, D. F. (1995). "Thirty years on: the development of Nile perch fishery in Lake Victoria." In Pitcher, T. J., and Hart, P. J. (eds.) *The Impacts of Species Chandes in African Lakes*: 181-214. London: Chapman and Hall.

Rhodes, D. H. (1966). FAO/UNDP Report to the Government, Kenya, Fisheries Development Possibilities. FAO/UNDP Report TA 2144, FAO, Rome.

Rhymer, Judith M. and Simberloff, Daniel. (1996). "Extinction by hybridization and introgression." *Annual Review of Ecology and Systematics,* 27, 83-109.

Rice, Exan, (2008). "Wildlife and livelihoods at risk in Kenyan wetlands biofuel project," (*The Guardian* (London), June 24.

Robbins, Paul. (2004). *Political ecology: a critical introduction.* Malden, Mass.: Blackwell Publishers.

Robinson, Mark. (2007). "The political economy of governance reform in Uganda." *Commonwealth and Comparative Politics*, 45 (4), 452-474.

Rockstrom, Johan et al. (September 24, 2009). "A safe operating space for humanity." Nature, 461: 472-476.

Rosenthal, Elizabeth, (2008) "In Europe the catch of the day is often illegal," *The New York Times,* January 15, p. 1.

Rotberg, Robert I. (2007). *Africa's Successes: Evaluating Accomplishments.* Cambridge, MA. BELSER-WPS Report 43, Program on Interstate Conflicts.

Rubinstein, Carl, (2009). "Speaking freely: China's eye on African agriculture," *Asia Times*, October 2.

Sachs, Jeffrey D. (2008). *Commonwealth: Economics for a crowded Planet.* New York: Penguin Group.

Sachs, Jeffrey D. et al. (2004). "Ending Africa's poverty trap," *Brookings Papers on Economic Activity*, 1204 (1), 117-216.

Salzburger, Walter et al. (2002). "Phylogeny of the Lake Tanganyika cichlid species flock and its relationship to the Central and East African haplochromine cichlid fish faunas," *Systematic Biology*, 51 (1), 113-135.

Sauper, Hubert. (2004). *Darwin's Nightmare*, a documentary film.

Schindler, Daniel et al. (1998). "Ecological consequences of alternative gillnet fisheries for Nile perch in Lake Victoria." *Conservation Biology*, 12 (1), 56-64.

Scudder, Thayer and Conelly, Thomas. (1985). Management Systems for Riverine Fisheries. FAO Technical Paper No. No. 263, FAO, Rome.

Schumpeter, Joseph. (2006). *Capitalism, Socialism and Democracy*. Cheltenham, U.K.: Edward Elgar Publishing.

Schuyt, K. D. (April 2005). "Economic consequences of wetland degradation for local populations in Africa," *Ecological Economics*, 53 (2), 177-190.

Scullion, J. (2007). *Inland Fisheries Co-management in East Africa*. LVFO, Jinja, Uganda.

Searchinger, Timothy et al. (2008). "Use of U.S. croplands for biofuels increases greenhouse gases through emissions from land use change," *Science*, 319 (5867), 1238-1240.

Seehausen, Ole et al, (2008). "Speciation through sensory drive in cichlid fish." *Nature*, 455 (7213), 620-627.

Seehausen, Ole et al. (1997). "Cichlid fish diversity threatened by eutrophication that curbs sexual selection." *Science*, 277 (5333), 1808-1811.

Sen, Amartya. (1981). *Poverty and famines: an essay on entitlement and deprivation*. Oxford, England: Clarendon Press.

Shacinda, Shapi, (2009). "African nations create fund to pool investment," Reuters, April 8.

Shackle, GLS. (1972). *Epistemics and Economics: a Critique of Economic Doctrines*. New York: Cambridge University Press.

Shaw, Timothy. (1976). "Zambia: dependence and underdevelopment." *Canadian Journal of African Studies*, 10 (1), 3-22.

Sijaona, M. E. R. (2002). Assessment of the situation and development prospects for the cashew nut sector, available at www.unctad.org/infocomm/francais/anacarde/Doc/tanzania.pdf

Simms, Andrew, (2009). "What is yours is mine," *New Statesman* (London), October 19. pp. 38-39.

Smith, Robert N., and Smith, Thomas M. (2001). *Ecology and Field Biology* 6th ed. San Francisco, California: Benjamin Cummings.

Smolker, Rachel et al. (2007)."The real cost of agrofuels: food, forest and the climate," *WRM Bulletin* No. 122, September, accessed January 15, 2010 from: *www.globalforestcoalition.org/img/userpics/File/publications/Therealcostofagrofuels.pdf*

SOMO. (2008). Sustainability issue in the tea sector: a comparative analysis of six leading producing countries. Center for Research on Multinational Corporations, Amsterdam, available at www.somo.nl

Spiegel, Samuel J. (2009). "Occupational health, mercury exposure, and environmental justice: learning from experiences in Tanzania." *American Journal of Public Health*, 99 (3), 550-558.

Steinfeld, Henning et al. (2006). *Livestock's Long Shadow: Environmental Issues and Options.* Rome: FAO.

Straaten, Peter van. (2000). "Mercury contamination associated with small-scale gold mining in Tanzania and Zimbabwe." *The Science of the Total Environment*, 259 (2-3),

Swyngedow, Erik. (2004). *Social power and the urbanization of water: flows of power.* New York: Oxford Press.

Syren, Per. (1990). "Estimated transport of the suspended load of the River Malewa, Kenya, between 1931-1959." *Geografiska Annaler. Series A Physical Geography*, 72 (3-4), 285-299.

Tangri, Roger and Mwenda, Andrew. (2008). "Elite corruption and politics in Uganda." *Commonwealth and Comparative Politics*, 46 (2), 177-194.

Tanzania Daily News (Dar Es Salaam). (2010). "Tanzania: modernized animal husbandry given priority under new policy." October 14.

Tanzania Coffee Board. (2007). "Coffee from the land of Kilimanjaoro," available at tanzaniacoffeeboard.org/products.asp

The East African (Nairobi). (2004). "Tanzania: Botswana ranch model won't work in Tanzania." March 4.

Tauli-Corpuz, Victoria and Pamang, Parshuram. (2007). "Oil palm and other commercial tree plantations, monocropping: impacts on indigenous people's land tenure and resource management system and livelihoods," paper presented to the U.N. Permanent Forum on Indigenous Issues, New York.

Taylor, Ian. (2010). *The International Relations of Sub Saharan Africa.* London: Continuum International Publishing Group LTD.

Taylor, Christopher S. (1978). "The struggle for economic control of Uganda, 1919 to 1922: formulation of an economic policy," *The International Journal of African Historical Studies*, 11 (1), 1-31.

Thado, Deborah P. (1998). *The Economic Sociology of Kenya's Lake Victoria Beach Communities: the Intersection of the Economy, Communal Social Relations and Gender.* Unpublished Dissertation, Michigan State University.

The East African (Nairobi), (2009). "East Africa: new viral disease threatens region's cassava farming." July 13.

The East African. (2008). "Tanzania: fruits of the soil." March 24.

The East African (Nairobi). (2004) "Tanzania: Botswana ranch model won't work in Tanzania." March 4.

The East African Standard (Nairobi). (2008). "Multi-billion project stuck in murky politics." April 27.

The East African Standard (London), (2008). "Kenya: horticulture takes lead as foreign exchange earner." March 26.

The Citizen (Dar Es Salaam). (2009). "Tanzania: miners scramble to dig up gold." December 5.

The Economist. (2010). "A bonanza beckons." April 3.

The Economist. (2003). "Untapped power." March 29.

The Economist. (1978). "Peru's anchovy: going, going." March 11. p. 82.

The Monitor (Kampala). (2010). "Uganda: increase in population causes decline of fish stocks." August 31.

The Monitor (Kampala). (2008). "Uganda: country's bet on rice begins to payoff." October 28.

The Nation (Nairobi), (2011). "Kenya: farmers hard hit by rising cost of feed." February 26.

The Nation (Nairobi). (2010). "Why herders are turning to the bee." April 17.

The Nation (Nairobi). (2007). "Kenya: Turkwel white elephant that weighs heavily on residents" July 13.

The Nation (Nairobi). (2001). "Kenya: BAT action shocks tobacco farmers." August 1.

The Nation (Nairobi), (1999). "Poor crop husbandry threatens tobacco." July 21.

Thieme, Michele L. et al. (2005). *Freshwater Ecoregions of Africa and Madagascar: A Conservation* Assessment. Washington D.C.: World Wildlife Fund.

Throup, David. (1985). "The origins of Mau-Mau, "*African Affairs*, 84 (3336), 399-433.

Thurow, Andrew. (2011). "The fertile continent, Africa: agriculture's final frontier." *Foreign Affairs,* November/December 2010, 102-110.

Tierney, Jessica E. et al. (2010). "Late-Twentieth-Century warming in Lake Tanganyika unprecedented since AD 500." *Nature*, 3, 422-425.

Tijaja, Julia. (2011). "The impact of China's demand on domestic value chain: lessons from the cassava value chains in Thailand." Paper prepared for the annual conference of Chinese Economic Association (UK), 12-13 July, 2010, University of Oxford, UK.

Tilman, David. (1997). "Community invisibility, recruitment limitation, and grassland biodiversity." *Ecology*, 78 (1), 81-92.

Tilman, David et al. (2001). "Forecasting agriculturally driven global environmental change." *Science: New Series*, 292 (5515), 281-284.

Tofani, Loretta. (2007). "Chinese Workers Lose their Lives Producing Goods for America," *Salt Lake Tribune*, October 25, 2007.

Toowoomba, T. P. (2000) *Interactions between Fish and Aquatic Macrophytes in Inland Waters: a Review*. FAO Technical Paper 396. Rome: FAO.

Tudge, Colin. (2004). "It's a meat Market." *New Scientist*, 181 (2438), 9.

Tumwebaze, R. et al, (2007). "Spatial and temporal changes in the distribution of Rastrineobola argentea in Lake Victoria." *Aquatic Ecosystem Health and Management*, 10 (4), 398-406.

Turpie, Jane K. (2004). "The valuation of riparian fisheries in Southern and Eastern Africa," accessed November 12, 2009 from: http://aquacomm.ffcla.edu/1274/1/SR1836_3.pdf

UCDA. (2009). "Coffee campaign starts to pay off." www.ugandacoffee.org/ accessed June 17, 2010.

UN Chronicle. (2007). "How to avoid the unmanageable and to manage the unavoidable climate changes," Vol XLIV No. 2, June UN Scientists Report.

UN IRIN Africa. (2011). "Analysis: subsidies not the only answer for Malawi's farmers," April 19.

UN IRIN. (2010). "Uganda: cassava disease threatens food supplies." May 20.

UN IRIN. (2010). "Uganda: health fears follow deadly mudslides." March 10.

UN IRIN Africa. (2008). "Malawi: subsidizing agriculture is not enough." February 5.

UN IRIN. (2007). "Kenya: pastoralists lose 100,000 animals to livestock disease." July 24.

UN IRIN. (2006). "East Africa: deforestation exacerbates droughts, floods." November 10.

UNESCO Courier, (1997). "Kenya becomes leading world tea exporter." December. p. 43.

UNEP. (2009). *Kenya: Atlas of our changing* environment. Nairobi, Kenya: United Nations Environment Program, Division of Early Warning and Assessment.

UNEP. (2002). *Africa Environment Outlook 2002: Past, Present and Future Perspectives.*

UN IRIN, "And then there were no fish," November 2, 2007.

United Republic of Tanzania. (December 2007). Country Profile, General Geographic and Economic Data. Food and Agriculture Organization of the United Nations, Rome.

Upton, Maureen. (2009). *The Golden Building Block: Gold Mining and the transformation of Developing Economies*, World Gold Council Report, www.gold.org/assets/file/pub.../WGC_Golden_Building_Block.pdf

U.S. Fed News Service (Washington D.C.). (2008). "USTDA awards grant for biofuels regulation in Uganda." August 28.

Vandenberg, Paul. (2003). "Ethnic sectoral cleavages and economic development: reflections on the second Kenyan debate," *Journal of Modern African Studies*, 41 (3), 437-455.

Van der Knaap, Martin. (1994). "Status of Fish Stocks and Fisheries in Thirteen Medium-sized African Reservoirs." Committee on Inland Fisheries in Africa, CIFA Technical Paper No. 26.

Verburg, Piet et al. (2003). "Ecological consequences of a century of warming in Lake Tanganyika," *Science*, July 25, vol. 301 No. 5632, pp. 5.5-507.

Vidal, John. (2009). "World: Electric fences to stop invasion of Kenya nature parks." *The Observer* (England), May 10, p. 32.

Wadhams, Nick. (2009). "How a biofuel 'miracle' ruined Kenyan farmers," AFP, October 4.

Wallenren, Maja. (1997). "Tanzania: cotton revival," *African* Business (London), July/August.

Wanink, Van H., and Witte, Frans. (2000). "Rapid morphological changes following niche shift in the zooplanktivorous cyprinid Rastrineobola argentea from Lake Victoria." *Journal of Zoology,* 50 (3), 365-372.

Waterbury, John. (2002). *The Nile Basin: National Determinants of Collective Action.* New Haven: Yale University Press.

Waters, Alan. (1972). "Change and evolution in the structure of the Kenyan coffee industry," *African Affairs*, 71 (283), 163-175.

Watkins, Eric. (2010). "Watching the world: Uganda opts for Russian jets." *Oil and Gas Journal*, April 12, p. 30.

Welcomme, R. L. et al. (2010). "Inland capture fishery." *Philosophical Transactions: Biological Sciences*, 365, 2881-2897.

Welcomme, R. L. (2003). "River fisheries in Africa: their relationship to flow regimes." *World Fish Center Quarterly*, 26 (3), 35-38.

Welcomme, R. L. (1988). *International Introductions of Inland Aquatic Species.* FAO Fisheries Technical Paper No. 294. Rome: FAO.

Welcomme, R. L. (1979). The Inland Fisheries of Africa, CIFA Occasional Paper No. 7, FAO Fisheries and Aqualculture Department, Rome.

Westaway, Elizabeth. (2007). "Feckless and reckless or forbearing and resourceful?" *African Affairs,* 106 (425), 663-679.

Westerkamp, Meike, and Houdret, Annabelle. (2010). "Peacebuilding across Lake Albert: reinforcing environmental cooperation between Uganda and the Democratic Republic of Congo." Retrieved August 19 2010 from www.initiativeforpeacebuilding.eu/pdf/peacebuilding _lake_albert.pdf

Westrup, Hugh. (1998). *"El Nino strikes again." Current Science*, 83 (9), 4-7.

Wetlands International. (2009). "Convention on biological diversity." February 2.

Weyl, I. O. (ed.) (2001). "Fisheries management and uncertainty: the causes and consequences of variability in inland fisheries in Africa with special reference to Malawi." Proceedings of Lake Malawi Fisheries Management Symposium, June 5-9, 2001. Lilnngwe, Malawi.

Wilson, Douglas C. et al. (1999). "The implication for participatory fisheries management of intensified commercialization on Lake Victoria." *Rural Sociology*, 64 (4), 554-572.

Wilson, J. D. K. (2004). "Fiscal Arrangements in Tanzanian Fisheries Sector". FAO Fisheries Circular No. 1000, FAO Rome.

Wilson, Edward O. (2002). *The Future of Life*. New York: Knopf Group Electronic Publishing.

Wilson, Edward O. (1998). *Consilience: The Unity of Knowledge*. New York: Knopf Doubleday.

Witte, Frans, et al. (2007). "Differential decline and recovery of haplochromine trophic groups in the Mwanza Gulf of Lake Victoria." *Aquatic Ecosystem Health and Management*, 10 (4), 416-433.

Witte, Frans, et al. (1992). "The destruction of an endemic species flock: quantitative data on decline of the haplochromine cichlids of Lake Victoria." *Environmental Biology of Fishes*, 1-2: 3-30.

World Bank. (September 2009). *World Bank Development Report 2010: Development and Climate Change*. Washington D.C.: The World Bank Group.

World Bank. (February 2009). Africa's infrastructure: a time for transformation, available at http://go.worldbank.org/R056W1GUB0

World Bank. (2008). *A World Bank Country Study: putting Tanzania's hidden economy to work, reform, management and protection of its natural resources sector*. Washington D.C.: The World Bank.

World Bank. (2007). "Increased Cotton Production in Uganda", available at http://go.worldbank.org/EZDHPI8D30

World Bank. (2005). "Where is the Wealth of Nations? Measuring capital for the 21st century".

World Bank. (2004). *Institutionalizing traditional community-based natural resources management*, Malawi. IK Notes, No. 64, available at www.worldbank.org/afr/ik/default.htm

World Bank. (1993). *A World Bank Country Study, Uganda: Growing out of* Poverty. Washington D.C.: World Bank.

World Bank Group. (1991). "Our Words: the Lawrence Summers Memo." Available at www.worldbank.org/ourwords/summers/.html

World Rainforest Movement. (2008). "Oil palm and rubber plantations in Western and Central Africa: an overview." Available at www.wrm.org.uy/publications/briefings/Western_Central_Africa.pdf

World Resources Institute. (2009). "Mapping a better future: how spatial analysis can benefit wetlands and reduce poverty in Uganda". Washington D.C.

World Wildlife Fund. (2003). *Managing Risk and Uncertainty in Deep Sea Fisheries: Lessons from Orange Roughy*. Available at www.worldwildlife.org/oceans/pdfs/orange_roghy

Worthington, S. and Worthington, E. B. (1933). *The Inland Waters of Africa: the Result of Two Expeditions to the Great Lakes of Kenya and Uganda, with Accounts of their Biology, Native Tribes and Development.* London: Macmillan and Co. LTD.

Wrong, Michela. (2009). *It's Our Turn to Eat: the Story of a Kenyan Whistle Blower.* New York: HarperCollins.

Xinhua General News Service. (2006). "Yangtze River Cancerous with Pollution." May 30.

Yohannes, Okbazghi. (2010). "The new oil Gulf of Africa: global geopolitics and enclave oil-producing economies." In Soyinka-Airewele, Peyi, and Edozie, Kiki Rita eds. *Reframing Contemporary Africa: Politics, Economics, and Culture in the Global* Era: 241-254. Washington D.C.: CQ Press.

Yohannes, Okbazghi. (2009). "Hydropolitics in the Nile Basin: in Search of Theory Beyond Realism and Neoliberalism." *Journal of East African* Studies, 3 (1), 74-93.

Yohannes, Okbazghi. (2008). *Water Resources and Inter-riparian Relations in the Nile Basin: the Search for an Integrative Discourse.* Albany, N.Y.: SUNY Press.

York, Geoffrey. (2009). "Land: Africa's last great treasure," *The Globe and Mail," (Canada),* May 6, p. A-16.

Zaret, Thomas M. and Paine, R. T. (1973). "Species introduction in a tropical lake." *Science: New Series*, 182 (411), 449-455.

Zwerdling, Daniel. (2009). "Green revolution trapping India's farmers in debt," *National Public Radio* (Morning Edition), April 14.

Index

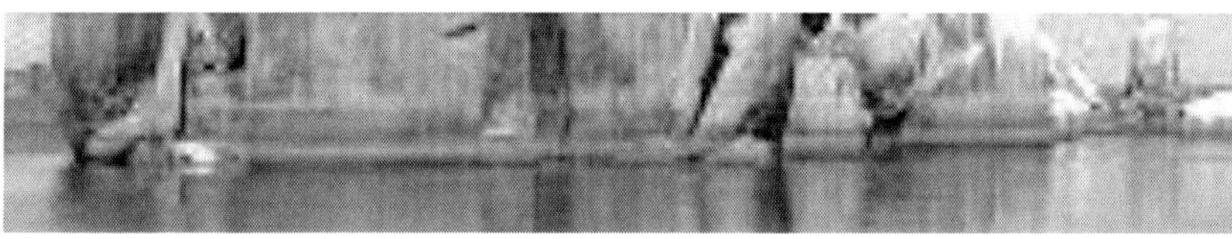

171114-20-1-60W